MW01598647

INVERSE PROBLEMS IN GROUNDWATER MODELING

Theory and Applications of Transport in Porous Media

Series Editor:
Jacob Bear, *Technion – Israel Institute of Technology, Haifa, Israel*

Volume 6

The titles published in this series are listed at the end of this volume.

Inverse Problems in Groundwater Modeling

by

Ne-Zheng Sun

Environmental Science Center,
Shandong University, P.R. China

Presently, Civil Engineering Department,
University of California, Los Angeles, U.S.A.

KLUWER ACADEMIC PUBLISHERS

DORDRECHT / BOSTON / LONDON

Library of Congress Cataloging-in-Publication Data

Sun, Ne-Zheng.
     Inverse problems in groundwater modeling / by Ne-Zheng Sun.
       p.   cm. -- (Theory and applications of transport in porous
  media ; v. 6)
     Includes bibliographical references and index.
     ISBN 0-7923-2987-2 (alk. paper)
     1. Groundwater--Mathematical models. 2. Inverse problems
  (Differential equations)   I. Title. II. Series.
  GB1001.72.M35S86   1994
  551.49'01'5118--dc20
                                                              94-22252

ISBN 0-7923-2987-2

Published by Kluwer Academic Publishers,
P.O. Box 17, 3300 AA Dordrecht, The Netherlands.

Kluwer Academic Publishers incorporates
the publishing programmes of
D. Reidel, Martinus Nijhoff, Dr W. Junk and MTP Press.

Sold and distributed in the U.S.A. and Canada
by Kluwer Academic Publishers,
101 Philip Drive, Norwell, MA 02061, U.S.A.

In all other countries, sold and distributed
by Kluwer Academic Publishers Group,
P.O. Box 322, 3300 AH Dordrecht, The Netherlands.

Printed on acid-free paper

Printed in the Netherlands

To
Fang, Yi-Xing and Yi-shan

Table of Contents

Preface

During the past twenty years, the technique of mathematical modeling has been extensively used in the study of groundwater resources management, seawater intrusion, aquifer remediation and other problems related to groundwater.

To build a model for a real groundwater system, two problems, the forward problem (simulation) and its inverse (calibration), must be solved. The former predicts unknown system states by solving appropriate governing equations, while the latter determines unknown physical parameters and other conditions of the system by fitting observed system states. We must first solve the inverse problem to find appropriate model structure and model parameters, and then solve the forward problem to obtain required prediction results.

Unfortunately, studies on these two problems are not in balance. The study of forward problems has developed rapidly. At this time, we can accurately simulate three-dimensional multicomponent transport in multiphase flow without any essential difficulty. On the other hand, the study of inverse problems is still limited to consideration of very simple models. In some case studies, we may see a situation where a very fine groundwater model was calibrated only by the primary trail-and-error method and insufficient observation data. If large errors are included in model structure, model parameters, sink/source terms and boundary conditions, the model can not be expected to produce accurate prediction results. Clearly, the model quality can not be improved by increasing only the accuracy of forward solutions if the inverse problem is not well solved. Like a person, the technique of groundwater modeling has two legs: the forward solution and the inverse solution. If one leg is long and another is short, how can the technique walk well?

The progress of inverse solution techniques is blocked by several inherent difficulties. First, the inverse problem is often ill-posed, i.e., its solution may be non-unique and unstable with respect to the observation error. Second, the quantity and quality of observation data are usually insufficient. Third, the model structure error, which is difficult to estimate, often dominates other errors. If these difficulties are not considered, satisfactory inverse solutions can never be obtained by changing only performance criteria and optimization algorithms.

In recent years, some significant developments in this field have been reported.

Methods coming from stochastics and control theory were successfully applied in the study of inverse problems. The concept of extended identifiabilities and the method of optimal data collection strategy may help us to find reliable models for given model applications. To make the "short leg" longer, however, more effort is needed and more difficulties must be overcome.

This is the first book on this subject. It systematically introduces basic concepts, theories, and methods, as well as recent developments in this field. The inverse problem is defined and solved in both deterministic and stochastic frameworks. Various direct and indirect methods are discussed and compared. As a useful tool, the adjoint state method and its applications are given in detail. Difficulties with the problem of ill-posedness in the inverse solution are highlighted throughout the book and the importance of data collection strategy is emphasized. The study of inverse problems requires knowledge of forward solution methods, as well as a background in mathematical programming, variational analysis and stochastics. For the reader's convenience, basic mathematical tools used in this book are either introduced in the text or given in Appendices. Exercises following each section are designed to help expand the reader's comprehension of the text.

Chapter 1 is an introduction to forward problems in groundwater modeling. Different types of models and their solution methods are reviewed. Chapter 2 is an introduction to inverse problems, in which the ill-posedness of inverse problems is explained using various examples related to groundwater modeling. From this chapter, the reader will obtain an overall understanding of the nature of inverse problems. Chapter 3 presents a general definition of inverse problems from the view-point of solving operator equations. The well-posedness of quasi-solutions and approximate solutions of an operator equation is discussed. If the reader is not interested in mathematical arguments, he or she can skip this Chapter. Various indirect methods of inverse solution are given in Chapter 4. Besides the general least squares criterion, other Lp-norm criteria and regularization methods are considered. Chapter 5 discusses direct methods of inverse solution. The quasi-linearization technique is presented as a special case of using mathematical programming methods. In Chapter 6, adjoint state equations are derived for flow problems, mass transport problems and general coupled problems in groundwater modeling. The adjoint state method is then used for parameter identification and sensitivity analysis. In Chapter 7, the inverse problem is defined and solved in the stochastic framework. The maximum likelihood estimate and other estimators are introduced. Kriging algorithms are derived for both stationary and unstationary stochastic fields, and co-kriging estimates are developed for both steady state and transient flow fields. In Chapter 8, different criteria of experimental design are presented for parameter identification, model prediction and decision making. After defining various extended identifiabilities, experimental designs based on these extended identifiabilities are discussed. In the last section of Chapter 8, criteria of experimental design for model structure identification are considered. A parameterization method that can directly incorporate geological structure into parameter identification of three- dimensional models is introduced.

There is a short conclusion at the end of the text, in which a step by step procedure of building mathematical models for real systems is summarized. This procedure is different from the conventional approach in that relationships between data collection, parameter identification, and model application are systematically considered. Additionally, some open problems in this area are presented that may help readers to develop original research projects.

A FORTRAN program developed by the author is given in Appendix C. By changing a few input parameters, it can solve either forward or inverse problems of coupled groundwater flow and mass transport in steady or transient flow fields. This program is able to deal not only with hypothetical but also practical problems.

This book is recommended as a text for graduate level courses or seminars. It should also provide a useful reference for hydrogeologists, geochemists, petroleum engineers, environmental engineers and applied mathematicians.

I wish to thank professor Jacob Bear, the editor of the book series, who suggested that I write a book on this topic and reviewed the outline. I wish to thank professor William W-G. Yeh for the support that he has provided for my work at the University of California, Los Angeles. Some research results quoted in this book were completed under his support. I wish to thank Dr. C. Wang, the senior engineer of Municipal Water District of Southern California, for his cooperation in the use of lately developed approaches to solve practical inverse problems of groundwater modeling. I wish to thank my friend, professor Julius Glater, who helped in the modification of the manuscript. The whole manuscript was typed by my son, Yi-shan Sun. He also drew figures and corrected all equations for the book. I thank Kluwer editors for their unfailing cooperation.

Forward Problems in Groundwater Modeling

1.1 Mathematical Models

What is a *mathematical model*? Why do we need mathematical models in planning and management of groundwater resources? How do we construct a model for groundwater systems and solve it with a computer? These problems have been expounded in detail by many authors over the past two decades (Domenico, 1972; Bear, 1972, 1979; Sun, 1981; Huyakorn and Pinder, 1983; Van der Heijde *et al.*, 1985; De Marsily, 1986; Kinzelbach, 1986; Bear and Veruijt, 1987; Willis and Yeh, 1987; Sun, 1989a, 1994a).

Intuitively, a mathematical model of a system is an equation or set of equations that can approximately simulate the excitation-response relation of the system. A mathematical model generally combines certain common physical and/or experimental rules along with particularities of the system.

1.1.1 EXAMPLES

Following are some examples that are often met within groundwater modeling.

Example 1.1.1 When the well loss is considered, a relationship between the stable drawdown S in a well and the pumping rate Q may be approximately represented by an algebraic equation

$$S = aQ + bQ^2, \tag{1.1.1}$$

where a and b are constant coefficients which depend on practical conditions of the aquifer and on the well structure. Equation (1.1.1) is an experimental model, in which pumping rate Q is the excitation and drawdown S the response. The particularities of the system are described by parameters a and b.

Example 1.1.2 The relationship between precipitation $p(t)$ and spring current $g(t)$ in a basin may be represented by a *convolution integral*

$$g(t) = \int_0^t K(t - \tau)p(\tau)\,\mathrm{d}\tau, \tag{1.1.2}$$

where *transfer function* $K(\cdot)$ is the system parameter which lumpily characterizes the particularities of the precipitation-spring system. In model (1.1.2), $p(t)$ is the excitation, $g(t)$ the response.

Example 1.1.3 The average concentration $C(t)$ of an aquifer is governed by an *Ordinary Differential Equation* (ODE)

$$\frac{\mathrm{d}(VC)}{\mathrm{d}t} = A(NC_N + RC_R - PC), \tag{1.1.3}$$

where V = volume of water in the aquifer (L^3);
A = surface area of the aquifer (L^3);
N = natural replenishment rate for unit area of aquifer (L/T);
R = artificial recharge rate for unit area of aquifer (L/T);
P = pumping rate for unit area of aquifer (L/T);
C_N = average concentration of natural replenishment water (M/L^3);
C_R = average concentration of artificial recharge water (M/L^3).

In order to determine a solution for Equation (1.1.3), an initial condition $C(0) = C_0$ must be given, where C_0 is a known constant. In this example, the model consists of an ODE and an initial condition. *State variable* (concentration) $C(t)$ is the response; *control variables* P, R and C_N can be seen as the excitation; V, A, N and C_R are parameters of the system.

Example 1.1.4 Two-dimensional groundwater flow in an isotropic and confined aquifer is governed by a partial differential equation (PDE)

$$S\frac{\partial \phi}{\partial t} = \nabla \cdot (T\nabla \phi) - Q, \quad (x, y) \in (\Omega), \quad t \geq 0, \tag{1.1.4}$$

subject to initial and boundary conditions

$$\phi(x, y, t)|_{t=0} = f_0(x, y), \tag{1.1.5a}$$

$$\phi(x, y, t)|_{(\Gamma_1)} = f_1(x, y, t), \tag{1.1.5b}$$

$$T\nabla \phi \cdot \mathbf{n}|_{(\Gamma_2)} = f_2(x, y, t), \tag{1.1.5c}$$

where
$\phi(x, y, t)$ = piezometric head (L);

$S(x, y)$ = storage coefficient (dimensionless);
$T(x, y)$ = transmissivity (L^2/T);
$Q(x, y, t)$ = sink/source term (L/T);
$\nabla = (\partial/\partial x, \partial/\partial y)$ the gradient operator vector;
(Ω) = flow region in horizontal plan (x, y) with boundary sections (Γ_1) and (Γ_2);
$\mathbf{n}$ = unit normal vector of (Γ_2);
f_1, f_2, f_3 = given functions;
x, y = horizontal coordinates (L);
t = time (T).

In this example, T and S are system parameters, Q is the control variable (excitation), ϕ the state variable (response).

Example 1.1.5 Consider a *leaky system* consisting of an *unconfined aquifer* and a *confined aquifer*. The two aquifers are separated by a semipervious layer. If the Dupuit's assumptions are used for the unconfined aquifer and storage in the semipervious layer is ignored, groundwater flow in the leaky system is governed by the following partial differential equations (PEDs):

$$S_y \frac{\partial h}{\partial t} = \nabla \cdot [K(h - b)\nabla h] - (h - \phi)/\sigma - Q_1, \tag{1.1.6a}$$

$$S \frac{\partial \phi}{\partial t} = \nabla \cdot [T\nabla \phi] + (h - \phi)/\sigma - Q_2, \tag{1.1.6b}$$

subject to initial and boundary conditions

$$h|_{t=0} = g_0, \qquad h|_{(\Gamma_1)} = g_1, \qquad -K(h - b)\nabla h \cdot \mathbf{n}|_{(\Gamma_2)} = g_2, \tag{1.1.7a}$$

$$\phi|_{t=0} = f_0, \qquad \phi|_{(\Gamma_3)} = f_1, \qquad -T\nabla \phi \cdot \mathbf{n}|_{(\Gamma_4)} = f_2, \tag{1.1.7b}$$

where
$h(x, y, t)$ = water level (L);
$S_y(x, y)$ = specific yield (dimensionless);
$K(x, y)$ = hydraulic conductivity (L/T);
$\sigma(x, y)$ = leakage coefficient (L/T);
Q_1, Q_2 = sink/source terms for unconfined and confined aquifers, respectively (L/T);
g_0, g_1, g_2 = given functions;
$b(x, y)$ = bedrock elevation (L).
Other symbols are the same as those in Example 1.1.4.

In this example, there are six system parameters $(S_y, K, b, S, T$ and $\sigma)$, two control variables $(Q_1$ and $Q_2)$, and two state variables $(h$ and $\phi)$.

Example 1.1.6 When density (ρ) and dynamic viscosity (μ) of groundwater are constant, two-dimensional *coupled groundwater flow-contaminant transport problem* is governed by PDEs

$$S\frac{\partial \phi}{\partial t} = \nabla \cdot (Km\nabla \phi) - Q, \tag{1.1.8a}$$

$$\frac{\partial (\theta C)}{\partial t} = \nabla \cdot (\theta \mathbf{D}\nabla C) - \nabla \cdot (\theta \mathbf{V}C) - M(C) \tag{1.1.8b}$$

subject to initial and boundary conditions

$$\phi|_{t=0} = f_0, \qquad \phi|_{(\Gamma_1)} = f_1, \qquad -Km\nabla \phi \cdot \mathbf{n}|_{(\Gamma_2)} = f_2, \tag{1.1.9a}$$

$$C|_{t=0} = g_0, \qquad C|_{(\Gamma_3)} = g_1, \qquad -(\theta \mathbf{D}\nabla C - \theta \mathbf{V}C) \cdot \mathbf{n}|_{(\Gamma_4)} = g_2, \tag{1.1.9b}$$

where
$C(x, y, t)$ = solute concentration (M/L^3);
$\mathbf{D}$ = hydrodynamic dispersion coefficient tensor (L^2/T);
$\mathbf{V}$ = average velocity vector (L/T);
θ = porosity (dimensionless);
$M(C)$ = sink/source terms (M/L^2T), the effects of ion exchange, chemical reaction, radioactive decay, adsorption, pumping and recharging of wastewater, etc. may be included in this term.

The boundary of the flow region (Ω) consists of either (Γ_1) and (Γ_2) or (Γ_3) and (Γ_4). Other symbols are the same as before. For an isotropic aquifer, components of the hydrodynamic dispersion coefficient tensor D can be expressed as

$$D_{xx} = \alpha_L \frac{V_x^2}{V} + \alpha_T \frac{V_y^2}{V} + D_d^*, \tag{1.1.10a}$$

$$D_{xy} = D_{yx} = (\alpha_L - \alpha_T)\frac{V_x V_y}{V}, \tag{1.1.10b}$$

$$D_{yy} = \alpha_T \frac{V_x^2}{V} + \alpha_L \frac{V_y^2}{V} + D_d^*, \tag{1.1.10c}$$

where
α_L = longitudinal dispersivity of the isotropic aquifer (L);
α_T = transverse dispersivity of the isotropic aquifer (L);
D_d^* = molecular diffusivity in porous media (L^2/T);
V = magnitude of the average velocity (L/T).

The distribution of average velocity is determined by Darcy's Law

$$\mathbf{V} = -\frac{K}{\theta}\nabla \phi. \tag{1.1.11}$$

The procedure of solving the above system includes the following steps: first, solve groundwater flow equation (1.1.8a) with (1.1.9a) to find the head distribution; second, use (1.1.11) to find the velocity distribution and use (1.1.10) to calculate the dispersion coefficients; and third, solve advection-dispersion-reaction equation (1.1.8b) with (1.1.9b) to obtain the concentration distribution.

Example 1.1.7 Two-dimensional steady flow in an isotropic and confined aquifer is governed by

$$\nabla \cdot (T\nabla\phi) = Q, \quad (x,y) \in (\Omega), \tag{1.1.12}$$

$$\phi|_{(\Gamma_1)} = f_1, \quad -T\nabla\phi \cdot \mathbf{n}|_{(\Gamma_2)} = f_2, \tag{1.1.13}$$

where the transmissivity $T(x,y)$ may be regarded as a *random function*. Freeze (1975) discovered that the logarithm of transmissivity ($\log T$) is often a normally distributed field. In addition, the sink/source term Q, functions f_1 and f_2 in the boundary conditions may also be random. As a result, Equation (1.1.12) becomes a *stochastic partial differential equation* (SPDE) and head $\phi(x,y,t)$ becomes a stochastic function.

All examples mentioned above can be found in Bear (1979), Huyakorn and Pinder (1983), De Marsily (1986), and Sun (1989b).

1.1.2 CATEGORIES OF MATHEMATICAL MODELS

Mathematical models mentioned above may be classified according to their mathematical forms as follows:

- *Deterministic models* and *stochastic models*, depending on whether random variables appear in the model;

- *Linear models* and *nonlinear models*, depending on the whether model equations are linear or nonlinear;

- *Stationary models* and *dynamic models*, depending on whether the time variable is included;

- *Lumped parameter models* and *distributed parameter models*, depending on whether the space variables are included.

In groundwater modeling, we prefer distributed parameter models, because this kind of model is more general, more accurate, and more suitable for the purpose of planning and management of groundwater resources. A distributed parameter model is described by a PDE or a set of PDEs. It may be deterministic or stochastic, linear or nonlinear, steady or unsteady. For instance, Example 1.1.5

is a deterministic, nonlinear and unsteady state model, while Example 1.1.7 is a stochastic, linear, and steady state model.

1.1.3 GENERAL FORM

Besides the models listed in Section 1.1.1, more complicated models have been considered by the hydrogeologist. Following are some examples: the models of three-dimensional groundwater flow in unsaturated-saturated zones; the models of *seawater intrusion* with or without the assumption of existing abrupt interfaces; the models of *chemical reaction* and *biodegradation* in groundwater; the coupled models of groundwater flow and *land subsidence, heat and mass transport* in *multiphase flow*, and etc. (Bear, 1972; Huyakorn and Pinder, 1983)

Generally, a distributed parameter model involves the following components:

· A space region and a time interval.

· One or more *system parameters* that characterize the geometry and/or physical nature of the system, such as the volume of the considered aquifer, porosity, hydraulic conductivity, and etc.

· One or more *subsidiary conditions* that define the initial state of the system and the relation to exchanging mass and energy with its neighboring systems.

· One or more *control variables* that represent the excitation to the system, such as pumping, artificial recharge, and etc. Sometimes control variables may appear in the subsidiary conditions, such as the controllable boundary inflow and boundary water level.

· One or more *state variables* that describe the state of the system, such as head, concentration, temperature and etc.

Henceforth, We will use vector functions

$$\mathbf{u} = (u_1, u_2, \ldots, u_n)^T, \tag{1.1.14a}$$
$$\mathbf{r} = (r_1, r_2, \ldots, r_m)^T, \tag{1.1.14b}$$
$$\mathbf{q} = (q_1, q_2, \ldots, q_k)^T, \tag{1.1.14c}$$

to denote state variables, system parameters and control variables, respectively, and G the time-space region. A variety of mathematical models then can be represented in a general form:

$$\mathbf{L}(\mathbf{u}; \mathbf{r}; \mathbf{q}; \mathbf{x}, t) = \mathbf{0}, \quad (\mathbf{x}, t) \in G, \tag{1.1.15}$$

with appropriate subsidiary conditions. In Equation (1.1.5),

$$\mathbf{L} = (L_1, L_2, \ldots, L_n)^T \tag{1.1.16}$$

is a *vector operator*. Its components may be algebraic equations, integral equations, ordinary differential equations, or partial differential equations.

Example 1.1.8 Considering the leaky flow model in Example 1.1.5 and assuming that the aquifer system is isotropic, then we have

state variables: $\quad u_1 = h, \quad u_2 = \phi;$

system parameters: $\quad r_1 = S_y, \quad r_2 = K, \quad r_3 = S, \quad r_4 = T, \quad r_5 = \sigma;$

control variables: $\quad q_1 = Q_1, \quad q_2 = Q_2, \quad q_3 = g_1, \quad q_4 = g_2;$

operators: $\quad L_1 = r_1 \dfrac{\partial u_1}{\partial t} - \nabla \cdot [r_2 (u_1 - b) \nabla u_1] + \dfrac{(u_1 - u_2)}{r_5} + q_1,$

$$L_2 = r_3 \dfrac{\partial u_2}{\partial t} - \nabla \cdot [r_4 \nabla u_2] - \dfrac{(u_1 - u_2)}{r_5} + q_2;$$

subsidiary conditions: $u_1|_{t=0} = g_0, \quad u_1|_{(\Gamma_1)} = q_3,$

$$-r_4 \nabla u_1 \cdot n|_{(\Gamma_2)} = q_4,$$

$$u_2|_{t=0} = f_0, \quad u_2|_{(\Gamma_1)} = f_1,$$

$$-r_2 (u_1 - b) \nabla u_1 \cdot \mathbf{n}|_{(\Gamma_2)} = f_2,$$

where water level q_3 on (Γ_1) and inflow q_4 on (Γ_2) of the unconfined aquifer are controllable.

Exercise 1.1.1 Write the one-dimensional model of mass transport in groundwater.

Exercise 1.1.2 Classify the models given in Example 1.1.1 to 1.1.7 according to the categories presented in section 1.1.2.

Exercise 1.1.3 Use the general form presented in section 1.1.3 to rewrite the coupled groundwater flow and mass transport model in Example 1.1.6.

Exercise 1.1.4 Write a mathematical model of seawater intrusion in a vertical plane with the abrupt interface assumption. Then, denote the state variables, system parameters and control variables.

1.2 The Solution of Forward Problems

Solving a *forward problem* (FP) for a system means to find the state of the system when the time-space region, system parameters, subsidiary conditions and control variables are known. In groundwater modeling, Equation (1.1.15) is often a PDE or a set of PDEs. The methods of solution may be divided into two kinds: *analytical solutions* and *numerical solutions*. We assume that the reader already has basic knowledge on both of them. Therefore, only a simple review of forward solutions is given in this section.

1.2.1 ANALYTICAL SOLUTIONS

An analytical solution, if available, is an accurate solution of the FP and can be expressed explicitly by *known functions*, or through some operations on them (infinite series or definite integrals). Superposition of fundamental solutions, separation of variables, Laplace transformation, Fourier transformation and other integral transformations are the often used techniques to obtain analytical solutions.

The main limitation of analytical solutions is that they are available only for very simple problems. For these methods, the shape of flow region must be regular and the parameters must be constant. Due to the rapid development and universal use of digital computers, analytical solutions are no longer used to any great extent in groundwater modeling. Presently, analytical solutions are mainly used to test the accuracy of numerical methods and simulate experiments when experimental conditions are relatively simple. In this book, several simple analytical solutions will be used for the purpose of explanation. Classical analytical solutions that have been obtained for groundwater flow and mass transport problems can be found in Li (1972) and Bear (1979). Following are several examples.

Example 1.2.1 *Steady flow to a well in a confined aquifer.* Consider the following model written in a polar coordinate system:

Governing equation: $\quad \dfrac{\partial^2 \phi}{\partial r^2} + \dfrac{1}{r}\dfrac{\partial \phi}{\partial r} = 0, \quad r_w \leq r \leq R,$ $\qquad$ (1.2.1)

Boundary conditions: $\quad 2\pi r_w T \dfrac{\partial \phi}{\partial r}\Big|_{r=r_w} = Q_w,$ $\qquad$ (1.2.2a)

$$\phi|_{r=R} = H,$$ $\qquad$ (1.2.2b)

where r_w is the radii of the well, Q_w the pumping rate, R a given distance, H a constant head. Integrating Equation (1.2.1) two times and using conditions (1.2.2a) and (1.2.2b), the solution of the model can be obtained as follows (Bear, 1979):

$$\phi(r) = H - \frac{Q_w}{2\pi T} \ln \frac{R}{r}, \quad r_w \le r \le R. \tag{1.2.3}$$

Example 1.2.2 *Unsteady flow to a well in a confined aquifer.* Its mathematical model may consist of

governing equation $\quad S\dfrac{\partial \phi}{\partial t} = T \left[\dfrac{\partial^2 \phi}{\partial r^2} + \dfrac{1}{r}\dfrac{\partial \phi}{\partial r} \right], \quad 0 < r < \infty, \ t \ge 0; \quad$ (1.2.4)

initial condition $\quad\quad \phi(r,t)\,|_{t=0} = \phi_0, \quad r > 0;$ (1.2.5a)

boundary conditions $\quad \lim\limits_{r \to \infty} \phi(r,t) = \phi_0, \quad t > 0;$ (1.2.5b)

$$\lim\limits_{r \to 0} 2\pi r T \frac{\partial \phi}{\partial r} = Q_w, \tag{1.2.5c}$$

where ϕ_0 is a given head. The solution of this model is known as the *Theis formula* (Bear, 1979):

$$\phi(r,t) = \phi_0 - \frac{Q_w}{4\pi T} W \left(\frac{Sr^2}{4Tt} \right), \tag{1.2.6}$$

where

$$W(u) = \int_u^\infty \frac{e^{-x}}{x}\, dx = -0.577 - \ln u + u - \frac{u^2}{2 \cdot 2!} + \frac{u^3}{3 \cdot 3!} - \frac{u^4}{4 \cdot 4!} + \cdots \tag{1.2.7}$$

is called the *well function* or *exponential integral*.

Example 1.2.3 *One-dimensional advection-dispersion-reaction in a semi-infinite column.* This problem may be described by the following mathematical model:

governing equation $\quad \dfrac{\partial C}{\partial t} = D\dfrac{\partial^2 C}{\partial x^2} - V\dfrac{\partial C}{\partial x} - \lambda C, \quad x \ge 0, \ t \ge 0; \quad$ (1.2.8)

initial condition $\quad\quad C(x,t)\,|_{t=0} = 0, \quad x > 0;$ (1.2.9a) *boundary conditions* $C\left(x,t\right)|_{x=0} = C_0, \quad t \geq 0;$ (1.2.9b)

$$C\left(x,t\right)|_{x=\infty} = 0, \quad t \geq 0,$$ (1.2.9c)

where D is the dispersion coefficient, V the velocity, λ the reaction (or radioactive decay) coefficient, C_0 is a given boundary concentration. Using Laplace transformation, the following solution of the model is obtained (Bear, 1979)

$$C\left(x,t\right) = \frac{C_0}{2} \exp\left(\frac{Vx}{2D}\right) \times$$

$$\times \left[\exp\left(-\frac{x}{2D}\sqrt{V^2 + 4\lambda D}\right) \mathrm{erfc}\left(\frac{x - \sqrt{V^2 + 4\lambda D}\,t}{2\sqrt{DT}}\right)\right.$$

$$\left. + \exp\left(\frac{x}{2D}\sqrt{V^2 + 4\lambda D}\right) \mathrm{erfc}\left(\frac{x + \sqrt{V^2 + 4\lambda D}\,t}{2\sqrt{Dt}}\right)\right].$$

(1.2.10)

Example 1.2.4   *An instantaneous injection at the origin into a uniform steady flow in an infinite plan.* Its mathematical model consists of:

governing equation $\dfrac{\partial C}{\partial t} = D_L \dfrac{\partial^2 C}{\partial x^2} + D_T \dfrac{\partial^2 C}{\partial y^2} - V \dfrac{\partial C}{\partial x}$

$$+ \frac{M}{\theta}\delta\left(x - 0, y - 0, t - 0\right),$$ (1.2.11)

initial condition $C\left(x,y,t\right)|_{t=0} = 0, \quad (x,y) \neq (0,0),$ (1.2.12a)

boundary conditions $C\left(x,y,t\right)|_{x=\pm\infty} = 0, \quad t \geq 0,$ (1.2.12b)

$$C\left(x,y,t\right)|_{y=\pm\infty} = 0, \quad t \geq 0,$$ (1.2.12c)

where D_L, D_T are longitudinal and transverse dispersion coefficients, respectively; V the velocity; θ the porosity; M the mass of the slug injected at $x = 0$, $y = 0$, $t = 0$; $\delta(\cdot,\cdot,\cdot)$ the Dirac δ-function. Through the superposition of fundamental solutions, the following solution can be found (Bear, 1979)

$$C\left(x,y,t\right) = \frac{M/\theta}{4\pi t\sqrt{D_L D_T}} \exp\left[-\frac{(x - Vt)^2}{4D_L t} - \frac{y^2}{4D_T t}\right].$$ (1.2.13)

1.2.2 Numerical Solutions

Using a numerical method to solve the FP is called *numerical simulation*. Although numerical methods can only generate approximate solutions to the FP, they provide the most powerful and general tool for solving the problems encountered in practice. In a numerical simulation, the time-space region is divided into elements (subdomains) with each element associated to one or several nodes. The governing equation (PDE or PDEs) is then discretized and replaced at these nodes by a system of algebraic equations. The solution of the algebraic system provides nodal values of unknown states. Several different approaches may be used to complete this procedure. Methods of discretizing time-space region and methods of deriving discretization equations may be different in different approaches. Finite Difference Methods (FDM), Finite Element Methods (FEM), Boundary Element Methods (BEM), and their variants and hybrids, including the hybrids of numerical and analytical solutions, have been extensively applied in groundwater modeling during the past two decades (Remson, 1971; Pinder and Gray, 1977; Sun, 1981; Huyokorn and Pinder, 1983; Bear and Verruijt, 1987; Kinzelbach, 1986, Sun, 1989b)

Although difficulties still exist in solving some special problems (for example, problems with sharp fronts), numerical simulations have been very successful in the field of groundwater modeling.

Numerical methods of solving transient groundwater flow and/or contaminant transport problems involve the following three common steps:

Step 1. Through the discretization of space variables and use of boundary conditions, to transform the PDE into a system of ODEs

$$\mathbf{A}\mathbf{u} + \mathbf{B}\frac{d\mathbf{u}}{dt} = \mathbf{f}, \tag{1.2.14}$$

where coefficient matrices $\mathbf{A}$, $\mathbf{B}$ and the right-hand side vector $\mathbf{f}$ depend on discretized system parameters, system state (for non-linear problems), control variables, and boundary conditions. They also depend on the numerical method to be used.

Step 2. Through the discretization of time variable, to transform the system of ODEs (1.2.14) into a system of algebraic equations

$$\mathbf{K}\mathbf{u}_{t+\Delta t} = \mathbf{e}_t, \tag{1.2.15}$$

where matrix $\mathbf{K}$ is determined by matrices $\mathbf{A}$ and $\mathbf{B}$. The right-hand side vector $\mathbf{e}$ depends on $\mathbf{B}$, $\mathbf{f}$, and the state at time t.

Step 3. To solve equations (1.2.15) for each time step. The initial condition is used to obtain a solution for the first time step.

For readers of this book, it is only require to know the basic FDM or FEM. Some programs (written in BASIC) for modeling groundwater flow and contaminant transport can be found in Bear and Verruijt (1987.) In Appendix C of this book, a subroutine MCB2D (written in FORTRAN) is given. This program can simulate two-dimensional coupled groundwater flow and mass trasport in confined and/or unconfined aquifers. It is based on the *Multiple Cell Balance* method (MCB) developed by Sun and Yeh (1983). This subroutine is ready for incorporation into a program of inverse solution.

1.2.3   WELL-POSEDNESS OF THE FORWARD PROBLEM

A *well-posed mathematical-physical problem*, according to Hardmard, must satisfy the following requirements:

(1) *Existence.* There exists a function which satisfies governing equations and subsidiary conditions.

(2) *Uniqueness.* There is no another solution which is different from **u**.

(3) *Stability.* The variation of solution **u** can be arbitrarily small, provided the variations of input data (e.g. physical parameters, control variables, initial and boundary conditions) are sufficiently small.

If any one of these requirements is not satisfied, the problem is ill-posed.

Forward problems (FP) in groundwater modeling are naturally well-posed. This fact has been strictly proved in mathematics. As a result, all examples mentioned in Section 1.1 are well-posed.

When an FP solution is obtained by a numerical method, the computer only gives nodal values of the solution and which may vary with grid size and time-step size. Thus, we may have numerous approximate solutions for an FP. This fact, however, is not in contradiction with the well-posedness of the FP, because numerical schemes often used in groundwater modeling have been proved to be convergent and stable (Lapidus and Pinder, 1982). When grid size and time-step size tend to zero, all approximate solutions of the FP must tend to the true solution of the problem which is unique.

For the general model (1.1.15), the solution of FP may be represented as

$$u = M\left(r; q; f; x, t\right),\tag{1.2.16}$$

where vector function **f** denotes that solution **u** depends on initial and boundary conditions, **M** can be thought as a subroutine for solving the FP either by an analytical method or by a numerical method with input **r;q;f** and output **u** (Figure 1.2.1).

We will use vector function **p** to represent not only physical parameters but also the parameters defining control variables and subsidiary conditions, i.e.,

$$p = (r; q; f),\tag{1.2.17}$$

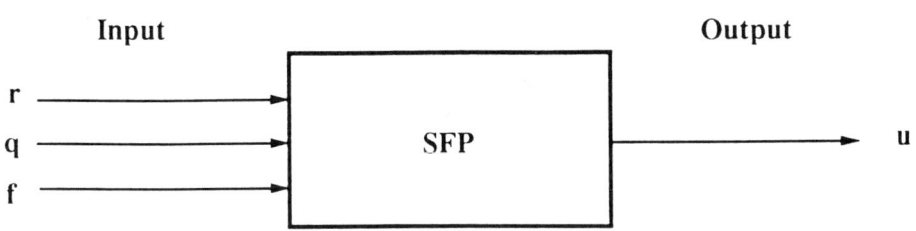

Fig. 1.2.1. The solution of the forward problem.

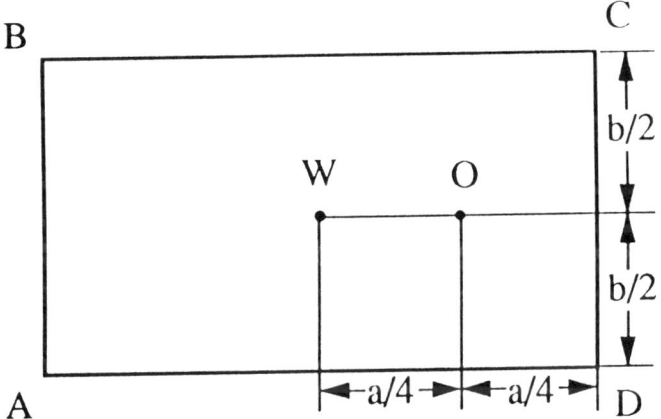

Fig. 1.2.2. Aquifer configuration for Exercise 1.2.2.

where **p** is called the *vector of model parameters*, or simply, the *model parameter*. Thus, (1.2.16) can be rewritten in a compact form

$$\mathbf{u} = \mathbf{M}(\mathbf{p}). \tag{1.2.18}$$

Exercise 1.2.1   Using Formula (1.2.3) to write a program with input parameters R, H, Q_w, r_w and output $\phi(r_w)$.

Exercise 1.2.2   Figure 1.2.2 shows a rectangular aquifer, where AB and CD are given head ($\phi = \phi_0$) boundaries and BC and DA are no flow boundaries.

A pumping well is located at the center of the aquifer. At $t = 0$, head $\phi = \phi_0$ everywhere. Storage coefficient S and transmissivity T are assumed to be constants.

(1) Write a program based on the method of image wells and Theis formula (1.2.6) to solve this problem.

(2) Change the values of physical parameter to observe the change of drawdown at the observation well (point O in Figure 1.2.2).

(3) Change the value of pumping rate Q_w to observe the change of drawdown at the observation well (point O in Figure 1.2.2).

(4) Change the values of geometric parameters a and b in Figure 1.2.2, to observe the change of drawdown at the observation well.

Exercise 1.2.3   Prepare a program based on a numerical method that can simulate two-dimensional groundwater flow in a confined aquifer (or learn how to use the program given in Appendix C for solving this problem).

Exercise 1.2.4   Solve the same problems presented in Exercise 1.2.2 with the prepared numerical model in Exercise 1.2.3 and compare their results.

An Introduction to Inverse Problems

2.1 Basic Concepts

2.1.1 MODEL STRUCTURE AND MODEL PARAMETERS

Once a simulation model, as shown in Figure 1.2.1, is built for a groundwater system, the *forecast problem* can be solved, i.e., we can forecast the response (states **u**) of the system for different excitation (Control variables **q**). As a result, different management decisions can be compared and an optimal decision can be selected based on certain criteria. Thus, the *management problem* can be solved by incorporating a simulation model with an optimization program (Bear, 1979).

The prerequisite of applying this procedure to a real problem is that the input-output relation of the simulation model must be consistent with the excitation-response relation of the original system. An accurate simulation model should satisfy the following requirements:

· The *model structure* given by Equation (1.1.15) is an exact description of the system.

· The *model parameters* $\mathbf{p} = (\mathbf{r}, \mathbf{q}, \mathbf{f})$ are exactly known.

In practice, however, it is very difficult to construct an accurate simulation model for a groundwater system. First, the governing equations of the constructed model may not be a suitable description of the original system. Second, the particularities of the system, such as geometry of the flow region, hydrogeological parameters, inflow or outflow boundary conditions, sink and source terms, are difficult to measure accurately in the field.

Fortunately, state variables of a groundwater system can be easily measured. These measurements, such as water level, solute concentration and temperature, are usually obtained from observation wells. Observation data can be collected either from historical records or from field experiments (pumping or tracer tests).

15

Since the input-output relation of a correct simulation model must fit any observed excitation-response relation, it is possible to indirectly determine the particularities of the system by using these observation data.

2.1.2 MODEL CALIBRATION AND INVERSE PROBLEM

Model calibration involves adjustment of the model structure and model parameters of a simulation model simultaneously or sequentially so as to make the input-output relation of the model fit any observed excitation-response relation of the real system.

Example 2.1.1 Consider one-dimensional flow in a saturated porous medium, the relationship between hydraulic gradient J and flux q is usually described by a linear model (*Darcy's Law*):

$$q = KJ, \tag{2.1.1}$$

where model parameter K is the hydraulic conductivity. When K is known, model (2.1.1) can forecast flux q for any J. When K is unknown, we can measure the value of flux q many times for different values of J. Assume that we obtain $q_1, q_2, \ldots, q_n$ for $J_1, J_2, \ldots, J_n$. Let

$$K_i = \frac{q_i}{J_i}, \quad i = 1, 2, \ldots, n. \tag{2.1.2}$$

If above ratio is close to a constant, then we believe that a linear model (2.1.1) is correct and the model parameter K can be estimated, e.g., let K be the mean value of all K_i.

Example 2.1.2 When K_i, $i = 1, 2, \ldots, n$, in (2.1.2) are not close to a constant, the model structure itself needs to be calibrated first. We can try, for example, to use the following nonlinear model to fit the observations:

$$J = aq + bq^2, \tag{2.1.3}$$

where model parameters a and b are determined by the *least squares method*. With the objective of minimizing

$$I = \sum_{i=1}^{n} (J_i - aq_i - bq_i^2)^2, \tag{2.1.4}$$

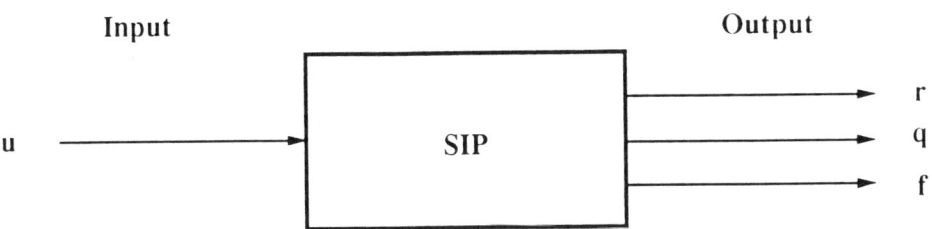

Fig. 2.1.1. The solution of the inverse problem.

constants a and b can be obtained by solving the following equations:

$$\frac{\partial I}{\partial a} = -2 \sum_{i=1}^{n} \left(J_i - aq_i - bq_i^2 \right) q_i = 0, \qquad (2.1.5)$$

$$\frac{\partial I}{\partial b} = -2 \sum_{i=1}^{n} \left(J_i - aq_i - bq_i^2 \right) q_i^2 = 0. \qquad (2.1.6)$$

If model (2.1.3) still does not fit the observations, other nonlinear models may be used. In the example above, both model structure and model parameters are calibrated by observed data.

If the model structure is determined, the problem of only determining some model parameters from the observed system states and other available information is called *parameter identification* (Example 2.1.1). In a certain sense, parameter identification is an inverse of the forward problem (FP). System state **u**, the output of FP, now becomes the input, while model parameter **p** from the input becomes the output (Figure 2.1.1). Therefore, parameter identification problems are often called *inverse problems* (IP).

From the general form of FP solution, (1.2.16), the following inverse problems may be presented:

Type 1 – Identify physical parameters **r**.

Type 2 – Identify sink and/or source terms **q**.

Type 3 – Identify initial and/or boundary conditions **f**.

Type 4 – Identify more than one of **r, q, f** simultaneously.

These cases are shown in Figure 2.1.2.

The importance of solving inverse problems is obvious. Although we can use an analytical or a numerical method to solve a forward problem with high levels of accuracy, if physical parameters, initial and boundary conditions used in the model

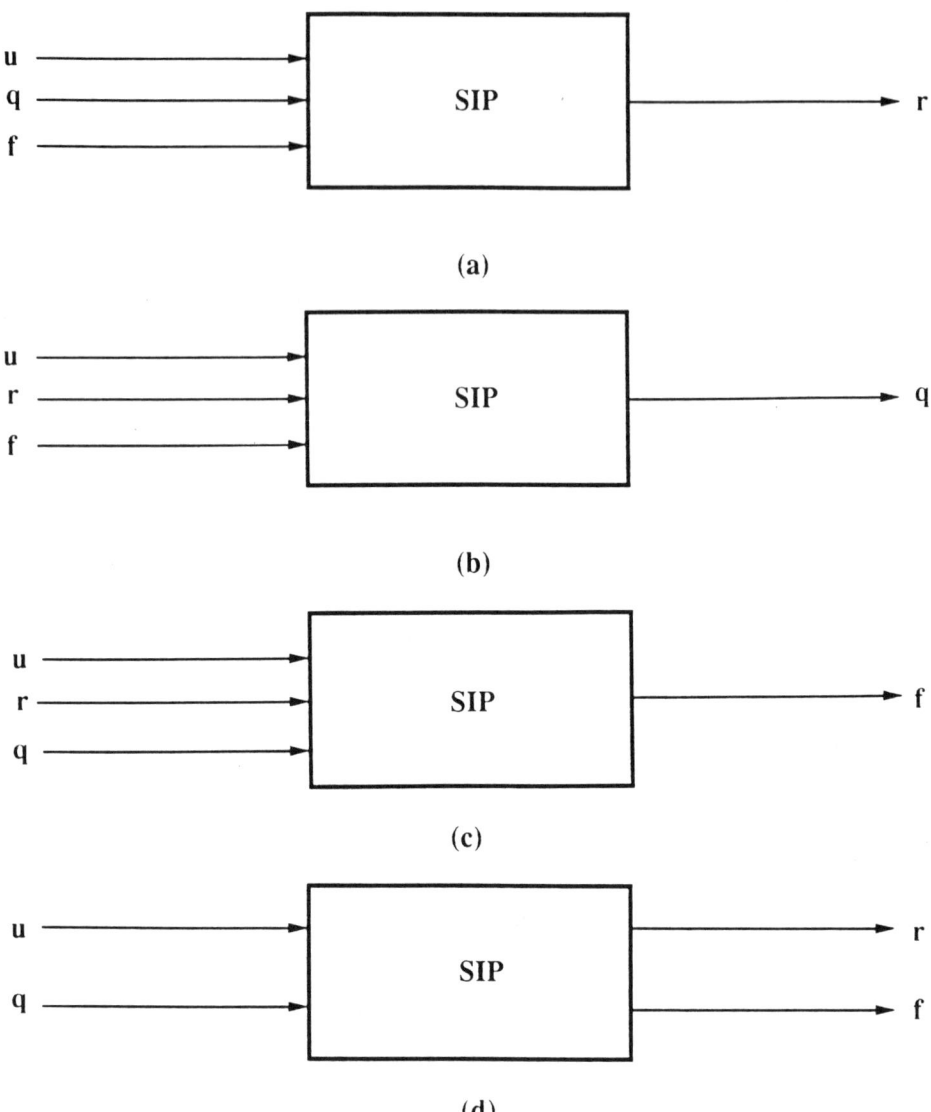

Fig. 2.1.2. Types of inverse problems. (a) Type 1 (b) Type 2 (c) Type 3 (d) Type 4.

are not correct, results obtained by this model are unreliable. When a simulation model has not been carefully calibrated by field observations, it should not be used to the purpose of groundwater management.

Therefore, solution of inverse problems is indeed a key in groundwater modeling. Unfortunately, as discussed in the following sections, there are several essential difficulties related to inverse problems.

Exercise 2.1.1.
(1) Explain the relationship between forecast and management problems.
(2) Explain the relationship between model calibration and parameter identification.
(3) Explain the relationship between forward and inverse problems.

Exercise 2.1.2. Assume that the following observation data were obtained in an experiment:

J	0.001	0.002	0.003	0.004	0.01	0.02	0.03	0.04
q	0.019	0.038	0.056	0.076	0.211	0.384	0.572	0.762

(1) Use linear model (2.1.1) to fit these data to determine K.
(2) Use nonlinear model (2.1.3) to fit these data to determine a and b.

Exercise 2.1.3. What types of inverse problems can be presented for Example 1.1.6?

2.2   Ill-Posedness of the Inverse Problem

2.2.1   EXISTENCE

Intuitively, existence of an inverse solution appears to be no problem at all, since the physical reality must be a solution. In practice, however, the observation error (or noise) of state variables cannot be avoided. As a result, an accurate solution of an inverse problem may not exist.

Example 2.2.1   In Example 1.2.2, if T and S are known, then drawdown $s(r,t)$ can be calculated for any r and t by

$$s\left(r,t\right) = \frac{Q_w}{4\pi T} W\left(\frac{Sr^2}{4Tt}\right). \qquad (2.2.1)$$

Therefore, the FP is well-posed. Now, assume that there are n noisy observations of drawdown:

$$\tilde{s}(r_0, t_i) = s(r_0, t_i) + \varepsilon_i, \quad (i = 1, 2, \ldots, n), \tag{2.2.2}$$

where r_0 is the distance between an observation well and the pumping well; $t_1, t_2, \ldots, t_n$ are observation times; $\varepsilon_1, \varepsilon_2, \ldots, \varepsilon_n$ are observation noises.

Obviously, there are no pairs of T and S that can exactly satisfy the following equations:

$$\frac{Q_w}{4\pi T} W\left(\frac{Sr_0^2}{4Tt_i}\right) = \tilde{s}(r_0, t_i), (i = 1, 2, \ldots, n). \tag{2.2.3}$$

From the example given above we see that an accurate inverse solution may not exist when noisy observation data are used. However, existence is not a major difficulty in the solution of inverse problems, because the physical reality at least is an approximate solution. In this example, if the noise is subject to a normal distribution, then parameters S and T can be determined by the least squares method and defined as the solution of Equations (2.2.3).

2.2.2 UNIQUENESS

Different combinations of hydrogeologic conditions may lead to similar observations of water level and solute concentration. It is thus impossible to uniquely determine the particularities of an aquifer by only observing the state variables, i.e., the *non-uniqueness* of inverse solutions is often observed.

Example 2.2.2 Considering a distributed parameter model of one-dimensional steady flow in a confined aquifer governed by

$$\frac{\mathrm{d}}{\mathrm{d}x}\left[T(x)\frac{\mathrm{d}\phi}{\mathrm{d}x}\right] = 0, \quad x_1 \le x \le x_2; \tag{2.2.4}$$

$$\phi(x_1) = \phi_1; \tag{2.2.5a}$$
$$\phi(x_2) = \phi_2. \tag{2.2.5b}$$

Assume that the head distribution $\phi(x)$ is known, consequently, $\phi'(x)$ is also known. Now, we want to find $T(x)$. The integration of (2.2.4) yields

$$T(x) = \frac{C}{\phi'(x)} . \tag{2.2.6}$$

Due to an arbitrary constant C included in (2.2.6), $T(x)$ is not unique, although the head is observed at each point x of the interval $[x_1, x_2]$. Therefore, this kind of non-uniqueness can not be removed by increasing the number of observation wells.

It is interesting to note that if anyone of the boundary conditions in (2.2.5) is replaced by a given flux boundary condition, for example,

$$T\left(\frac{d\phi}{dx}\right)\Big|_{x_1} = -q, \qquad (2.2.7)$$

then $C = -q$ can be obtained from (2.2.6). Thus,

$$T(x) = -\frac{q}{\phi'(x)} \qquad (2.2.8)$$

is uniquely determined.

Another way to obtain a unique $T(x)$ is to supplement a measurement of $T(x)$ at any point x_0 ($x_1 \le x_0 \le x_2$), say $T(x_0)$, then

$$T(x) = T(x_0)\frac{\phi'(x_0)}{\phi'(x)}. \qquad (2.2.9)$$

In this case, the non-uniqueness is removed by the supplementary information.

Example 2.2.3   Again, considering one-dimensional steady flow governed by (2.2.4), but assuming that transmissivity $T(x)$ is a piecewise constant:

$$T(x) = \begin{cases} T_1, & \text{when } x_1 \le x < a; \\ T_2, & \text{when } a \le x \le x_2, \end{cases} \qquad (2.2.10)$$

and boundary condition (2.2.5a) is replaced by

$$T_1\frac{d\phi}{dx}\Big|_{x_1} = -q, \quad (q > 0). \qquad (2.2.11)$$

We want to find out how many observation wells are enough to uniquely determine T_1 and T_2 (Figure 2.2.1.).

First, suppose that there is only one observation well located at point c_1 and $a \le c_1 \le x_2$, then we have

$$T_2 = q\frac{x_2 - c_1}{\phi(c_1) - \phi_2}. \qquad (2.2.12)$$

But T_1 is undetermined and the inverse solution is not unique. This kind of non-uniqueness, however, can be removed by increasing the number of observation wells. In fact, if there is another observation well located at c_2, and $x_1 \le c_2 < a$, then we have

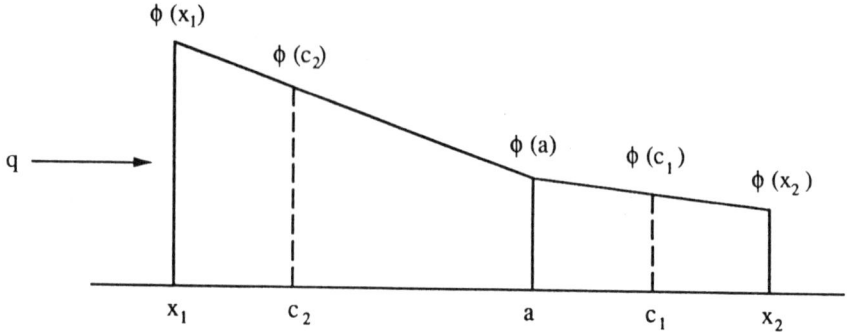

Fig. 2.2.1. The identification of $T(x)$, when it is a piecewise constant.

$$T_1 = q \frac{a - c_2}{\phi(c_2) - \phi(a)},$$

(2.2.13)

where

$$\phi(a) = \phi_2 + \frac{q(x_2 - a)}{T_2}.$$

(2.2.14)

Example 2.2.4 Two-dimensional steady flow in a confined aquifer is governed by

$$\frac{\partial}{\partial x}\left(T\frac{\partial \phi}{\partial x}\right) + \frac{\partial}{\partial y}\left(T\frac{\partial \phi}{\partial y}\right) = Q.$$

(2.2.15)

The associated inverse problem is to find $T(x, y)$ when $\phi(x, y)$ is known, where $(x, y) \in (\Omega)$. Equation (2.2.15) can be rewritten as

$$a\frac{\partial T}{\partial x} + b\frac{\partial T}{\partial y} + cT - Q = 0,$$

(2.2.16)

where

$$a = \frac{\partial \phi}{\partial x}, \qquad b = \frac{\partial \phi}{\partial y}, \qquad c = \frac{\partial^2 \phi}{\partial x^2} + \frac{\partial^2 \phi}{\partial y^2}.$$

(2.2.17)

Since $\phi(x, y)$ is known, a, b and c are all known functions.

Equation (2.2.16) is a first order PDE with respect to transmissivity $T(x,y)$. Its general solution involves an arbitrary function. For instance, if $T_1(x, y)$ is a solution of the equation, and $T_2(x, y)$ is a solution of the following equation:

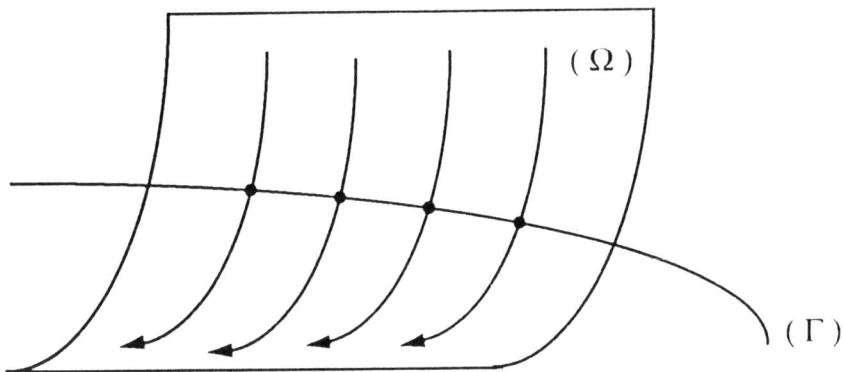

Fig. 2.2.2. Determine $T(x, y)$ by Cauchy data.

$$a\frac{\partial T}{\partial x} + b\frac{\partial T}{\partial y} + cT = 0, \tag{2.2.18}$$

then $T_1 + T_2$ is also a solution of (2.2.16). Therefore, the solution of the inverse problem is non-unique, except that additional information is given.

According to the basic theory of first order PDEs, assume that (Γ) is a curve in (Ω) and the values of $T(x, y)$ are known on it (these values are named *Cauchy data*), then $T(x, y)$ can be determined along all streamlines passing through (Γ) as shown in Figure 2.2.2.

This fact is easy to prove. On the one hand, *equipotentials* $\phi(x, y) = $ constant of Equation (2.2.15) satisfy

$$\frac{\partial \phi}{\partial x}\, dx + \frac{\partial \phi}{\partial y}\, dy = 0, \tag{2.2.19}$$

and *streamlines* satisfy

$$\frac{\partial \phi}{\partial y}\, dx - \frac{\partial \phi}{\partial x}\, dy = 0 \quad \text{or} \quad \frac{dx}{a} = \frac{dy}{b}. \tag{2.2.20}$$

On the other hand, the *characteristics* of Equation (2.2.16) satisfy

$$\frac{dx}{a} = \frac{dy}{b} = \frac{dT}{(Q - cT)}. \tag{2.2.21}$$

i.e., streamlines are just the projects of characteristics on the (x, y)-plane. If $T(x_0, y_0)$ is known at any point (x_0, y_0), then $T(x, y)$ can be obtained along a streamline passing through the point by solving Equation (2.2.21).

Therefore, if all streamlines passing through curve (Γ) fill the whole domain (Ω) and the Cauchy data are known on (Γ), then the inverse solution $T(x, y)$ can be uniquely determined in (Ω).

Cauchy data may be obtained from the boundary condition. If (Γ) is a given flux boundary, i.e. $T(\partial\phi/\partial n)|_{(\Gamma)} = q$, then $T = q/(\partial\phi/\partial n)|_{(\Gamma)}$ can serve as Cauchy data. However, if (Γ) is only a part boundary of (Ω), the Cauchy data may not be sufficient.

Suppose that there is a pumping well in (Ω) with known pumping rate Q, and (Γ) is a curve around the well, then we have

$$\int_{(\Gamma)} T \left(\frac{\partial\phi}{\partial x} \, dy - \frac{\partial\phi}{\partial y} \, dx \right) = q. \tag{2.2.22}$$

Generally, the values of $T(x, y)$ on (Γ) can not be determined from (2.2.22). However, if we supplement an assumption that $T(x, y)$ is a constant on (Γ), then we will have Cauchy data

$$T = q \left/ \int_{(\Gamma)} \left(\frac{\partial\phi}{\partial x} \, dy - \frac{\partial\phi}{\partial y} \, dx \right) \right. . \tag{2.2.23}$$

If all streamlines pass through (Γ), the inverse solution $T(x, y)$ can be determined uniquely in (Ω).

Summarizing, the above examples show that the inverse solution in groundwater modeling is often non-unique. However, the non-uniqueness may be removed in some cases by supplementary information, e.g., the Cauchy data in Examples 2.2.2 and 2.2.4, or appropriate limitations on unknown parameters, e.g., the assumption of piecewise constants in Example 2.2.3.

2.2.3 STABILITY

The requirement of *stability* is also very important for a well-posed problem. A solution, although it is existent and unique, can not be accepted, if it does not continuously depend upon the input data. It is well known that forward solutions in groundwater modeling are always stable. This fact has been seriously proved in mathematics and is easy to be understood from physics. For example, when hydraulic parameters and/or boundary conditions change slightly, the water level should be slightly affected. Unfortunately, inverse solutions in groundwater modeling are often unstable.

Example 2.2.5 Again, consider the one-dimensional steady flow model

$$\frac{d}{dx} \left[T(x) \frac{\partial\phi}{\partial x} \right] = 0, \quad x_1 \le x \le x_2, \tag{2.2.24}$$

$$\phi(x_1) = \phi_1, \tag{2.2.25a}$$

$$T\frac{\mathrm{d}\phi}{\mathrm{d}x}\bigg|_{x_2} = -q, \quad (q > 0). \tag{2.2.25b}$$

The forward problem is defined as finding $\phi(x)$ when $T(x)$ is known. The forward solution can be obtained directly by the integration of (2.2.24):

$$\phi(x) = \phi_1 - q \int_{x_1}^{x_2} \frac{\mathrm{d}x}{T(x)}. \tag{2.2.26}$$

Suppose that $T(x)$ is replaced by $T^*(x) = T(x) + \varepsilon(x)$, where $\varepsilon(x)$ is an error, then the solution changes to

$$\phi^*(x) = \phi_1 - q \int_{x_1}^{x_2} \frac{\mathrm{d}x}{T^*(x)}. \tag{2.2.27}$$

Assume that $T(x) > 0$ and $\varepsilon(x)$ is so small that

$$T(x) - |\varepsilon(x)| \geq \lambda > 0, \tag{2.2.28}$$

then we have

$$|\phi(x) - \phi^*(x)| = q \left| \int_{x_1}^{x_2} \frac{\varepsilon(x)}{T(x)\,[T(x) + \varepsilon(x)]}\,\mathrm{d}x \right| \leq M \max |\varepsilon(x)|, \tag{2.2.29}$$

where

$$M = \frac{q}{\lambda} \int_{x_1}^{x_2} \frac{\mathrm{d}x}{T(x)}. \tag{2.2.30}$$

Equation (2.2.29) shows that the forward problem is stable.

Now, turn to the inverse problem, i.e., to find $T(x)$ when $\phi(x)$ is known. Because a Cauchy data is already given by boundary condition (2.2.25b), the inverse problem has a unique solution

$$T(x) = -\frac{q}{\partial \phi/\partial x}. \tag{2.2.31}$$

In fact, the true distribution of $\phi(x)$ can never be known. We can only have a noisy observation head $\phi^*(x) = \phi(x) + \eta(x)$, where $\eta(x)$ is an unknown observation error, and the cooresponding inverse solution must be

$$T^*(x) = -q \bigg/ \left(\frac{\partial \phi}{\partial x} + \frac{\partial \eta}{\partial x}\right). \tag{2.2.32}$$

From (2.2.31) and (2.2.32), we can obtain the following absolute error estimation:

$$|T(x) - T^*(x)| = \frac{q|\partial\eta/\partial x|}{|\partial\phi/\partial x \, (\partial\phi/\partial x + \partial\eta/\partial x)\,|}.$$ (2.2.33)

It is well known that although $|\eta(x)|$ is very small, $|\partial\eta/\partial x|$ may be very large. Thus, solution (2.2.31) is unstable with respect to the observation error.

In practice, the number of observation wells for measuring hydraulic head is usually limited. Distributions of head and its derivatives can be obtained only by interpolation processes. As a result, observation and computation errors can never be avoided and that often cause the inverse solution to be unstable.

2.2.4 THE ILL-POSEDNESS OF NUMERICAL INVERSE SOLUTIONS

When a mathematical model described by a PDE or a set of PDEs is discretized by a numerical method, such as the FDM, FEM or other alternative methods, it reduces to a set of algebraic equations. The state variables and parameters all reduce to finite-dimensional vectors. Each component is associated with a node (or an element) in the time-space domain.

Let us consider the two-dimensional transient flow given in Example 1.1.4. If a square grid is used to discretize Equation (1.1.4), the *implicit finite difference equation* for an inner block (i,j) at the kth time step can be written as

$$\frac{1}{2}(T_{i-1,j} + T_{i,j})\left(\phi_{i-1,j}^k - \phi_{i,j}^k\right) + \frac{1}{2}(T_{i+1,j} + T_{i,j})\left(\phi_{i+1,j}^k - \phi_{i,j}^k\right)$$

$$+\frac{1}{2}(T_{i,j-1} + T_{i,j})\left(\phi_{i,j-1}^k - \phi_{i,j}^k\right) + \frac{1}{2}(T_{i,j+1} + T_{i,j})\left(\phi_{i,j+1}^k - \phi_{i,j}^k\right)$$

$$= S_{i,j}\phi_{i,j}^k - \frac{\phi_{i,j}^{k-1}}{\Delta t}(\Delta x)^2 + Q_{i,j}(\Delta x)^2.$$ (2.2.34)

In the first term on the left-hand side, $\frac{1}{2}(T_{i-1,j} + T_{i,j})$ represents the transmissivity between blocks $(i-1,j)$ and (i,j). The terms $\phi_{i-1,j}^k$ and $\phi_{i,j}^k$ are heads of the two blocks at time $t_k = k\Delta t$, where Δt is the time step. Other terms on the left-hand side of the equation have similar meaning. On the right-hand side, $S_{i,j}$ and $Q_{i,j}$ are block values of S and Q associated with block (i,j); $\phi_{i,j}^{k-1}$ the head of block (i,j) at time t_{k-1}; $(\Delta x)^2$ the block area. Block (i,j) and its neighboring blocks are shown in Figure 2.2.3.

Let the block values of parameters T, S and sink term Q be:

$$\mathbf{T} = (T_1, T_2, \ldots, T_N)^T,$$ (2.2.35a)
$$\mathbf{S} = (S_1, S_2, \ldots, S_N)^T,$$ (2.2.35b)
$$\mathbf{Q} = (Q_1, Q_2, \ldots, Q_N)^T,$$ (2.2.35c)

and let the head distributions for all time steps be

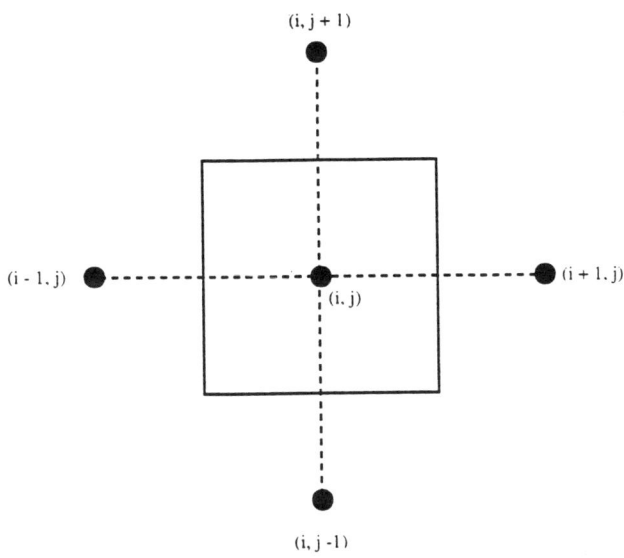

Fig. 2.2.3. Grid (i, j) and its neighboring grids.

$$\phi^k = \left(\phi_1^k, \phi_2^k, \ldots, \phi_N^k\right)^T, \quad (k = 1, 2, \ldots, K). \tag{2.2.36}$$

In (2.2.35) and (2.2.36), N is the total number of blocks and K the total number of time steps. In solution of the forward problem, $\mathbf{T}$, $\mathbf{S}$ and $\mathbf{Q}$ are given. After incorporating initial and boundary conditions, we can obtain a set of algebraic equations for the unknown ϕ^k

$$\mathbf{A}\phi^k = \mathbf{b}^k, \quad (k = 1, 2, \ldots, K), \tag{2.2.37}$$

where $\mathbf{A}$ is a positive definite and symmetric square matrix depending on parameters T and S. Vector $\mathbf{b}^k$ depends on boundary conditions, sink term Q and ϕ^{k-1}. Since the coefficient matrix is positive definite, there is no difficulty in solving (2.2.37).

Suppose that the numerical inverse problem presented is identification of parameters $\mathbf{T}$ and $\mathbf{S}$, while ϕ^k $(k = 1, 2, \ldots, K)$ and $\mathbf{Q}$ are given. To solve $\mathbf{T}$ and $\mathbf{S}$, Equations (2.2.37) may be rearranged into the following form:

$$\mathbf{H}\begin{pmatrix} \mathbf{T} \\ \mathbf{S} \end{pmatrix} = \mathbf{r}, \tag{2.2.38}$$

where $\mathbf{H}$ is a matrix depending on all ϕ^k $(k = 1, 2 \ldots, K)$. Vector $\mathbf{r}$ depends on $\mathbf{Q}$ and also on boundary conditions. Can we obtain the unknown $\mathbf{T}$ and $\mathbf{S}$ by solving

Equation (2.2.38)? Since **H** in (2.2.38) may not be a square matrix, the following three cases should be considered.

(1) *Underdetermination.* The number of equations is less than the number of unknown parameters. For example, if the number of blocks is six ($N = 6$) and the number of time steps is one ($K = 1$), then we only have 6 finite differential equations but with 12 unknown parameters ($T_1, T_2, \ldots, T_6, S_1, S_2, \ldots, S_6$). As a result, the number of solutions of (2.2.38) is infinite. The requirement of uniqueness is not satisfied.

(2) *Superdetermination.* The number of equations is larger than the number of unknown parameters. For example, if ten time steps ($K = 10$) are considered for six blocks ($N = 6$), then we have 60 finite difference equations but the number of unknown parameters is still 12. In this case, the requirement of existence is not satisfied.

(3) *Determination.* The number of equations is just equal to the number of unknown parameters. If 2 time steps are considered for the 6 block problem, the number of equations and the number of unknown parameters are both 12. The solution of (2.2.38) will be existent and unique, provided matrix **H** is not singular. Unfortunately, the value of the determinant of **H** is generally very small. A very small error in head value may cause a large error in the solution of parameters. Thus, the requirement of stability is not satisfied. The following example explains this situation.

Example 2.2.6 Suppose that a steady flow field is divided into 4 blocks as shown in Figure 2.2.4. The left side of block #1 and the right side of block #4 are given head boundary with transmissivity T_1 and T_4, respectively. Other boundary sections are assumed to be impermeable.

We have 4 finite difference equations corresponding to the 4 blocks

$$
\begin{cases}
T_1 (\phi_0 - \phi_1) + \frac{1}{2} (T_1 + T_2)(\phi_2 - \phi_1) + \frac{1}{2} (T_1 + T_3)(\phi_3 - \phi_1) = Q_1 (\Delta x)^2, \\
\frac{1}{2} (T_1 + T_2)(\phi_1 - \phi_2) + \frac{1}{2} (T_2 + T_4)(\phi_4 - \phi_2) = Q_2 (\Delta x)^2, \\
\frac{1}{2} (T_1 + T_3)(\phi_1 - \phi_3) + \frac{1}{2} (T_3 + T_4)(\phi_4 - \phi_3) = Q_3 (\Delta x)^2, \\
\frac{1}{2} (T_2 + T_4)(\phi_2 - \phi_4) + \frac{1}{2} (T_3 + T_4)(\phi_3 - \phi_4) + T_4 (\phi_5 - \phi_4) = Q_4 (\Delta x)^2.
\end{cases}
$$

$$(2.2.39)$$

If block heads $\phi_1, \phi_2, \phi_3, \phi_4$, boundary heads ϕ_0, ϕ_5 and block sinks Q_1, Q_2, Q_3, Q_4

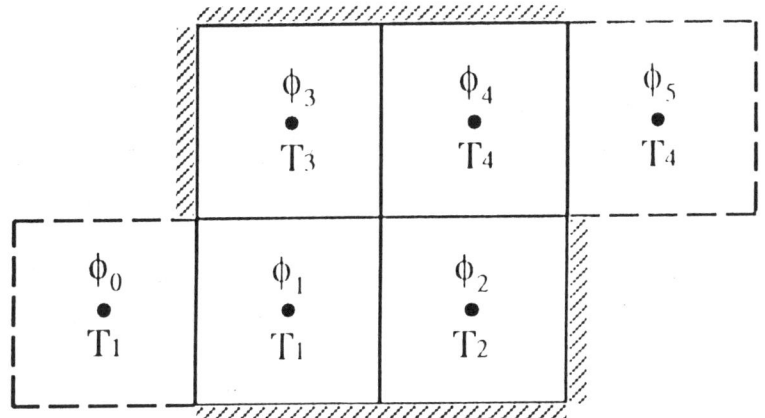

Fig. 2.2.4. The finite difference grids for Example 2.2.6.

are measured, Equation (2.2.39) can be rearranged as:

$$\begin{cases} (a + b + 2\delta) T_1 + aT_2 + bT_3 = Q_1 (\Delta x)^2, \\ -aT_1 + (c - a) T_2 + cT_4 = Q_2 (\Delta x)^2, \\ -bT_1 + (d - b) T_3 + dT_4 = Q_3 (\Delta x)^2, \\ -cT_2 - dT_3 + (-c - d + 2\gamma) T_4 = Q_4 (\Delta x)^2, \end{cases} \qquad (2.2.40)$$

where

$$\phi_0 - \phi_1 = \delta, \qquad \phi_5 - \phi_4 = \gamma, \qquad \phi_2 - \phi_1 = a,$$

$$\phi_3 - \phi_1 = b, \qquad \phi_2 - \phi_4 = c, \qquad \phi_4 - \phi_3 = d.$$

When $\delta = \gamma = 0$, we will have $\det \mathbf{H} = 0$, i.e., the coefficient matrix of (2.2.40) is singular. When $\delta \approx 0$, and $\gamma \approx 0$, or $a \approx b \approx c \approx d$, we will have $\det \mathbf{H} \approx 0$. In this case, the solution of (2.2.40) becomes very sensitive with respect to the observation error.

Therefore, when numerical models are used, the ill-posed nature of inverse problems does not change, only its manifestation changes. More discussions on instability of aquifer identification can be found in Yakowitz and Duckstein (1980).

Exercise 2.2.1. Present an example from the view of hydrogeology, in which solution of the inverse problem is non-unique.

Exercise 2.2.2. Extend Example 2.2.3 to the case that $T(x) = T_1$, when $x_1 \leq x < a_1$; $T(x) = T_2$, when $a_1 \leq x < a_2$; and $T(x) = T_3$, when $a_2 \leq x \leq x_2$.

Exercise 2.2.3.   Present inverse problems for Example 1.2.1 and consider their well-posedness.

Exercise 2.2.4.   Use a numerical example to test the results of Example 2.2.6.

2.3   Trial and Error Method

The *trial and error procedure* is the simplest method for solving inverse problems. It only needs:

· Some observation data on state variables;

· A subroutine SFP that solves the forward problem;

· An expert hydrogeologist who is familiar with the considered aquifer.

Let

$$\mathbf{u}^{obs} = \{u_1^{obs}, u_2^{obs}, \ldots, u_L^{obs}\} \qquad (2.3.1)$$

be a set of observation data and

$$\mathbf{u}^{cal}(\mathbf{p}) = \{u_1^{cal}(\mathbf{p}), u_2^{cal}(\mathbf{p}), \ldots, u_L^{cal}(\mathbf{p})\} \qquad (2.3.2)$$

be the corresponding output of subroutine SFP when $\mathbf{p}$ is used as the model parameter.

Figure 2.3.1 shows the trial and error procedure, where Steps 2, 4, 5, 6 are completed by the expert. This person can provide a good initial guess, p_0, for the unknown model parameters. After analyzing the model output, he or she knows how to modify the model parameters to generate a better fit between the observation data and the model output. This step is repeated until a fitting is judged to be satisfactory. The search procedure then can be ended and the unknown model parameters are found. If a satisfactory fitting can not be achieved, the modification of model structure should be considered.

The trial and error method has at least three advantages. First, it is unnecessary to write any new program for inverse problem solutions. Second, it can be used to solve any type of inverse problems (Actually, any uncertainty in a simulation model can be calibrated by this procedure). Third, the judgement of an expert can be incorporated into the search procedure. This may help to overcome the ill-posedness problem. For instance, when two sets of model parameters generate similar output of state variables, the expert can select one according to his or her judgement. Since these advantages, the primitive method is still extensively used in practice.

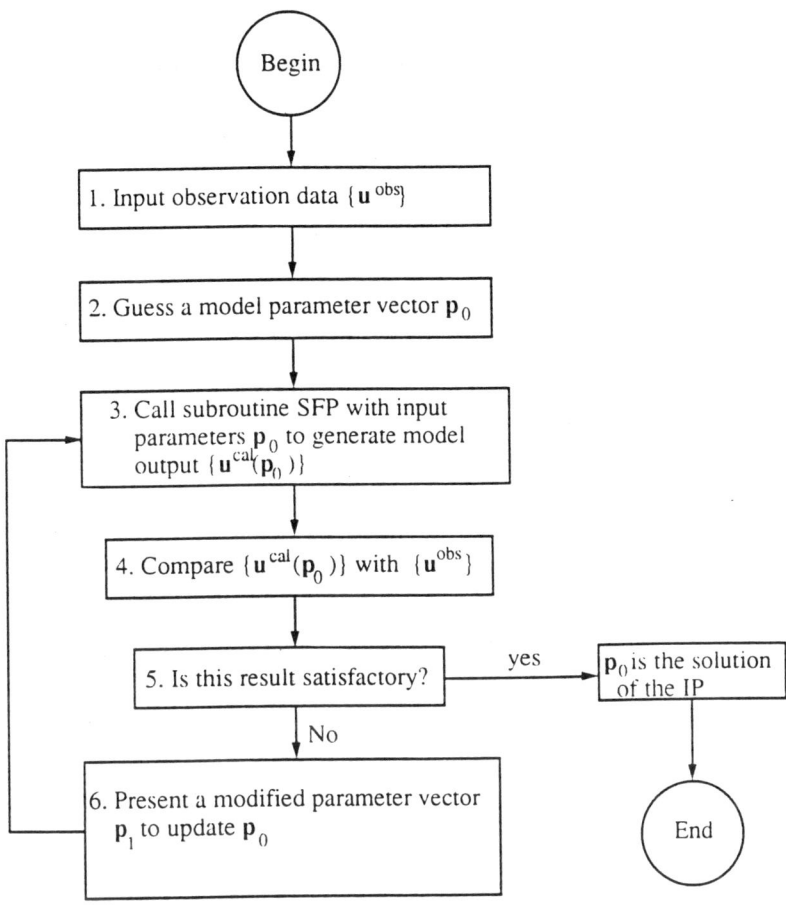

Fig. 2.3.1.   Flow chart of the trial and error method.

However, the trial and error procedure is time consuming, especially, when the number of unknown parameters is large. Moreover, accurate solutions of inverse problems can not be found in this procedure, and different results may be obtained by different users. Therefore, it is necessary to develop more sophisticated methods.

Exercise 2.3.1.   For the Theis model given in Example 1.2.2, assume that the pumping rate Q_w is equal to 1000 m^3/day and the distance r between the pumping well and the observation well is equal to 2 m. The drawdown in the observation well is observed at different observation times as shown in the following Table:

time	0.001	0.005	0.01	0.05	0.1	0.5
drawdown	0.982	1.607	1.880	2.519	2.795	3.435

(1) Use the trial and error method to identify parameters T and S.

(2) If pumping rate Q_ω is also unknown, can it be identified?

2.4 Transfer Inverse Problems into Optimization Problems

2.4.1 THE INDIRECT METHOD

Many computer programs have been developed to replace the human participation in the trial and error method. Let us return to Figure 2.3.1 and see what steps can be completed by computer.

To replace Step 4, one can use a criterion to measure the difference between $\{u^{obs}\}$ and $\{u^{cal}(p)\}$. The most common criterion is the *Output Least Squares* defined by

$$E(p) = \sum_{l=1}^{L} \left[u_l^{cal}(p) - u_l^{obs} \right]^2. \tag{2.4.1}$$

If $E(p_2) < E(p_1)$, then we conclude that p_2 is better than p_1.

To replace Step 5, one can set up a bound $\varepsilon > 0$. If $E(\hat{p}) < \varepsilon$, then the search procedure can be stopped and we accept $\hat{p}$ as the inverse solution.

Finally, in Step 6, the modified parameter p_1 can be generated in an optimization procedure. The "best" parameter $\hat{p}$ should be the minimizer of $E(p)$ in (2.4.1).

Thus, the trial and error procedure can be completed automatically as shown in Figure 2.4.1. With this approach, the inverse solution may be obtained by solving the following optimization problem: find $\hat{p} \in P_{ad}$, such that

$$E(\hat{p}) = \min E(p), \tag{2.4.2}$$

where P_{ad} is an admissible set of the unknown parameters. Problem (2.4.2) can be solved by a *nonlinear programming* subroutine (SNP). In this procedure, subroutine SFP is called for many times. In other words, the inverse problem is solved indirectly through the solution of the forward problem. As a result, it has been named the "*indirect method*" (Neuman, 1973).

An advantage of the indirect method is the capability of solving the inverse problem rapidly by computer without human participation. In addition, a set of "best fitting" parameters can be found by solving an optimization problem. However, when objective function (2.4.1) has more than one minimizer in the admissible set, the computer may select a local minimum other than the global one. A detailed discussion concerning the indirect method will be given in Chapter 4.

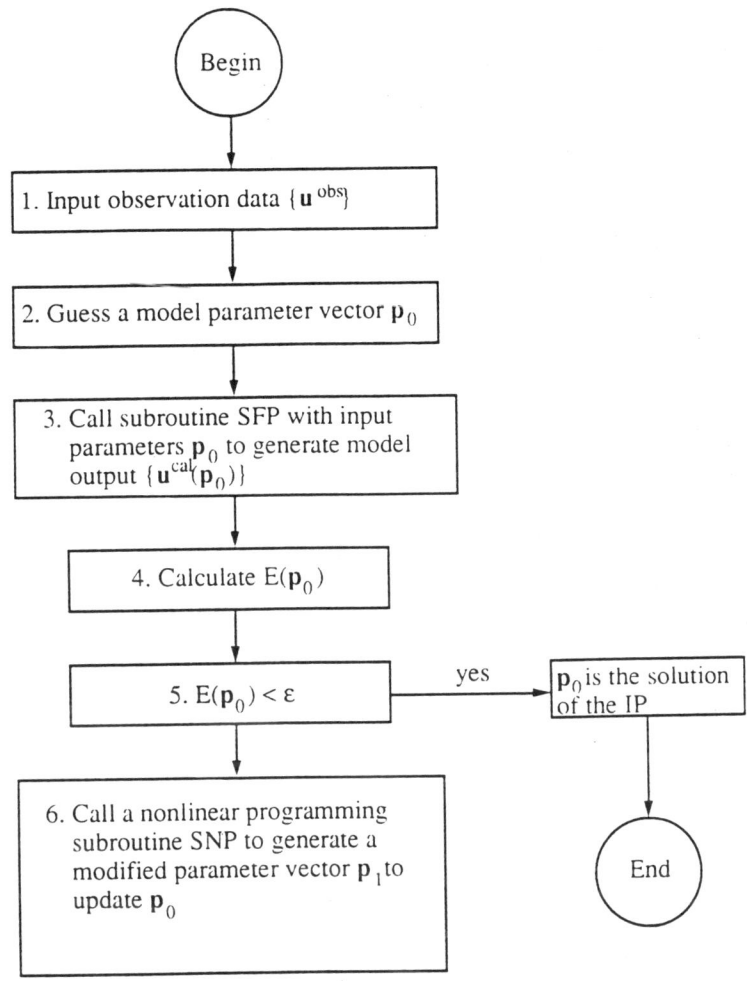

Fig. 2.4.1.   Flow chart of the indirect method.

The flow chart contains the following steps:

Begin

1. Input observation data $\{\mathbf{u}^{obs}\}$

2. Guess a model parameter vector $\mathbf{p}_0$

3. Call subroutine SFP with input parameters $\mathbf{p}_0$ to generate model output $\{\mathbf{u}^{cal}(\mathbf{p}_0)\}$

4. Calculate $E(\mathbf{p}_0)$

5. $E(\mathbf{p}_0) < \varepsilon$ yes → $\mathbf{p}_0$ is the solution of the IP

6. Call a nonlinear programming subroutine SNP to generate a modified parameter vector $\mathbf{p}_1$ to update $\mathbf{p}_0$

End

2.4.2   THE DIRECT METHOD

Another approach of inverse solution is the use of superdeterminate equations. As mentioned in section 2.2.4, after rearranging finite difference or finite element equations such that the parameters are considered as unknown variables, we obtain equations

$$\mathbf{Hp} - \mathbf{r} = 0, \qquad\qquad (2.4.3)$$

where $\mathbf{H}$ is an $L \times M$ matrix, $\mathbf{r}$ an L row-vector, L the number of observations, and M the number of unknown parameters. For the superdeterminate case, we have $L > M$. Matrix $\mathbf{H}$ and vector $\mathbf{r}$ in (2.4.3) depend on all nodal values of state variables.

Because Equations (2.4.3) is superdeterminate, there is no value of $\mathbf{p}$ that can satisfy all of them. Let the residual of lth equation be $\mu_l(\mathbf{p})$, i.e.,

$$\mu_l(\mathbf{p}) = H_{l1}p_1 + H_{l2}p_2 + \cdots + H_{lm}p_m - r_l, \quad (l = 1, 2, \ldots, L). \qquad (2.4.4)$$

An acceptable solution of the superdeterminate Equations (2.4.4) is such a vector $\hat{\mathbf{p}}$ that minimizes these residuals in a certain sense. For example, if $\hat{\mathbf{p}}$ is the minimizer of optimization problem

$$\min D(\mathbf{p}), \quad \mathbf{p} \in P_{\text{ad}}, \quad \text{where } D(\mathbf{p}) = \sum_{l=1}^{L} [\mu_l(\mathbf{p})]^2, \qquad (2.4.5)$$

then we accept $\hat{\mathbf{p}}$ as the solution of equations (2.4.4). Thus, the inverse problem is again transformed into an optimization problem.

If the admissible set P_{ad} is given by

$$\underline{p}_m \leq p_m \leq \bar{p}_m, \quad (m = 1, 2, \ldots, M), \qquad (2.4.6)$$

where $\underline{p}_m$ and $\bar{p}_m$ are the estimated lower bound and upper bound of parameter p_m, respectively, then problem (2.4.5) can be solved by a *quadratic programming* subroutine (SQP). In this procedure, we do not need any subroutine SFP. The inverse solution is obtained directly by solving a set of superdeterminate equations. Therefore, it is named the *"direct method"* (Neuman, 1973). The main steps of the direct method are shown in Figure 2.4.2.

A detail discussion of the direct method will be given in Chapter 5. Here, we only point out that data requirements in direct and indirect methods are very different. No matter how many observation data are collected, the indirect method can always be used. On the other hand, in order to form the matrix $\mathbf{H}$ in the direct method, all nodal values of the state variables should be given. In practice, however, state variables are measured only at a few observation wells. An interpolation process is always needed to generate their distributed values. As a result, extra interpolation error (noise) is introduced. Unfortunately, the inverse solution obtained by the direct method is very sensitive to any noise associated with the observation data.

2.4.3 Prior Information and Constraints

Despite inverse problems have been transformed into optimization problems in both direct and indirect methods, the ill-posed nature of inverse problems can not be shaken off. The non-uniqueness of inverse solution strongly displays itself in the indirect method through the existence of many local minimizers, while the solution of the direct method is often unstable.

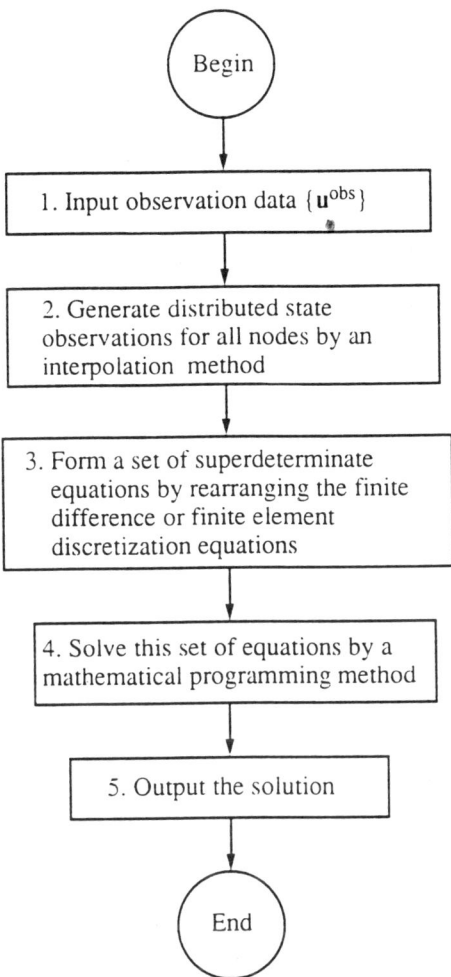

Fig. 2.4.2.   Flow chart of the direct method.

To overcome the ill-posedness, it is necessary to supplement with extra information that is independent of the measurement of state variables. Extra information may be produced by designating parameter values at some points, by limiting the admissible set P_{ad} in a small range, by assuming that the unknown parameter is piecewise constants, etc.

Generally, this supplementary information, often called *prior information*, can be imposed on an optimization problem presented either in the indirect method or in the direct method as *constraints*. With constraints, a unique and stable inverse solution may be found.

Exercise 2.4.1. Transform the problem presented in Exercise 2.3.1 into an optimization problem by the indirect method.

Exercise 2.4.2. If $\phi_0, \phi_1, \phi_2, \phi_3, \phi_4, \phi_5$ in Example 2.2.6 are observed three times, list the superdeterminate equations and transform the inverse problem of determining T_1, T_2, T_3, T_4 into an optimization problem by the direct method.

Exercise 2.4.3. If you have a subroutine that can be applied to solve non-linear programming problems, use it to get the results for Exercise 2.4.1 and 2.4.2.

2.5   Further Considerations

In this book, the inverse problem will be further considered from the following aspects.

(1) *Solution Methods.* Although the methods of transforming inverse problems into optimization problems have been presented, there are many problems that need to be further considered. For example:

  · Are there any other criteria better than (2.4.3) in the indirect method and (2.4.6) in the direct method?

  · Which numerical method is more effective for solving inverse problems?

  · How can we incorporate prior information into an optimization problem?

  · How can we verify the convergency and stability of the optimal solution?

  · How can we estimate the error of the identified parameters?

These problems will be discussed in Chapter 4 through Chapter 6. Some of them are still open.

(2) *Stochastic Framework.* It is absolutely necessary to study inverse problems in the framework of stochastics, since the observation error is always a random variable. Moreover, it is often convenient to treat physical parameters **p**, control variables **q** and subsidiary conditions **f** as random functions. For example, the infiltration caused by precipitation in a basin is difficult to be represented by a deterministic function. As mentioned in Example 1.1.7, hydraulic conductivity is often considered as a logarithm normally distributed random field. The problems associated with identifying random fields in groundwater modeling will be considered in Chapter 7.

(3) *Reliability Evaluation.* When a numerical model is used to predict the states of an aquifer, there are three kinds of errors that can not be avoided.

· *Model structure error* – the structure of the mathematical model does not coincide with that of the physical reality.

· *Forward solution error* – the numerical solution is not an accurate solution of the mathematical model.

· *Inverse solution error* – the identified parameters do not coincide with the true parameters.

If one kind of error is large, it is meaningless to decrease other kinds of errors. For instance, when the inverse solution error is large, the use of high order numerical methods to solve the forward problem does not improve the model quality. In many cases, the total error is determined mainly by the model structure error.

We believe that the quantitative analysis on the reliability of a simulation model can not be separated from its applications. We will define several kinds of extended identifiabilities in Chapter 8 which can help us make a detour from the difficulty caused by the ill-posedness of inverse problems.

(4) *Data Collection Strategy*. The reliability of an inverse solution is closely dependent on the quantity and quality of observation data. The study of *optimal data collection strategies* allows us either to decrease the expense of data collection or increase the reliability of model applications. We could say that the fundamental way out for solving inverse problems lies in the sufficiency of data. A methodology presented in Chapter 8 systematically considers experimental design, parameter identification, model prediction and decision making. The optimal experimental design for model structure identification will be also considered.

Classical Definition of Inverse Problems

3.1 The Solution of Operator Equations

It is known that various mathematical models in groundwater modeling can be represented commonly by operator Equations relating state variables and parameters. In forward problems we solve for state variables when parameters are given, while in inverse problems we solve parameters when state variables are measured. Thus, the general theory of inverse problems should be based on the solution of operator Equations. In this section, we will introduce accurate and approximate solutions of operator Equations and discuss their well-posedness. Introducing the classical definition of inverse problems not only makes these concepts more general and clear, but also helps to understand their origin and developments in depth. If the reader is not familiar with terminologies used in this section, such as mapping, vector space, norm and etc., it is suggested to read Appendix A of the book first. In this section, we assume the model to be free of structure error. Therefore, it is unnecessary to differentiate a *"real system"* and a *"model"*.

3.1.1 CLASSICAL WELL-POSEDNESS OF OPERATOR EQUATIONS

If the problem of interest only involves one state variable u and one parameter p, its distributed parameter model can be represented by an operator Equation

$$L(u, p, \mathbf{x}, t) = 0, \tag{3.1.1}$$

subject to appropriate subsidiary conditions. Using different values of p in (3.1.1) to solve the forward problem, different values of u are obtained. The dependence of u on p may be represented by

$$u = M(p), \tag{3.1.2}$$

where M can be seen as a *mapping* which transfers each *preimage* $p \in P$ into an unique *image* $u \in U$. Here, P and U are function spaces called *parameter space* and *state space*, respectively.

Generally, parameter p can vary only in a subset $P_{ad} \subset P$, which is called the admissible set of parameter p. For all $p \in P_{ad}$, the corresponding range of $M(p)$ is denoted by $M(P_{ad})$, which is a subset of state space U.

If there is an output system or an observation system for observing the model state u, the system output or the observed state can be represented as

$$u_D = D(u), \qquad\qquad (3.1.3)$$

where mapping D, with preimage $u \in U$ and image $u_D \in U_D$, describes characteristics of the observation system. U_D is called *system output space* or *observation space*. Hereinafter we assume that P, U and U_D are all *Banach spaces*. Thus, all of these spaces are furnished by norms. The distance between two elements in P, the distance between two elements in U, and the distance between two elements in U_D are all defined. As a result, the statement "u_1 tends to u_2 when p_1 tends to p_2" now does make sense. The definition of Banach space is given in Appendix A.

Example 3.1.1. Consider a 2-dimensional groundwater flow model of a confined aquifer and assume that the initial and boundary conditions are given. Let state u be the head distribution $h(x, y, t)$, parameter p be the transmissivity distribution $T(x, y)$. Parameter space P may consist of all non-negative functions, and admissible set P_{ad} may consist of all piecewise continuous functions. State space U consists of functions that have continuous second order derivatives. If the output system involves a continuous observation with respect to time at a point (x_0, y_0), then the model output is a function $h(x_0, y_0, t)$ and the observation space U_D is a function space. On the other hand, if the model output system includes only three observation points $(x_1, y_1), (x_2, y_2), (x_3, y_3)$ and two observation times t_1 and t_2, then the resulting observations consist of six numbers: $h(x_1, y_1, t_1)$, $h(x_2, y_2, t_1)$, ..., $h(x_3, y_3, t_2)$. Consequently, the observation space becomes a finite-dimensional space, R^6.

The combination of (3.1.2) and (3.1.3) yields a complex mapping (from parameter to observation)

$$u_D = DM(p), \qquad\qquad (3.1.4)$$

which transfers each preimage $p \in P_{ad}$ into an image u_D. The range of u_D is often denoted by $DM(P_{ad})$. It is a subset of space U_D (Figure 3.1.1).

The inverse problem of simulation involves finding model parameter p from the model output u_D, which is equivalent to solving operator Equation

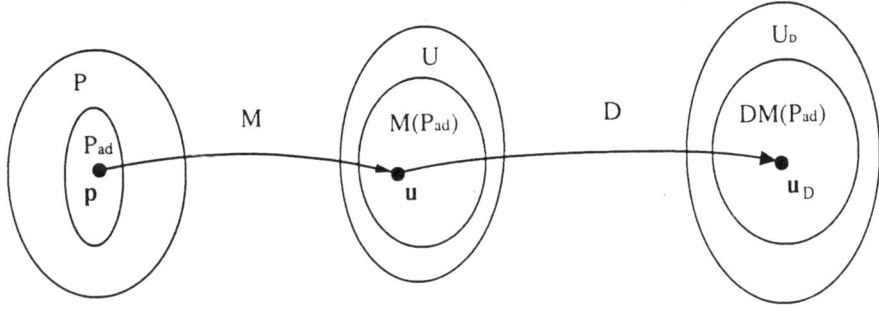

Fig. 3.1.1. The complex mapping $u_D = DM(p)$.

$$DM(p) = u_D. \tag{3.1.5}$$

An operator Equation is said to be well-posed, if the following conditions are satisfied:

(1) *Existence.* For each observation $u_D \in DM(P_{ad})$, there is a parameter $p \in P_{ad}$, such that $DM(p) = u_D$.

(2) *Uniqueness.* If $DM(p_1) = u_D$ and $DM(p_2) = u_D$, we must have $p_1 = p_2$.

(3) *Stability.* If $DM(p) = u_D$ and $DM(p') = u'_D$, where $p, p' \in P_{ad}$ and $u_D, u'_D \in DM(P_{ad})$, then $u'_D \to u_D$ implies $p' \to p$.

Thus, when operator Equation (3.1.5) is well-posed, there must exist a continuous mapping $(DM)^{-1}$ from observation space to parameter space, i.e.,

$$p = (DM)^{-1} u_D. \tag{3.1.6}$$

It is an inverse mapping of mapping DM. Unfortunately, the problem of solving an operator Equation is often ill-posed. Some examples of ill-posedness have been given in Chapter 2.

3.1.2 EXTENDED WELL-POSEDNESS

For a practical problem, the existence of a parameter satisfying Equation (3.1.5) can often be judged from physics, and the range of the parameter can often be estimated by prior information. Instead of the classical well-posedness defined above, a kind of extended well-posedness was suggested by Tikhonov (1963).

The problem of solving operator Equation (3.1.5) is considered as *well-posed in the Tikhonov sense*, if

(1) It is previously known that a solution p of Equation (3.1.5) exists and belongs to a set P_0, which is a subset of P_{ad}.

(2) The solution is unique in P_0, i.e. when $P_1 \in P_0$ and $P_2 \in P_0$, then $DM(p_1) = DM(p_2)$ implies $p_1 = p_2$.

(3) The solution is stable, that is, for any $\varepsilon > 0$, there is a $\delta > 0$, such that $\|DM(p) - DM(p')\|_{U_D} < \delta$ implies $\|p - p'\|_P < \varepsilon$, where $p \in P_0$ and $p' \in P_0$.

In the above definition, the uniqueness often implies stability. One can prove that if the solution of Equation (3.1.5) is unique, the inverse mapping $(DM)^{-1}$ must be continuous on P_0, provided that mapping DM is continuous and P_0 is compact (Lavrent'ev *et al.*, 1980).

The well-posedness in the Tikhonov sense provides a theoretical background for the *trial-and-error method* of parameter identification. For example, when the trial-and-error method (Section 2.3) is used to identify the hydraulic conductivity of an aquifer, the flow region is often divided into several homogeneous zones based on certain prior information. In this procedure, we have indeed assumed that the identified parameter (vector) exists and belongs to a subset P_0 which consists of all positive and piecewise constant functions. Uniqueness is also assumed in the trial-and-error procedure, because only one guessed parameter is tested in each iteration. According to the Tikhonov's theory, the identified parameter must be stable. In fact, when we use the trial-and-error method to calibrate a groundwater model, we could have the experience that small variations in the head observations cause only small variations in the identified hydraulic conductivity.

3.1.3 APPROXIMATE SOLUTION

Suppose that Equation (3.1.5) is well-posed in the Tikhonov sense, but the right-hand side u_D of the Equation contains some error. In this case, the true model parameter p^* and its corresponding model output u_D^* are both unknown. We only have an approximate model output $\tilde{u}_D$ and know that the "distance" between $\tilde{u}_D$ and u_D^* does not exceed a positive number δ, that is,

$$\|\tilde{u}_D - u_D^*\|_{U_D} < \delta, \tag{3.1.7}$$

where $\| \cdot \|_{U_D}$ is a norm defined in the observation space U_D. When observations have noises, it is impossible to find the exact solution, but we can define an approximate solution.

For each $p \in P_0$, we can find its image $DM(p)$ in space U_D and calculate the distance between $DM(p)$ and $\tilde{u}_D$. Thus, we have a function $E(p)$ defined on P_0,

$$E(p) = \|DM(p) - \tilde{u}_D\|_{U_D}. \tag{3.1.8}$$

A parameter $\tilde{p} \in P_0$ is called a *quasi-solution* of the equation

$$DM(p) = \tilde{u}_D, \tag{3.1.9}$$

if it is a minimizer of functional $E(p)$ on P_0, i.e.

$$E(\tilde{p}) = \min_{p \in P_0} \|DM(p) - \tilde{u}_D\|_{U_D}. \tag{3.1.10}$$

Since the true parameter p^* is also in P_0, we must have

$$\|DM(\tilde{p}) - \tilde{u}_D\|_{U_D} \leq \|DM(p^*) - \tilde{u}_D\|_{U_D} = \|u_D^* - \tilde{u}_D\|_{U_D} < \delta. \tag{3.1.11}$$

From the triangle inequality, we obtain

$$\|DM(\tilde{p}) - DM(p^*)\|_{U_D} < 2\delta. \tag{3.1.12}$$

Thus, we can conclude that the image of quasi-solution $\tilde{p}$ is close to the image of the true parameter p^*. According to the stability condition of well-posedness, $\tilde{p}$ must be close to p^*, i.e.,

$$\|p^* - \tilde{p}\|_{U_P} < \varepsilon. \tag{3.1.13}$$

where $\| \cdot \|_{U_P}$ is the norm defined in the parameter space P. In (3.1.13), $\varepsilon > 0$ can be arbitrarily small, provided that $\delta > 0$ in (3.1.7) is small enough (Figure 3.1.2).

Therefore, the quasi-solution $\tilde{p}$ defined by (3.1.10) is indeed an approximate solution of Equation (3.1.9). When observation error tends to zero, the quasi-solution must tend to the true parameter. The concept of quasi-solution is the common background of all *indirect methods* for parameter identification (Section 2.4.1).

Exercise 3.1.1. Define parameter space P, state space U, observation space U_D, as well as mappings M, D and DM for Exercise 2.3.1.

Exercise 3.1.2. Consider a steady flow in a confined aquifer. Let M be its numerical model obtained by the finite difference method, u and p be discretized head and transmissivity distributions, respectively. Write operator Equation (3.1.5) for this case and discuss its well-posedness.

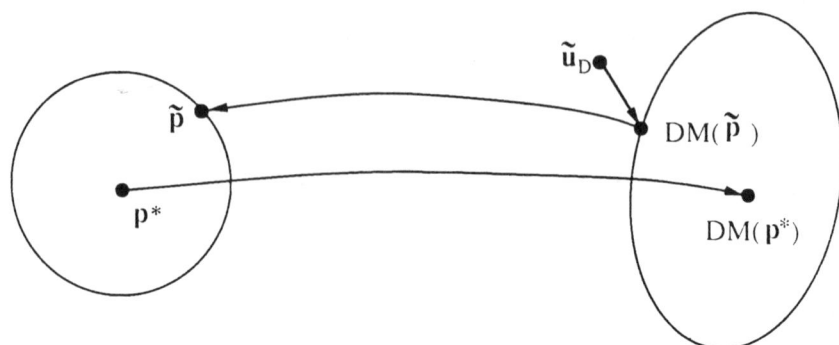

Fig. 3.1.2. The determination of a quasisolution $\tilde{p}$.

Exercise 3.1.3. What are the differences between the classical definition of well-posedness given in Section 3.1.1 and the well-posedness in the Tikhonov sense given in Section 3.1.2?

Exercise 3.1.4. Define the quasi-solution when observation space U_D is a finite-dimensional space and the l_2-norm is used.

3.2 Identifiability

In the solution of inverse problems, a most important consideration is that whether the observation data contain sufficient information for identifying the unknown parameters. This constitutes the *identifiability problem* that will run through the whole book. In this section, however, we deal only with certain classical definitions of identifiability and explain the relationship between identifiability and well-posedness.

3.2.1 CLASSICAL DEFINITION OF IDENTIFIABILITY

When there is no observation error, the *classical identifiability* requires that each preimage $p \in P_0$ can be found uniquely from its image $u_D \in DM(P_0)$, that is,

$$\|DM(p_1) - DM(p_2)\|_{U_D} = 0 \text{ implies } \|p_1 - p_2\|_{U_P} = 0, \qquad (3.2.1)$$

where $p_1, p_2 \in P_0$. Equation (3.2.1) is equivalent to the requirement that mapping DM be an injection, or the inverse mapping $(DM)^{-1}$ exists on set $DM(P_0) \subset U_D$.

Therefore, the identifiability defined above requires the uniqueness of inverse solution. When P_0 is compact, it is equivalent to the well-posedness in the Tikhonov sense. An interesting and difficult problem is to determine under what conditions the unknown parameter is identifiable. Up to now, only a few results have been obtained for distributed parameter models. Kitamera and Nakagiri (1977) discussed the identifiability of coefficients in a linear, one-dimensional parabolic PDE. The necessary and sufficient conditions for identifiability obtained in that paper provide important insights into parameter identification of groundwater modeling.

Roughly speaking, if we want to identify a distributed hydraulic conductivity by head observations, the gradient of head, i.e., Darcy's velocity, must be unequal to zero almost anywhere in the flow region. In fact, if there is a subregion where Darcy's velocity is zero, then hydraulic conductivity can take any value in that subregion.

As denoted by Chavent (1987), however, the identifiability definition (3.2.1) is not practical. It is very difficult to prove that a mapping is injective, except for a few particular cases. Moreover, this definition does not consider observation error, which, of course, can never be avoided in real problems.

3.2.2 IDENTIFIABILITY FOR QUASISOLUTIONS

When model output contains error, true parameters can never be found. In this case, it is natural to define the identifiablity in terms of a quasi-solution.

If a quasi-solution $\tilde{p} \in P_0$ can be determined uniquely by the noisy model output $\tilde{u}_D$, and it is continuously dependent on $\tilde{u}_D$, then the model parameter is called *quasi-identifiable*. In other words, if $\tilde{p} \in P_0$ and $\hat{p} \in P_0$ are quasi-solutions corresponding to $\tilde{u}_D$ and $\hat{u}_D$, respectively, where $\tilde{u}_D$ and $\hat{u}_D$ are in a neighborhood of $DM(P_0)$, then

$$\|\tilde{u}_D - \hat{u}_D\|_{U_D} = 0 \text{ implies } \|\tilde{p} - \hat{p}\|_{D_P} = 0, \tag{3.2.2}$$

and when $\hat{u}_D \to \tilde{u}_D$, we must have $\hat{p} \to \tilde{p}$.

We have shown that the image $DM(\tilde{p})$ of a quasi-solution $\tilde{p} \in P_0$ is a point in set $DM(P_0)$ which is closest to the observation $\tilde{u}_D$. Therefore, quasi-identifiability include two requirements: there exists an unique point in set $DM(P_0)$ which is closest to $\tilde{u}_D$, and mapping DM is injective on P_0. Generally, the uniqueness of a quasi-solution can not be guaranteed when $DM(P_0)$ is not a convex set. Figure 3.2.1 shows two quasi-solutions corresponding to an observation $\tilde{u}_D$. Only a few results have been obtained for the quasi-identifiability (Chavent, 1987b, 1990). Roughly speaking, it requires model output u_D to be sensitive enough with respect to the perturbation of model parameters. In the next chapter, we will consider a special case in which the l_2-norm is used in the definition (3.2.2).

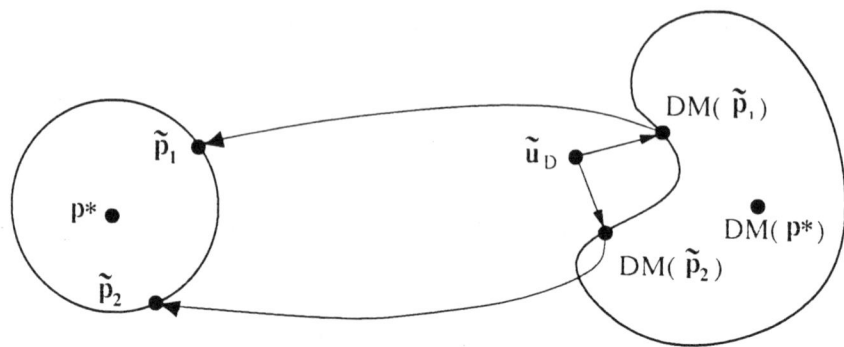

Fig. 3.2.1. The nonuniqueness of quasi-solutions.

Exercise 3.2.1. Explain that the classical definition (3.2.1) of identifiability is impractical.

Exercise 3.2.2. Discuss the relationship between the well-posedness of operator Equation (3.1.5) and the identifiability of parameter p when observations are made with or without error.

3.3   Discretization and Parameterization

For any real problem in groundwater modeling, the quantity and quality of available data are very limited. It is impossible to identify a distributed parameter with infinite or very large freedom. When $u = M(p)$ is a numerical model, however, both parameter and state spaces are finite dimensional spaces. In this section, we will explain how to represent a unknown distributed parameter with a finite dimensional vector and consider its identifiability.

3.3.1   PARAMETERIZATION

Parameterization is a technique used to simplify the structure of an infinite or N-dimensional parameter space such that it can be represented approximately by a M-dimensional space ($M \ll N$), where M is the *dimension of parameterization*.

When a finite difference method or a finite element method is used to solve the forward problem $u = M(p)$, parameter p is discretized according to the discretization of space variables. Each node (or element) is associated with a value of this parameter. Let the number of nodes be N. The distributed parameter p is now approximately represented by a N-dimensional vector

$$\mathbf{p}_N = (p_1, p_2, \ldots, p_N), \tag{3.3.1}$$

which can be regarded as a point in space R^N, where $p_1, p_2, \ldots, p_N$ are nodal values of parameter p. Operator equation (3.1.5) now becomes

$$DM(\mathbf{p}_N) = u_D, \tag{3.3.2}$$

where model M is either in its continuous form or in its discrete form. The number of nodes (N) of the simulation model, however, is usually so large that the information provided by observation u_D is not sufficient for uniquely determining so many components of $\mathbf{p}_N$. In other words, $\mathbf{p}_N$ is not identifiable.

There are several approaches in groundwater modeling for decreasing the dimension of an unknown parameter:

(1) *Piecewise Constant or Zonation Method*

From geological analysis, the flow region may be divided into M homogeneous zones: $(\Omega_1), (\Omega_2), \ldots, (\Omega_M)$. In each zone, the unknown parameter p (hydraulic conductivity, storage coefficient and so forth) can be seen as a constant. Let the parameter value in the mth zone be k_m, then we have

$$p(\mathbf{x}) = k_m, \quad \text{when } \mathbf{x} \in (\Omega_m), \quad m = 1, 2, \ldots, M. \tag{3.3.3}$$

The identified parameter now becomes a M-dimensional vector

$$\mathbf{k} = (k_1, k_2, \ldots, k_m), \quad \mathbf{k} \in K_{\text{ad}}, \tag{3.3.4}$$

where K_{ad} is an admissible subset of $\mathbf{k}$ in R^M.

Zonation is the most popular method for parameterization. However, there is a difficulty associated with the use of this method: prior information is usually insufficient to determine the geometry of zones. As denoted by Sun and Yeh (1985), an incorrect geometry of zones always leads to an incorrect estimate of parameter values. The problem of zonation pattern identification will be considered in Chapter 8.

(2) *Polynomial approximation method*

In this method, the unknown parameter is replaced approximately by a low order (linear, quadratic and cubic) polynomial in the region of interest. For example, when quadratic polynomial is used in a two-dimensional model, we will have

$$p(x, y) = k_1 x^2 + k_2 y^2 + k_3 xy + k_4 x + k_5 y + k_6, \tag{3.3.5}$$

where k_m $(m = 1, 2, \ldots, 6)$ are undetermined coefficients. The identified parameter now becomes a 6-dimensional vector:

$$\mathbf{k} = (k_1, k_2, \ldots, k_6), \quad \mathbf{k} \in R^6. \tag{3.3.6}$$

Of course, $\mathbf{k}$ must belong to an admissible subset $K_{\mathrm{ad}} \subset R^6$, such that the physical constraint $p(x, y) \geq 0$ is satisfied. For practical problems, of course, the polynomial approximation method is neither accurate nor flexible.

(3) *Interpolation Method*

All *interpolation methods* can be used for the purpose of parameterization. For example, we can use the following finite element interpolation method to represent the unknown parameter

$$p(\mathbf{x}) = \sum_{m=1}^{M} k_m \phi_m(\mathbf{x}), \tag{3.3.7}$$

where $\phi_m(\mathbf{x})$ $(m = 1, 2, \ldots, M)$ are basis functions and M is the dimension of finite element subspace. Linear and bilinear basis functions are often used for this purpose (Distefano and Rath, 1975; Yoon and Yeh, 1976). Note that the dimension of finite element subspace for parameter discretization should be much smaller than that for state variable discretization, i.e., $M \ll N$ is required.

The identified parameter now becomes the coefficients in (3.3.7),

$$\mathbf{k} = (k_1, k_2, \ldots, k_M), \quad \mathbf{k} \in K_{\mathrm{ad}}. \tag{3.3.8}$$

The basis functions defined in (3.3.7) satisfy

$$\phi_m(\mathbf{x}_n) = \begin{cases} 0 & \text{if } n \neq m, \\ 1 & \text{if } n = m, \end{cases} \tag{3.3.9}$$

where $\mathbf{x}$ represents the coordinates of node n. From (3.3.7), we have $k_m = p(\mathbf{x}_m)$, i.e. k_m is just the parameter value at node m. From physics, all k_m must be non-negative.

Other interpolation methods, such as the cubic spline, may also be used to generate a parameterization.

(4) *Distance Weighting Method*

In the method presented by Sun and Yeh (1985), M points were selected from the flow region and called basis points. Let the value of parameter p at basis point k be $p^{(k)}$ $(k = 1, 2, \ldots, M)$. The distance between an arbitrary point $\mathbf{x}$ and a basis

k is denoted by $d_{\mathbf{x},k}$. We can then divide the flow region into M subregions S_k ($k = 1, 2, \ldots, M$) according to their relative distances. We define $\mathbf{x} \in S_k$, if

$$d_{\mathbf{x},k} = \min(d_{\mathbf{x},1}, d_{\mathbf{x},2}, \ldots, d_{\mathbf{x},M}), \tag{3.3.10}$$

where k is the smallest number when (3.3.10) is satisfied by more than one basis points.

When $\mathbf{x} \in S_k$, $p(\mathbf{x})$ is defined as a linear combination of parameter values of all basis points,

$$p(\mathbf{x}) = p^{(k)} + \frac{\displaystyle\sum_{m=1}^{M} \omega_{\mathbf{x},m} \sigma_{\mathbf{x},m} (p^{(m)} - p^{(k)})}{\displaystyle\sum_{m=1}^{M} \omega_{\mathbf{x},m}}, \tag{3.3.11}$$

where

$$\sigma_{\mathbf{x},m} = \frac{(\gamma_{k,m} - d_{\mathbf{x},m})}{\gamma_{k,m}^{\alpha}}, \quad m \neq k, \tag{3.3.12a}$$

$$\sigma_{\mathbf{x},k} = 0;$$

$$\omega_{\mathbf{x},m} = \frac{1}{d_{\mathbf{x},m}^{\beta}}, \quad m \neq k, \tag{3.3.12b}$$

$$\omega_{\mathbf{x},k} = 1.$$

In (3.3.12), $\gamma_{k,m}$ is the distance between basis points k and m, $\alpha \geq 1$ and $\beta \geq 1$ are constant. Changing α and β, parameterization algorithm (3.3.11) can describe different types of parameters. When $\alpha > 2$, $p(\mathbf{x})$ is able to represent piecewise constant functions. When $\alpha = 1, \beta > 10$, $p(\mathbf{x})$ is able to represent continuous functions.

(5) Stochastic Field Method

In the stochastic frame of inverse problems, the unknown parameter is regarded as a random field. It is partially characterized by its first two moments, the mean (or drift) and the covariance function. Low order polynomials or other simple functions with a few undetermined coefficients can be used to represent the unknown drift and covariance function. Thus, the unknown parameter is parameterized by these coefficients. A detailed discussion of the stochastic field method will be given in Chapter 7.

(6) Geological Structure Method

This method of parameterization was presented recently by Sun *et al.* (1994b), in which the unknown parameters, such as hydraulic conductivity and porosity, are directly associated with different kinds of geological materials. Some local three-dimensional geological structures may be known from existing well-logs and other prior geological information. A distributed geological structure is then obtained by the geostatistical method or kriging.

Using the geological structure method for parameterization, all geological information can be incorporated into the solution of inverse problems, and the dimension of parameterization does not depend on the complexity of the simulation model. A detailed discussion of this method will be given in Chapter 8.

In the preceding discussion, we have introduced several parameterization methods. Generally speaking, parameterization can be seen as a mapping

$$p = P(\mathbf{k}),   \tag{3.3.13}$$

where P is a parameterization operator with preimage $\mathbf{k} \in K_{ad} \subset R^M$ and image $p \in P_{ad} \subset P$. Substituting (3.3.13) into (3.1.5) and using DMP to represent the complex mapping of mappings P and DM, we then have

$$DMP(\mathbf{k}) = u_D.   \tag{3.3.14}$$

Now the inverse problem becomes one of finding the model parameter $\mathbf{k} \in K_{ad}$ of parameterized model DMP from the model output u_D.

3.3.2   IDENTIFIABILITY OF DISCRETIZED MODELS

Because operator Equation (3.3.14) is always finite-dimensional in practical applications, it is necessary to study the identifiability of finite-dimensional operators. To date, however, only a few works have been concerned with this topic.

Kunisch and White (1987a,b) and Kunisch (1988) considered the identifiability of coefficients in one-dimensional parabolic and elliptic $PDEs$. They used the standard Galerkin approximation to discrete the state as well as the parameter spaces. Let

$$p^M = \sum_{j=1}^{M} k_j \phi_j \quad \text{and} \quad u^N = \sum_{i=1}^{N} u_i \psi_i   \tag{3.3.15}$$

be the approximate solutions for parameter p and state u, respectively, where ϕ_j and ψ_i are basis functions of M-dimensional finite element subspace V^M for the parameter and N-dimensional finite element subspace H^N for the state, respectively.

For each $p^M \in P_{ad}^M$, where P_{ad}^M is the admissible set of p^M in space V^M, image $u^N(p^M)$ can be found by solving the finite element equations. If $p_1^M \neq p_2^M$ implies $u^N(p_1^M) \neq u^N(p_2^M)$ for all $p_1^M \in P_{ad}^M$ and $p_2^M \in P_{ad}^M$, that is, preimage p^M

can be uniquely determined by its image $u^N(p^M)$, then the discretized unknown parameter is called identifiable.

This definition was given by Kunisch and White (1987a) and named by them as *identifiability under approximation.*

Certain sufficient conditions for identifiability under approximation have been obtained. Roughly speaking, the discretized state u^N must vary enough from node to node, the dimension of H^N must be much larger than the dimension of V^M, i.e. $M \ll N$, the order of basis functions ψ_i in H^N must be higher than the order of basis functions ϕ_j in V^M.

A weakness of the above definition is that the stability problem caused by observation error is not considered. Moreover, all nodal values of u^N be given for the identification of p^M. In practice, of course, it is impossible to obtain all nodal values of u^N which are noise free.

Studies in this direction may provide useful theoretical background for the *direct methods* of inverse solution (Section 2.4.2).

Exercise 3.3.1. Give the expression of unknown parameter p in each element, When triangle elements and linear basis functions are used in the two-dimensional finite element interpolation method for parameterization.

Exercise 3.3.2. In the program given in Appendix C, the zonation method is used for parameterization. By revising subroutine DATAP and PARZA, it can combine with other parameterization methods. Change these two subroutines such that the linear finite element interpolation method defined in Exercise 3.3.1 can be used to generate parameterization.

Exercise 3.3.3. Generally speaking, can we attain the identifiability under approximation by decreasing the dimension of parameterization?

Indirect Methods for the Solution of Inverse Problems

4.1 The Output Least Squares Criterion

In a special case of defining the quasisolution, the observation space is furnished with the L_2-norm. The corresponding performance is called the output least squares (OLS) criterion which is the most popular tool for parameter identification.

4.1.1 OUTPUT LEAST SQUARES FORMATION

Consider a physical system S, in which p^t is the system parameter. With an observation system D, the true system state $DS(p^t)$ will be smeared by *observation error* η_D^{obs}. When the physical system is simulated by a model M, model output $DM(p^M)$ is different from the system output. Thus, a *model error* or a *model structure error* η_D^M is introduced. Moreover, if model parameter p^M is parameterized by a finite-dimensional parameter vector k, a *parameterization error* η_D^P is generated. Therefore, the observed system state, $\tilde{u}_D$, can be represented as:

$$
\begin{aligned}
\tilde{u}_D &= DS(p^t) + \eta_D^{\text{obs}} \\
&= DM(p^M) + \eta_D^M + \eta_D^{\text{obs}} \\
&= DMP(\mathbf{k}) + \eta_D^P + \eta_D^M + \eta_D^{\text{obs}}.
\end{aligned}
\tag{4.1.1}
$$

We will use η_D to denote the *total error* between $\tilde{u}_D$ and $DMP(\mathbf{k})$, namely,

$$
\eta_D = \eta_D^P + \eta_D^M + \eta_D^{\text{obs}}.
\tag{4.1.2}
$$

For a real problem, we only know $\tilde{u}_D$. All terms on the right-hand side of (4.1.1) are unknown. Usually, we assume that the total error η_D is normally distributed with a zero mean. In Chapter 7, we will prove that under some assumptions, the best estimation of the unknown parameters, $\hat{\mathbf{k}}$, is the solution of the following optimization problem:

$$E(\hat{k}) = \min_{k \in K_{ad}} E(k), \tag{4.1.3}$$

where

$$E(k) = \|DMP(k) - \tilde{u}_D\|_{L_2}^2. \tag{4.1.4}$$

In Section 3.1.3, we have defined quasisolutions for operator equations. Obviously, (4.1.4) is a special case of (3.1.8) when the observation space U_D is furnished by L_2 norm. When observation system D is distributed, i.e., the system state is measured continuously on a time-space domain (Ω), the corresponding L_2 norm is defined by (A.35) in Appendix A, and we have

$$E(k) = \int_{(\Omega)} W_D^2(x,t)[DMP(k,x,t) - \tilde{u}_D(x,t)]^2 \, d\Omega, \tag{4.1.5}$$

where $W_D(x,t)$ is a weighting function. On the other hand, when observation system D is pointwise, i.e., only L observations are available in the time-space domain, the corresponding L_2 norm is defined by (A.22), and we have

$$E(k) = \sum_{l=1}^{L} W_{D,l}^2[DMP(k)_l - \tilde{u}_{D,l}]^2, \tag{4.1.6}$$

where $W_{D,l}$ is a set of weighting coefficients. When (4.1.5) or (4.1.6) is used as objective function for parameter identification, optimization problem (4.1.3) is called *the generalized output least squares criterion*.

An advantage of using the weighted objective function (4.1.6) is that the relative importance of each observation can be adjusted by selecting different weighting coefficients. In Chapter 7, we will show that when observation errors are independent of each other and normally distributed, the optimal weighting coefficients are given by

$$W_{D,l} = \frac{1}{\sigma_l}, \quad (l = 1, 2, \ldots, L), \tag{4.1.7}$$

where $\sigma_1, \sigma_2, \ldots, \sigma_L$ are standard deviations of observation errors. In practice, however, these standard deviations are usually unknown. Therefore, the following rules may be referenced:

· An observation, which is relatively accurate, should be associated with a large weighting coefficient, otherwise with a smaller one;

· An observation, which represents the system state of a large time-space subdomain, should be associated with a large weighting coefficient, otherwise with a small one;

· If an observation is important for determining a parameter component k_m and the parameter component is important for model applications, then it should be associated with a large weighting coefficient, otherwise with a small one.

4.1.2 THE OUTPUT LEAST SQUARES IDENTIFIABILITY

Using the Output Least Squares Criterion, inverse solution $\hat{k}$ is obtained by solving the optimization problem (4.1.3), where $E(k)$ is defined by (4.1.5) for distributed observations or (4.1.6) for pointwise observations. An important problem is whether optimization problem (4.1.3) is well-posed, or, whether the optimization problem have a global minimum which is continuously dependent on the observation $\tilde{u}_D$. For answering this question, Chavent (1979) defined a kind of *Output Least Squares Identifiability* (OLSI) as follows:

Parameter $k \in K_{ad}$ is said to be OLSI with respect to an observation system D, if the optimization problem (4.1.3)–(4.1.4) has a unique solution and the solution depends continuously on observation $\tilde{u}_D$. Obviously, OLSI is a special case of quasi-identifiability defined in Section 3.2.2. Here, the L_2 norm is used to furnish the observation space U_D. The OLSI can be achieved if and only if the following two conditions are satisfied:

(1) The projection $\overline{u}_D$ of $\tilde{u}_D$ onto $DMP(K_{ad})$ is unique and continuously dependent on $\tilde{u}_D$.

(2) Mapping $u_D = DMP(k)$ is injective and continuous on K_{ad}.

Chavent (1983) presented a set of sufficient conditions for OLSI. Roughly speaking, model output u_D must be sensitive enough to the perturbation of model parameters, and observation error must be small enough. Chavent (1987b, 1990) proposed another sufficient condition for OLSI based on a geometrical approach. These conditions, however, are difficult to verify directly. As denoted by Chavent (1987a), OLSI is seldomly realized in practice. In other words, optimization problem (4.1.3)–(4.1.4) often has more than one local minimum. Therefore, in order to obtain a unique inverse solution, we usually need additional prior information on the unknown parameters.

Exercise 4.1.1. Let $\hat{k}$ be a minimizer of problem (4.1.3)–(4.1.4) and $\sigma_D = \|DMP(\hat{k}) - \tilde{u}_D\|_{L_2}$, where σ_D is called the residual of least squares problems. Prove that inequality $\sigma_D \leq \overline{\eta}_D$ is always correct, where $\overline{\eta}_D$ is the norm of total error defined in (4.1.2).

Exercise 4.1.2. Does prior information help inverse solutions to be *OLS* identifiable?

4.2 Optimization Algorithms

This section discusses various numerical methods used in the solution of *OLS* problems. Since many books have contributed to this topic (Levenberg, 1977;

Dennis, 1983; Fletcher, 1987), it is unnecessary to describe various optimization algorithms in detail. Instead, we will emphasize their applications in the solution of inverse problems.

4.2.1 Unconstrained Nonlinear Optimization

Consider the following multi-dimensional optimization problem

$$\min E(\mathbf{k}), \qquad \mathbf{k} \in K_{\mathrm{ad}} \subset R^M. \tag{4.2.1}$$

If objective function $E(\mathbf{k})$ is second-order differentiable, the following are *necessary conditions* for $\hat{\mathbf{k}}$ being a *local minimum* of $E(\mathbf{k})$:

· *gradient* $g = \nabla E(\mathbf{k})$ vanishes at $\hat{\mathbf{k}}$, i.e.,

$$\left. \frac{\partial E}{\partial k_m} \right|_{\hat{\mathbf{k}}} = 0, \quad (m = 1, 2, \ldots, M). \tag{4.2.2}$$

· Hessian matrix $\nabla^2 E(\mathbf{k})$, i.e.,

$$G = \begin{bmatrix} \dfrac{\partial^2 E}{\partial k_1^2} & \dfrac{\partial^2 E}{\partial k_1 \partial k_2} & \cdots & \dfrac{\partial^2 E}{\partial k_1 \partial k_M} \\ \dfrac{\partial^2 E}{\partial k_1 \partial k_2} & \dfrac{\partial^2 E}{\partial k_2^2} & \cdots & \dfrac{\partial^2 E}{\partial k_2 \partial k_M} \\ \vdots & \vdots & \ddots & \vdots \\ \dfrac{\partial^2 E}{\partial k_1 \partial k_M} & \dfrac{\partial^2 E}{\partial k_2 \partial k_M} & \cdots & \dfrac{\partial^2 E}{\partial k_M^2} \end{bmatrix} \tag{4.2.3}$$

is a positive semi-definite matrix at $\hat{\mathbf{k}}$.

On the other hand, we have the following *sufficient conditions* for $\hat{\mathbf{k}}$ being a local minimum of $E(\mathbf{k})$:

· gradient $\nabla E(\hat{\mathbf{k}}) = 0$,

· Hessian $\nabla^2 E(\hat{\mathbf{k}})$ is a positive definite matrix.

If $E(\mathbf{k})$ is a differentiable convex function, then $\nabla E(\hat{\mathbf{k}}) = 0$ is the necessary and sufficient condition for $\hat{\mathbf{k}}$ being a local minimum of $E(\mathbf{k})$. The above conclusions can be found in any textbook on multi-variable calculus.

A point k satisfying Equations (4.2.2) is called a Stationary Point. It is a candidate of optimization solutions.

In the identification of model parameters, objective function $E(\mathbf{k})$ defined in (4.1.4) depends on model output $DMP(\mathbf{k})$ and the latter is obtained generally from a numerical model. Since we do not have an explicit expression for $\nabla E(\mathbf{k})$, it is impossible to obtain stationary points by solving equations (4.2.2) directly. For practical optimization problems, we can only use numerical methods to find their approximate solutions. A numerical method of optimization generally consists of three steps:

Step 1. Choose a starting point $\mathbf{k}_0$.

Step 2. Designate a way to generate a search sequence:

$$\mathbf{k}_0, \quad \mathbf{k}_1, \quad \mathbf{k}_2, \quad \ldots, \quad \mathbf{k}_n, \quad \ldots \tag{4.2.4}$$

such that $E(\mathbf{k}_{n+1}) < E(\mathbf{k}_n)$ for all n.

Step 3. Stipulate a convergence criterion. If it is satisfied, then end the search procedure, and a local minimum is approximately achieved.

Search sequence (4.2.4) has the following general form:

$$\mathbf{k}_{n+1} = \mathbf{k}_n + \lambda_n \mathbf{d}_n, \tag{4.2.5}$$

where vector $\mathbf{d}_n$ is called a *displacement direction*, λ_n is a *step size* along this direction. Different optimization methods may use different algorithms to generate $\mathbf{d}_n$ and λ_n.

In the study of numerical optimization, the most useful tool is the Taylor expansion of objective function $E(\mathbf{k})$ around a point $\mathbf{k}_0$. It can be represented in the following matrix form:

$$E(\mathbf{k}_0 + \Delta\mathbf{k}) = E(\mathbf{k}_0) + \mathbf{g}^T \Delta\mathbf{k} + \frac{1}{2}(\Delta\mathbf{k})^T \mathbf{G}(\Delta\mathbf{k}) + \mathbf{HOT}, \tag{4.2.6}$$

where gradient vector $\mathbf{g}$ and Hessian matrix $\mathbf{G}$ are evaluated at point $\mathbf{k}_0$, $\mathbf{HOT}$ represents the third and higher order terms.

Note that the condition of using (4.2.6) is that the third or higher order derivatives of $E(\mathbf{k})$ must exist in K_{ad}. We can sort the various optimization algorithms into three main categories:

· An optimization algorithm is called a *search method* if it only utilizes values of objective functions;

· An optimization algorithm is called a *gradient method* if it utilizes gradients of objective functions;

· An optimization algorithm is called a *second order method* if it utilizes second derivatives of objective functions.

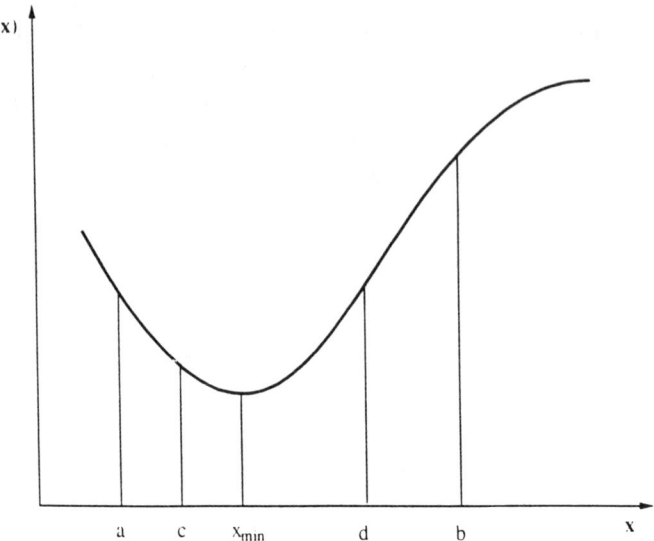

Fig. 4.2.1.   Single variable optimization.

In the solution of inverse problems, we often assume that both gradient vector **g** and Hessian matrix **G** do exist. However, we should remember the following features of *OLS* problems in groundwater modeling:

· Function evaluation of $E(\mathbf{k})$ can be completed through running a simulation model, but it is generally time consuming.

· Although the gradient of $E(\mathbf{k})$ can be calculated by a numerical method, it requires running the simulation model many times.

· It is impractical to calculate the Hessian matrix of $E(\mathbf{k})$ using a simulation model.

4.2.2   SEARCH METHODS

Let us begin with the one-dimensional minimization problem:

$$\min f(\lambda), \qquad a \leq \lambda \leq b, \tag{4.2.7}$$

where objective function $f(\lambda)$ only depends on a single variable λ. If we know that $f(\lambda)$ has a unique minimum in an interval $[a, b]$ (see Figure 4.2.1), how can we determine this minimum through function evaluaton?

Let us introduce a general *bracket method* to solve the above problem. A bracket is defined by three points $a < c < b$, when condition $f(c) \leq \min[f(a), f(b)]$ is satisfied. Obviously, the minimum must take place in a bracket. In Figure 4.2.1,

$[a, c, b]$ is an initial bracket. If we take a new point d in the bracket and calculate $f(d)$, then either $f(c) \le f(d)$ or $f(c) > f(d)$. For the former, $[a, c, d]$ forms a new bracket, that is, we can cut interval $[d, b]$ without missing the minimum. For the latter, $[a, c]$ can be cut and $[c, d, b]$ can be taken as the new bracket. This procedure is then repeated until the length of the maintained bracket is so small that a convergence test is satisfied.

Golden section search is a way to determine the new search points, in which

$$c = b - \tau(b - a), \tag{4.2.8a}$$

$$d = a + \tau(b - a), \tag{4.2.8b}$$

where $\tau = 0.618$. *Fibonacci section search* is a similar method in which τ depends on the Fibonacci sequence.

The *quadratic interpolation method* is another way for determining a new search point, in which c is the middle point of the bracket and d is the minimizer of a quadratic polynomial defined by three points $(a, f(a))$, $(c, f(c))$ and $(b, f(b))$, and we have

$$d = c + \frac{s[f(a) - f(b)]}{2[f(b) - 2f(c) + f(a)]}, \tag{4.2.9}$$

where $s = (b - a)/2$.

Detailed discussions about the golden section search, Fibonacci section search and the quadratic interpolation method can be found in any textbook on numerical optimization. Corresponding subroutines can be found in *Numerical Recipes* (Press *et al.*, 1992). It is suggested that the reader prepare a one-dimensional minimizatoin subroutine using any method mentioned above and write it in the following form:

$$\textit{Subroutine} \quad LINESH(FUNCT, A, B, EPS), \tag{4.2.10}$$

where *LINESH* (*Line Search*) is the name of this subroutine, *Funct* is an external function, i.e., the objective function; $[A, B]$ is the initial search interval, and *EPS* is a criterion for convergence test. When the length of the current bracket is smaller than *EPS*, end the subroutine.

Now, let us turn back to the multiple variable minimization problem (4.2.1). The simplest search method consists of adjusting parameter components, one at a time, with a one-dimensional search approach. In each iteration, all components are adjusted one by one. This procedure is repeated until no further improvement can be obtained. Using the notations of (4.2.5), coordinate directions $e_1, e_2, \ldots, e_M$ are now taken as the displacement directions d_n alternatively, while optimal step size λ_n is determined by solution of the following one-dimensional minimization problem:

$$f(\lambda_n) = \min_{\lambda \in K_{ad,n}} f(\lambda), \qquad f(\lambda) = E(\mathbf{k}_n + \lambda \mathbf{d}_n), \tag{4.2.11}$$

where $K_{ad,n}$ is the projection of K_{ad} on $\mathbf{d}_n$.

The *one at a time* search method is often used for parameter identification when the dimension of parameterization is small, for example, less than 10. The most sensitive parameter for observations should be adjusted first.

Although it is very easy to use the *one at a time* search method, its convergence speed is generally very slow. In the following modified method, M mutually *conjugate directions* are used as search directions instead of the axes directions. If we assume that $E(\mathbf{k})$ is convex in a neighborhood of $\mathbf{k}_0$, Hessian matrix $\mathbf{G}$ in (4.2.6) must be symmetric and non-negative definite. Thus, we can define conjugate directions as follows:

A set of M vectors $\mathbf{u}_1, \mathbf{u}_2, \ldots, \mathbf{u}_M$ is said to be mutually conjugate with respect to $\mathbf{G}$, if

$$\mathbf{u}_i^T \mathbf{G} \mathbf{u}_j = 0 \quad \text{for } i \neq j. \tag{4.2.12}$$

When $E(\mathbf{k})$ is a quadratic function, conjugate directions become the principle axes of the quadratic surface. As a result, its minimum can be found in M search steps along these directions.

If $E(\mathbf{k})$ can be replaced approximately by a quadratic function as shown in (4.2.6), to search along conjugate directions defined above may increase the convergent speed. Powell (1964) proposed a method for generating conjugate directons without calculating the derivatives of $E(\mathbf{k})$. A subroutine based on the Powell's method can be found in *Numerical Recipes* (Press *et al.*, 1992).

4.2.3 GRADIENT METHODS

The basic gradient method is steepest descent method. It uses the negative gradient direction as the search direction in each iteration, i.e., let

$$\mathbf{d}_n = -\mathbf{g}_n, \quad (n = 1, 2, \ldots), \tag{4.2.13}$$

where $\mathbf{g}_n = \nabla E(\mathbf{k}_n)$. Direction $\mathbf{d}_n$ is the steepest descent direction of $E(\mathbf{k})$ at point $\mathbf{k}_n$. The optimal step size along this direction can be determined by a line search as described by (4.2.11). Since we can use the simulation model to calculate $E(\mathbf{k})$ for any parameter k, gradient $\nabla E(\mathbf{k}_n)$ may be obtained by *the forward finite difference approximation*

$$\left. \frac{\partial E}{\partial k_i} \right|_{\mathbf{k}_n} \approx \frac{E(\mathbf{k}_n + \Delta k_i \mathbf{e}_i) - E(\mathbf{k}_n)}{\Delta k_i}, \quad (i = 1, 2, \ldots, M), \tag{4.2.14}$$

where e_i is the unit vector of ith coordinate, Δk_i is an increment of ith component $k_{n,i}$. $E(\mathbf{k}_n)$ and $E(\mathbf{k}_n + \Delta k_i e_i)$ can be obtained by running the simulation model with parameter $\mathbf{k}_n$ and the perturbed parameter $\mathbf{k}_n + \Delta k_i e_i$, respectively.

From a theoretical point of view, (4.2.14) is more accurate if the increment Δk_i is small. When Δk_i is too small, however, the round-off error may control the results. Usually, Δk_i is taken to be directly proportional to k_i obtained in the nth iteration. That is, let

$$\Delta k_i = \alpha k_i, \quad (i = 1, 2, \ldots, M), \tag{4.2.15}$$

where coefficient α can be determined by a trial-error procedure. Its value is generally in the range $10^{-5} \sim 10^{-2}$.

If using (4.2.14) to calculate $\nabla E(\mathbf{k}_n)$, we have to run the simulation model $M + 1$ times in each iteration. In Chapter 6, we will introduce a variational method which allows us to obtain $\nabla E(\mathbf{k}_n)$ with only two simulation runs.

The *steepest descent method* was adopted by many authors for parameter identification in the past (Vemuri *et al.*, 1969; Chavent *et al.*, 1975, among others). This method, however, usually terminates far from the solution due to round-off effects. Therefore, it is generally inefficient and unreliable in practice (Fletcher, 1987).

A conjugate gradient method presented by Fletcher and Reeves (1964) uses the following successive search directions in M iterations:

$$\mathbf{d}_1 = -\mathbf{g}_0,$$

$$\mathbf{d}_n = -\mathbf{g}_{n-1} + \frac{\mathbf{g}_{n-1}^T \mathbf{g}_{n-1}}{\mathbf{g}_{n-2}^T \mathbf{g}_{n-2}} \mathbf{d}_{n-1}, \quad (n = 1, 2, \ldots, M). \tag{4.2.16}$$

After each cycle of M iterations, the first search direction is reset to the steepest descent direction of the current point. Along these search directions, exact line searches are required. Since the search directions defined by (4.2.16) are conjugate directions, the Fletcher–Reeves method allows us to find the minimum for a quadratic function in M iterations. This method is different from the Powell method because it requires evaluation of derivatives. Carrera (1984) and Carrera and Neuman (1986b) compared the Fletcher–Reeves method with the other conjugate gradient algorithms. Their synthetic test cases showed that the Fletcher–Reeves method is superior for identifying the transmissivity of 2-dimensional groundwater flow models.

4.2.4 QUASI-NEWTON METHODS

Let $\mathbf{k}_n$ be a point of search sequence (4.2.4). In the neighborhood of $\mathbf{k}_n$, gradient vector $\mathbf{g}(\mathbf{k})$ can be expressed approximately by

$$g(k) \approx g_n + G_n \Delta k, \tag{4.2.17}$$

where $\Delta k = k - k_n$, $g_n = g(k_n)$, $G_n = G(k_n)$. Equation (4.2.17) is obtained through the differentiation of (4.2.6). According to the necessary condition (4.2.2) for a minimum, we expect that $g(k_{n+1}) \approx 0$. From (4.2.17), this requirement may be satisfied, if we take $k_{n+1} = k_n + \Delta k_n$, where Δk_n is the solution of the following equation

$$g_n + G_n \Delta k = 0. \tag{4.2.18}$$

Thus, we have

$$k_{n+1} = k_n - G_n^{-1} g_n. \tag{4.2.19}$$

When $E(k)$ is a quadratic function, there will be no error in (4.2.17). As a result, its minimum can be found after solving (4.2.18) only once. For the general case, however, an iteration procedure is necessary. The method given above is called *the Newton's method*, which is the fundamental form of second order methods. Comparing (4.2.19) with (4.2.5), we can find that the search direction d_n now is $-G_n^{-1} g_n$, and the optimal step size λ_n is always equal to 1.

The major advantage of Newton method is that it may rapidly converge when $E(k)$ is near a quadratic function. However, there are several disadvantages associated with this method. First, it is impractical to calculate the inverse of Hessian matrix, G^{-1}, in each iteration, when $E(k)$ is obtained from a numerical model. Second, when k_n is not close to the minimum, G_n is not necessarily positive definite. As a result, there is no guarantee that we could obtain $E(k_{n+1}) < E(k_n)$. In other words, the Newton's search sequence (4.2.19) may not be convergent. In practice, the Newton's method is never used for parameter identification in groundwater modeling.

In order to overcome these disadvantages, several *quasi-Newton methods* have been proposed. Their common idea is replacement of G_n^{-1} approximately by a symmetric positive definite matrix H_n, which is updated from iteration to iteration. Generally, the nth iteration includes the following steps:

· Set search direction $d_n = -H_n g_n$.

· Define the next search point $k_{n+1} = k_n + \lambda_n d_n$ as usual, where λ_n is determined by a line search.

· Update H_n with H_{n+1}.

The initial matrix H_1 can be any symmetric positive definite matrix. Taking $H_1 = I$, where I is the unit matrix, is the simplest choice.

The following updating fomula was suggested by Davidon, Fletcher and Powell (see Fletcher, 1987) which is known as the DFP method:

$$\mathbf{H}_{n+1} = \mathbf{H}_n + \frac{\Delta \mathbf{k}_n \Delta \mathbf{k}_n^T}{\Delta \mathbf{k}_n^T \Delta \mathbf{g}_n} - \frac{\mathbf{H}_n \Delta \mathbf{g}_n \Delta \mathbf{g}_n^T \mathbf{H}_n}{\Delta \mathbf{g}_n^T \mathbf{H}_n \Delta \mathbf{g}_n}, \tag{4.2.20}$$

where

$$\Delta \mathbf{k}_n = \mathbf{k}_{n+1} - \mathbf{k}_n, \qquad \Delta \mathbf{g}_n = \mathbf{g}_{n+1} - \mathbf{g}_n. \tag{4.2.21}$$

Another updating formula was suggested by Broyton, Fletcher, Goldfarb and Shanno (see Fletcher, 1987), which is known as the BFGS method:

$$\mathbf{H}_{n+1} = \mathbf{H}_n + \left(1 + \frac{\Delta \mathbf{g}_n^T \mathbf{H}_n \Delta \mathbf{g}_n}{\Delta \mathbf{k}_n^T \Delta \mathbf{g}_n}\right) \frac{\Delta \mathbf{k}_n \Delta \mathbf{k}_n^T}{\Delta \mathbf{k}_n^T \Delta \mathbf{g}_n}$$
$$- \left(\frac{\Delta \mathbf{k}_n \Delta \mathbf{g}_n^T \mathbf{H}_n + \mathbf{H}_n \Delta \mathbf{g}_n \Delta \mathbf{k}_n^T}{\Delta \mathbf{k}_n^T \Delta \mathbf{g}_n}\right). \tag{4.2.22}$$

A detailed discussion of *DFP* and *BFGS* formulae and more updating formulae can be found in Fletcher (1987) or other textbooks.

From (4.2.21) and (4.2.22), we can easily find that only the evaluation of gradient is required for updating matrix $\mathbf{H}$ in each iteration. Therefore, these algorithms may be used for parameter identification in groundwater modeling (Neuman, 1980; Carrera and Neuman, 1986b; Cheng and Yeh, 1992).

4.2.5 STOPPING CRITERIA

There are several stopping criteria used to test the convergency of an iteration sequence.

Criterion 1. The norm of displacement $\Delta \mathbf{k}_n$ becomes very small after n iterations:

$$\|\mathbf{k}_n - \mathbf{k}_{n-1}\| < \varepsilon_1. \tag{4.2.23}$$

Criterion 2. The norm of gradient $\Delta E(\mathbf{k}_n)$ becomes very small after n iterations, i.e.,

$$\|\nabla E(\mathbf{k}_n)\| < \varepsilon_2. \tag{4.2.24}$$

Criterion 3. The value of objective function $E(\mathbf{k})$ has no significant change after n iterations, i.e.,

$$|E(\mathbf{k}_n) - E(\mathbf{k}_{n-1})| < \varepsilon_3. \tag{4.2.25}$$

Criterion 4. The value of objective function $E(\mathbf{k})$ becomes so small that the following inequality is satisfied after n iterations:

$$E(\mathbf{k}_n) \leq \|\eta_D\|^2, \tag{4.2.26}$$

where $\|\cdot\|$ is the l_2 norm defined in the observation space.

In (4.2.23)–(4.2.25), ε_1, ε_2 and ε_3 are given positive numbers.

In practice, if Criterion 1 and Criterion 2 or Criterion 1 and Criterion 3 are satisfied simultaneously, it seems reasonable to stop the iteration procedure. However, we often find that Criteria 2 and 3 are satisfied, but Criterion 1 is not. This means that the quantity of observation data is probably insufficient for identifying the unknown parameters.

If the lower bound of $\|\eta_D\|$ can be estimated, and Criterion 4 is satisfied but others are not, we can probably conclude that the quality of observation data is insufficient for identifying the unknown parameters.

Since the quantity and quality of observations are often limited in practice, we should be not surprised that iteration sequences are not convergent. Generally, incorporating upper and lower bounds of unknown parameters or other constraints into the formulation of inverse problems may enable the iteration procedure to become convergent (See Section 4.4).

4.2.6   PROGRAM AND EXAMPLE

Presently, it is not difficult to find appropriate documented subroutines for numerical optimization. For example, in *Numerical Recipes* (Press *et al.*, 1992), the following subroutines can be found.

(1) Golden section search in one-dimension;

(2) Parabolic interpolation in one-dimension;

(3) Direction set (Powell's) method in multidimensions;

(4) Conjugate gradient methods in multidimensions;

(5) Variable metric (quasi-Newton) methods in multidimensions.

The general form of a numerical optimization subroutine can be represented as

Subroutine MINIM (M, VAR0, VAR1, FUNCT0, FUNCT1, GRAD0, GRAD1, EPS, ...),

where M is the number of variables, $VAR0(M)$ the initial variables, $VAR1(M)$ the updated variables, *FUNCT0* and *FUNCT1* the values of objective function at

VAR0 and *VAR1*, respectively, *GRAD0(M)* and *GRAD1(M)* are the corresponding gradients. These variables are necessary only for conjugate gradient methods and quasi-Newton methods. *EPS* is a stopping criterion.

When Subroutine *MINIM* is used for the purpose of parameter identification, we have to prepare the following data and subroutines:

· the dimension of parameterization (M);

· the initial guess of unknown parameters (VAR_0);

· observation data $\tilde{u}_{D,L}$ and weighting coefficients $W_{D,L}$ $(l = 1, 2, \ldots, L)$;

· a function procedure *FUNCT* for the function evaluation of $E(\mathbf{k})$;

· a subroutine *GRAD* for the gradient evaluation of $E(\mathbf{k})$ if it is necessary;

· one or more stopping criteria *(EPS)*.

Subroutine *FUNCT* is easy to generate by incorporating the observation data into the simulation model. Using current variables $\mathbf{k}$ as model parameters, the simulation model can give the model state distribution $u(\mathbf{x}, t)$ for any $\mathbf{x}$ and t. Thus, we can find $DMP(\mathbf{k})_l$ and calculate $E(\mathbf{k})$ according to (4.1.6).

The forward finite difference approximation (4.2.14) may be used to write Subroutine *GRAD*. To generate all components of the gradient vector, we have to call *FUNCT* for $M + 1$ times. When the central finite difference approximation is used to replace (4.2.14), $2M + 1$ simulation runs are required.

In appendix C, readers can find a subroutine "MCB2D" for simulating 2-dimensional coupled flow and mass transport problems which is ready to combine with any optimization program for parameter identification.

Example 4.2.1. Consider a horizontal, 2-dimensional confined aquifer *ABCD* in Figure 4.2.2, where $\overline{AD}$ is a constant head boundary ($h_B = 100$ m), other boundaries ($\overline{AB}$, $\overline{BC}$ and $\overline{CD}$) are impervious; The length of $\overline{AB}$ is 6500 m, $\overline{BC}$ is 4500 m, and the aquifer is heterogeneous. It can be divided into 3 zones (*AEGD*, *EFHG*, *FBCH*) with transmissivities T_1, T_2, T_3, respectively, but their storage coefficient S are the same. Assume there is a pumping well located at the third zone with constant pumping rate $Q = 2000$ m^3/day. At the beginning of pumping ($t = 0$), the head is a constant ($h_0 = 100$ m) everywhere in the aquifer.

Problem 1. If we know that $T_1 = 500$ m^2/day, $T_2 = 1000$ m^2/day, $T_3 = 2000$ m^2/day and $S = 0.0001$, find the head distributions at $t_1 = 0.5$, $t_2 = 1.0$ and $t_3 = 1.5$ (day).

This is a forward problem which can be solved by any numerical method. In this example, the finite difference method is used. The aquifer is divided into 3 (in the

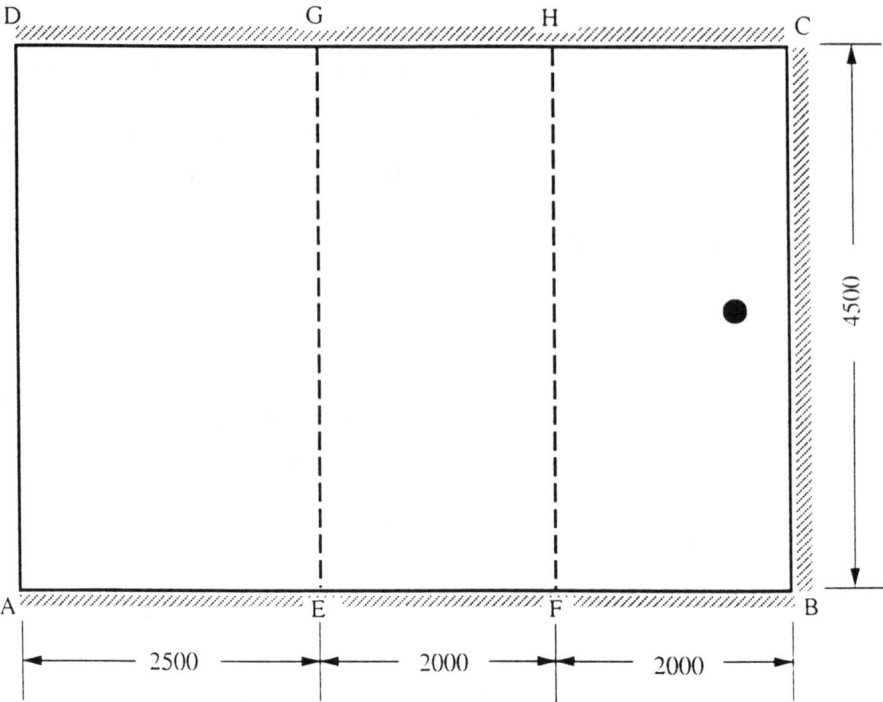

Fig. 4.2.2. A 2-dimensional confined aquifer used in Example 4.2.1, where • indicates the location of pumping well.

AD direction) by 6 (in the *AB* direction) blocks. The calculated head distributions are listed in Table 4.2.1.

Problem 2. Using the data listed in Table 4.2.1 as the observations, find T_1, T_2, T_3 and S by solving the inverse problem.

This problem can be solved by any numerical optimization method, for example, by the one at a time search method. The upper and lower bounds of all parameters are given in Table 4.2.2. The optimal parameters obtained after six circles are also shown in Table 4.2.2. The relative errors of all parameters are already less than 5%. If more circles are used, these errors can be further decreased.

Problem 3. If we decrease the number of observation wells from 17 to 6 (two observation wells in each zone) and increase the observation times from 3 to 10 ($t_1 = 0.5$, $t_{i+1} = t_i + \Delta t$, $\Delta t = 0.1$, $i = 1, 2, \ldots, 9$), can the parameters be identified?

Still using the one at a time search method for six circles, the results are shown in Table 4.2.3. The relative errors are all less than 5%, except that for T_1.

In this example, it implies that the model structure error, observation error and parameterization error are all assumed to be zero. However, we have tested that

TABLE 4.2.1
Head distributions obtained by solving the Forward
Problem

Node No.	0.5 (day)	1.0 (day)	1.5 (day)
1	99.85	99.19	98.17
2	99.42	97.65	95.38
3	98.36	95.14	91.12
4	96.50	91.96	87.74
5	93.64	88.12	83.43
6	90.83	84.93	80.07
7	99.85	99.19	98.17
8	99.42	97.65	95.38
9	98.35	95.14	91.71
10	96.46	91.92	87.71
11	93.40	87.88	83.19
13	99.85	99.19	98.17
14	99.42	97.65	95.38
15	98.36	95.14	91.72
16	96.50	91.96	87.74
17	93.64	88.12	83.43
18	90.83	84.93	80.07

TABLE 4.2.2
Upper bounds, Lower bounds, True Values and Identified Values
of Parameters T_1, T_2, T_3 and S

Parameters	T_1	T_2	T_3	S
Upper Bound	1000	1500	3000	5×10^{-4}
Lower Bound	100	500	1000	5×10^{-5}
True Value	500	1000	2000	10^{-4}
Identified Parameter	525	972	2078	1.006×10^{-4}
Relative Error	5%	3%	4%	0.5%

when ± 5 cm error is added to the observed head distributions (Table 4.2.1), the
relative errors of identified parameters are still less than 5% after 10 circles.

Exercise 4.2.1. Derive the Taylor expansion (4.2.6) from its scalar form.

TABLE 4.2.3
The Identified Parameters for Problem 3 of Example 4.2.1.

Parameters	T_1	T_2	T_3	S
True Value	500	1000	2000	10^{-4}
Identified Parameter	574	991	2090	0.991×10^{-4}
Relative Error	15%	0.9%	4.5%	0.9%

Exercise 4.2.2. Give the representation of Hessian matrix $\mathbf{G}$ defined in (4.2.3) when objective function E is given by (4.1.6).

Exercise 4.2.3. Prepare a line search subroutine and then use the one at a time search method to identify transmissivity T and storage coefficient S when the observations are obtained from a Theis model. Test the stability of inverse solution with respect to the observation error.

Exercise 4.2.4. Write a program that can simulate the one-dimensional tracer test described in Example 1.2.3 and determine the longitudinal dispersivity by solving the inverse problem.

4.3 Gauss–Newton Method

The Gauss–Newton optimization method is specially designed for minimizing the objective function which has the form of sum of square functions, such as function $E(\mathbf{k})$ in (4.1.6). Some modified algorithms of this method with constraints are often used for parameter identification in groundwater modeling.

4.3.1 THE FUNDAMENTAL GAUSS–NEWTON METHOD

Using $\mathbf{u}_D(\mathbf{k})$ to represent model output $DMP(\mathbf{k})$, (4.1.6) can be rewritten as:

$$E(\mathbf{k}) = \sum_{l=1}^{L} w_{D,l}^2 [u_{D,l}(\mathbf{k}) - u_{D,l}^{obs}]^2 \tag{4.3.1}$$

or

$$E(\mathbf{k}) = \sum_{l=1}^{L} f_l^2(\mathbf{k}), \tag{4.3.2}$$

where

$$f_l(\mathbf{k}) = w_{D,l}[u_{D,l}(\mathbf{k}) - u_{D,l}^{\text{obs}}]. \tag{4.3.3}$$

The first order derivatives of $E(\mathbf{k})$ can be obtained directly from (4.3.2):

$$\frac{\partial E}{\partial k_i} = 2 \sum_{l=1}^{L} f_l \frac{\partial f_l}{\partial k_i} \quad (i = 1, 2, \ldots, M) \tag{4.3.4}$$

and thus we have second order derivatives

$$\frac{\partial^2 E}{\partial k_i \partial k_j} = 2 \sum_{l=1}^{L} \left[\frac{\partial f_l}{\partial k_i} \frac{\partial f_l}{\partial k_j} + f_l \frac{\partial^2 f_l}{\partial k_i \partial k_j} \right]$$

$$(i = 1, 2, \ldots, M; \ j = 1, 2, \ldots, M) \tag{4.3.5}$$

Since $f_l(\mathbf{k})$ is a residual, when $\mathbf{k}$ is not too far from the minimizer, it is reasonable to assume that the value of $f_l(\mathbf{k})$ is small and the second order terms on the right-hand side of (4.3.5) can be ignored. Thus,

$$\frac{\partial^2 E}{\partial k_i \partial k_j} \approx 2 \sum_{l=1}^{L} \frac{\partial f_l}{\partial k_i} \frac{\partial f_l}{\partial k_j}. \tag{4.3.6}$$

It is now convenient to define a new matrix

$$\mathbf{A} = \begin{bmatrix} \dfrac{\partial f_1}{\partial k_1} & \dfrac{\partial f_1}{\partial k_2} & \cdots & \dfrac{\partial f_1}{\partial k_M} \\[2ex] \dfrac{\partial f_2}{\partial k_1} & \dfrac{\partial f_2}{\partial k_2} & \cdots & \dfrac{\partial f_2}{\partial k_M} \\[2ex] \vdots & \vdots & \ddots & \vdots \\[2ex] \dfrac{\partial f_L}{\partial k_1} & \dfrac{\partial f_L}{\partial k_2} & \cdots & \dfrac{\partial f_L}{\partial k_M} \end{bmatrix}, \tag{4.3.7}$$

which consists of derivatives of functions $f_1(\mathbf{k})$, $f_2(\mathbf{k})$, ..., $f_L(\mathbf{k})$ with respect to the variation of each parameter components $k_1, k_2, \ldots, k_M$. Since L is usually much larger than M, it is not a square matrix. Using matrix $\mathbf{A}$ and considering (4.3.4) and (4.3.6), gradient ∇E can be represented in matrix form

$$\nabla E = 2\mathbf{A}^T \mathbf{f}, \tag{4.3.8}$$

where

$$\mathbf{f} = (f_1, f_2, \ldots, f_L)^T \tag{4.3.9}$$

and Hessian matrix $\mathbf{G}$ can be replaced approximately by

$$\mathbf{G} \approx 2\mathbf{A}^T \mathbf{A}. \tag{4.3.10}$$

Substituting (4.3.8) and (4.3.10) into Newton algorithm (4.2.19), we obtain:

$$\mathbf{k}_{n+1} = \mathbf{k}_n - \left(\mathbf{A}_n^T \mathbf{A}_n\right)^{-1} \mathbf{A}_n^T \mathbf{f}_n, \tag{4.3.11}$$

where the subscript n of $\mathbf{A}_n$ and $\mathbf{f}_n$ indicates that $\mathbf{A}$ and $\mathbf{f}$ are evaluated at $\mathbf{k}_n$. The iteration sequence generated by (4.3.11) is known as the Gauss–Newton sequence and

$$\Delta\mathbf{k}_n = -\left(\mathbf{A}_n^T \mathbf{A}_n\right)^{-1} \mathbf{A}_n^T \mathbf{f}_n \tag{4.3.12}$$

the Gauss–Newton direction.

Note that the Gauss–Newton direction is determined only by first order derivatives. When $f_1, f_2, \ldots, f_L$ are linear functions, it coincides with the Newtion direction because (4.3.10) becomes accurate. When $f_1, f_2, \ldots, f_L$ are nonlinear functions, the deviation of the Gauss–Newton direction from the Newton direction depends on the accuracy of (4.3.10). Therefore, for small residual problems (the observation error is small and $\mathbf{k}_n$ is close to the true parameter), the Gauss–Newton sequence will converge rapidly. Unfortunately, the parameter identification problems in groundwater modeling often belong to the case of large residuals. In the generation of a Gauss–Newton sequence for inverse solution, the following difficulties are often encountered:

· The search sequence does not converge, i.e., $E(\mathbf{k}_{n+1}) > E(\mathbf{k}_n)$ for some n.

· Matrix $\mathbf{A}^T \mathbf{A}$ is near singular, the Gauss–Newton direction in (4.3.11) can not be determined.

· The displacement vector $\Delta\mathbf{k}_n$ obtained by (4.3.11) is so large that $\mathbf{k}_{n+1}$ is no longer in the admissible set K_{ad}.

In order to avoid these difficulties, it is necessary to modify the Gauss–Newton algorithm.

4.3.2 MODIFIED GAUSS–NEWTON METHODS

There are several ways to modify the Gauss–Newton algorithm. The first is to add a line search in each iteration. Let

$$\mathbf{k}_{n+1} = \mathbf{k}_n + \lambda_n \mathbf{d}_n, \tag{4.3.13}$$

where $\mathbf{d}_n = \Delta\mathbf{k}_n$ is the Gauss–Newton direction in nth iteration, and λ_n is the optimal step size along this direction. By introducing λ_n, inequality $E(\mathbf{k}_{n+1}) \leq E(\mathbf{k}_n)$ can always be guaranteed. Of course, additional computational effort is needed for completing the line search in each iteration.

The second method involves modifying the displacement direction. Let

$$
\mathbf{d}_n = \begin{cases} \Delta \mathbf{k}_n, & \text{if } \mathbf{g}_n \cdot \Delta \mathbf{k}_n < 0, \\[2mm] -\mathbf{g}_n, & \text{if } \mathbf{g}_n \cdot \Delta \mathbf{k}_n \geq 0. \end{cases} \tag{4.3.14}
$$

Equation (4.3.14) means that if the Gauss–Newton direction $\Delta \mathbf{k}_n$ is agreeable with the steepst descent direction $-\mathbf{g}_n$, then $\Delta \mathbf{k}_n$ is used as the displacement direction, otherwise, $-\mathbf{g}_n$ must be used. In other words, when the Gauss–Newton step is unsuccessful, it becomes necessary to use a steepest descent step instead.

The Levenberg-Marquardt method is another way to guarantee that $\mathbf{k}_{n+1}$ is better than $\mathbf{k}_n$. In this case, the Gauss–Newton direction is replaced by

$$
\Delta \mathbf{k}_n = (\mathbf{A}_n^T \mathbf{A}_n + \lambda \mathbf{I})^{-1} \mathbf{A}_n^T \mathbf{f}_n, \tag{4.3.15}
$$

where $\mathbf{I}$ is the unit matrix, λ is a coefficient. When $\lambda = 0$, $\Delta \mathbf{k}_n$ reduces to the Gauss–Newton direction. On the other hand, when λ tends to infinity, $\Delta \mathbf{k}_n$ turns to the steepest descent direction and the size of $\Delta \mathbf{k}_n$ tends to zero. Therefore, $E(\mathbf{k}_{n+1}) < E(\mathbf{k}_n)$ can always be expected by increasing the value of λ. Marquardt suggested that if condition $E(\mathbf{k}_{n+1}) < E(\mathbf{k}_n)$ is not satisfied for an initial value of λ, then multiply λ by 10 and recompute $\Delta \mathbf{k}_n$ with (4.3.15) until the reduction condition is satisfied. Since a line search is not required in the Levenberg–Marquardt method, it may be more effective than using (4.3.13).

Up to now, we have not considered a way to limit the identified parameter into a designated admissible set K_{ad}. Suppose that each component k_i of the unknown parameter $\mathbf{k}$ is limited by a upper bound $\overline{k}_i$ and a lower bound $\underline{k}_i$, thus

$$
K_{ad} = \{ \mathbf{k} | \underline{k}_i \leq k_i \leq \overline{k}_i, \quad i = 1, 2, \ldots, M \}. \tag{4.3.16}
$$

When (4.3.13) is used to generate $\mathbf{k}_{n+1}$, the interval for line search must be limited in K_{ad}. When (4.3.15) is used to generate $\mathbf{k}_{n+1}$, the value of coefficient λ must be large enough so that $\mathbf{k}_{n+1} \in K_{ad}$ and $E(\mathbf{k}_{n+1}) < E(\mathbf{k}_n)$ are both satisfied.

A general inverse solution program for groundwater modeling is given in Appendix C. In this program, the Levenberg–Marquardt method is used to identify various hydrogeological parameters, such as hydraulic conductivities, storage coefficients, porosities and dispersivities.

4.3.3 SENSITIVITY COEFFICIENTS

Most computational effort in the Gauss–Newton method and its modified algorithms results in generating matrix $\mathbf{A}$ defined by (4.3.7) in each iteration.

Substituting (4.3.3) into (4.3.7), we obtain

$$
\mathbf{A} = \mathbf{W}_D \mathbf{J}_D, \tag{4.3.17}
$$

where $\mathbf{W}_D$ is a diagonal matrix with elements $w_{D,1}, w_{D,2}, \ldots, w_{D,L}$; $\mathbf{J}_D$ is the Jacobian of model output with respect to model parameters, i.e.,

$$
\mathbf{J}_D = \begin{bmatrix}
\dfrac{\partial u_{D,1}}{\partial k_1} & \dfrac{\partial u_{D,1}}{\partial k_2} & \cdots & \dfrac{\partial u_{D,1}}{\partial k_M} \\[2ex]
\dfrac{\partial u_{D,2}}{\partial k_1} & \dfrac{\partial u_{D,2}}{\partial k_2} & \cdots & \dfrac{\partial u_{D,2}}{\partial k_M} \\[2ex]
\vdots & \vdots & \ddots & \vdots \\[2ex]
\dfrac{\partial u_{D,L}}{\partial k_1} & \dfrac{\partial u_{D,L}}{\partial k_2} & \cdots & \dfrac{\partial u_{D,L}}{\partial k_M}
\end{bmatrix}. \tag{4.3.18}
$$

The elements of $[\mathbf{J}_D]$ are often called *sensitivity coefficients* and $[\mathbf{J}_D]$ the *sensitivity matrix*.

There are three approaches for obtaining sensitivity coefficients: the *finite difference method*, the *sensitivity equation method* and the *variational method*.

In the finite difference method, sensitivity coefficients $\partial u_{D,l}/\partial k_i$ are replaced by their finite difference approximations as we did in Section 4.2.3 for calculating the gradient $\nabla E(\mathbf{k})$. Using this method to get all sensitivity coefficients evaluated at a give parameter vector $\mathbf{k}$, $M+1$ simulation runs are needed.

In order to introduce the sensitivity equation method, let us consider the following *PDE*:

$$
\frac{\partial(au)}{\partial t} = \frac{\partial}{\partial x}\left(b_{11}\frac{\partial u}{\partial x} + b_{12}\frac{\partial u}{\partial y} - c_{11}u\right) + \frac{\partial}{\partial y}\left(b_{21}\frac{\partial u}{\partial x} + b_{22}\frac{\partial u}{\partial y} - c_{22}u\right)
$$

$$
+du + e \quad (x,y) \in (\Omega),\ t \geq 0 \tag{4.3.19}
$$

subject to

$$
u|_{t=0} = f_0, \tag{4.3.20a}
$$

$$
u|_{\Gamma_1} = f_1, \tag{4.3.20b}
$$

$$
-\left[\left(b_{11}\frac{\partial u}{\partial x} + b_{12}\frac{\partial u}{\partial y} - c_{11}u\right)n_x + \left(b_{21}\frac{\partial u}{\partial x} + b_{22}\frac{\partial u}{\partial y} - c_{22}u\right)n_y\right]\Bigg|_{\Gamma_2} = f_2,
$$

$$
\tag{4.3.20c}
$$

where f_0, f_1, f_2 are given functions, coefficients $a, b_{11}, b_{12}, b_{21}, b_{22}, c_{11}, c_{22}, d, e$ may be functions of x, y, t. They may also depend on u when the model is nonlinear. Various groundwater flow equations, mass transport equations, heat transport equations are all special cases of Equation (4.3.19).

Assume that all unknown coefficients $a, b_{11}, b_{12}, b_{21}, b_{22}, c_{11}, c_{22}, d, e$ have been parameterized by a M-dimensional vector $\mathbf{k} = (k_1, k_2, \ldots, k_M)$. By differentiation of Equations (4.3.19) and (4.3.20) with respect to a component k_i, we obtain

$$
\frac{\partial(a u_i')}{\partial t} = \frac{\partial}{\partial x}\left(b_{11}\frac{\partial u_i'}{\partial x} + b_{12}\frac{\partial u_i'}{\partial y} - c_{11} u_i' \right)
$$
$$
+ \frac{\partial}{\partial y}\left(b_{21}\frac{\partial u_i'}{\partial x} + b_{22}\frac{\partial u_i'}{\partial y} - c_{22} u_i' \right) + d u_i' + e'
$$
$$
(x, y) \in (\Omega),\ t \geq 0 \tag{4.3.21}
$$

and

$$
u_i'|_{t=0} = 0, \tag{4.3.22a}
$$
$$
u_i'|_{\Gamma_1} = 0, \tag{4.3.22b}
$$
$$
-\left[\left(b_{11}\frac{\partial u_i'}{\partial x} + b_{12}\frac{\partial u_i'}{\partial y} - c_{11} u_i' \right) n_x \right.
$$
$$
\left. + \left(b_{21}\frac{\partial u_i'}{\partial x} + b_{22}\frac{\partial u_i'}{\partial y} - c_{22} u_i' \right) n_y \right]\Bigg|_{\Gamma_2} = f_2', \tag{4.3.22c}
$$

where $u_i' = \partial u / \partial k_i$ is the sensitivity coefficient that we want to find, and

$$
e' = \frac{\partial}{\partial x}\left(\frac{\partial b_{11}}{\partial k_i}\frac{\partial u}{\partial x} + \frac{\partial b_{12}}{\partial k_i}\frac{\partial u}{\partial y} - \frac{\partial c_{11}}{\partial k_i}u \right)
$$
$$
+ \frac{\partial}{\partial y}\left(\frac{\partial b_{21}}{\partial k_i}\frac{\partial u}{\partial x} + \frac{\partial b_{22}}{\partial k_i}\frac{\partial u}{\partial y} - \frac{\partial c_{22}}{\partial k_i}u \right) + \frac{\partial d}{\partial k_i}u + \frac{\partial e}{\partial k_i} - \frac{\partial}{\partial t}\left(\frac{\partial a}{\partial k_i}u \right),
$$

$$
f_2' = \left[\left(\frac{\partial b_{11}}{\partial k_i}\frac{\partial u}{\partial x} + \frac{\partial b_{12}}{\partial k_i}\frac{\partial u}{\partial y} - \frac{\partial c_{11}}{\partial k_i}u \right) n_x \right.
$$
$$
\left. + \left(\frac{\partial b_{21}}{\partial k_i}\frac{\partial u}{\partial x} + \frac{\partial b_{22}}{\partial k_i}\frac{\partial u}{\partial y} - \frac{\partial c_{22}}{\partial k_i}u \right) n_y \right]\Bigg|_{\Gamma_2}.
$$

Equation (4.3.21) is called the sensitivity equation with respect to k_i. Equations (4.3.22a)–(4.3.22c) are its subsidiary conditions.

Note that $\partial a / \partial k_i$, $\partial b_{11} / \partial k_i$, $\ldots$, $\partial e / \partial k_i$ can be obtained from the parameterization formulae, and u, $\partial u / \partial x$, $\partial u / \partial y$ and $\partial u / \partial t$ can be obtained from the solution of simulation problem (4.3.19)–(4.3.20). Therefore, e' and f_2' are both known and sensitivity problem (4.3.21)–(4.3.22) may be determined.

Since the form of the sensitivity problem is exactly the same as that of the simulation problem, we can use the same subroutine to solve both of them.

Solving the simulation problem once and solving the sensitivity problem for each component of $\mathbf{k}$, we can obtain all sensitivity coefficients and form the sensitivity matrix $\mathbf{J}_D$. The total computational effort is the same as that of using the finite difference method. The difficulty of determining the size of perturbation increments can thus be avoided.

The variational method for calculating sensitivity coefficients will be given in Chapter 6. We will see that sensitivity matrix $\mathbf{J}_D$ plays an important role in the study of inverse problems, not only for parameter identification, but also for reliability analysis and observation design. In order to avoid $\mathbf{A}^T\mathbf{A}$ being close to a singular matrix, it is required that there exist at least one observation for each component k_i $(i = 1, 2, \ldots, M)$ which is sensitive enough with repect to this component.

In Appendix C, readers can find a numrical example which uses the Gauss–Newton–Levenberg–Marquardt method to identify hydraulic conductivities and storage coefficients for a confined aquifer. Both head and concentration observations are used for parameter identification. Since the presented inverse problem is well-posed, the Gauss–Newton sequence converges very fast. The true parameters are found in only five iterations. Uniqueness and stability of the inverse solution can also be observed.

Exercise 4.3.1.   Prove that when $f_1, f_2, \ldots, f_L$ in (4.3.2) are linear functions, the Gauss–Newton method can find the optimal solution in only one iteration.

Exercise 4.3.2.   Find the sensitivity equations and their subsidiary conditions for 2-dimensional groundwater flow in a confined aquifer. (The forward problem is given in Example 1.1.4).

Exercise 4.3.3.   Write a program that can simulate the two-dimensional tracer test described in Example 1.2.4 and identify the longitudinal and transverse dispersivities based on the Gauss–Newton method.

Exercise 4.3.4.   Test the uniqueness and stability of the inverse solution of the numerical example given in Appendix C.

Exercise 4.3.5   Using the inverse solution program given in Appendix C to solve an inverse problem designed by the reader.

4.4   The Regularization Method

We have shown that the OLSI is seldom achieved. When a parameter is not OLSI, minimizing the least squares criterion will be difficult. There may exist many local minima, the search sequence may be divergent, the Gauss–Newton direction may be undetermined due to the singularity of matrix $A^T A$, and so on. In order to avoid these difficulties, we must use other information besides state observations. Regularization is a way of alleviating the ill-posedness of inverse problems through incorporation of prior information into the objective function.

4.4.1 THE REGULARIZED *OLS* PROBLEM

The regularization method was first proposed by Tikhonov (1963) and other Russian mathematician-physicists and further developed by Banks and Daniel (1985), Kravaris and Seinfeld (1985), Kunisch and White (1986), and Chavent (1987b) among others.

For the parameter identification of distributed parameter systems, the regularized objective function $E_\alpha(\mathbf{k})$ often takes the following form:

$$E_\alpha(\mathbf{k}) = E_{LS}(\mathbf{k}) + \alpha^2 E_R(\mathbf{k}), \tag{4.4.1}$$

where

$$E_{LS}(\mathbf{k}) = \|\mathbf{u}_D(\mathbf{k}) - \mathbf{u}_D^{\text{obs}}\|_{U_D}^2 \tag{4.4.2}$$

is the *OLS* criterion used before, and

$$E_R(\mathbf{k}) = \|\mathbf{k} - \mathbf{k}_0\|_{U_K}^2. \tag{4.4.3}$$

In (4.4.1)–(4.4.3), $\alpha^2 E_R(\mathbf{k})$ is called the regularization term, $\alpha > 0$ the regularization coefficient, $\mathbf{k}_0$ is the initial guess of unknown parameter $\mathbf{k}$ obtained from prior information. $\|\cdot\|_{U_D}$ and $\|\cdot\|_{U_K}$ are weighted least square norms defined in observation space U_D and parameter space U_K, respectively. Thus, when the observation is pointwise, the regularization objective function $E_\alpha(\mathbf{k})$ in (4.4.1) can be represented as

$$E_\alpha(\mathbf{k}) = \sum_{l=1}^{L} w_{D,l}^2 [u_{D,l}(\mathbf{k}) - u_{D,l}^{\text{obs}}]^2 + \alpha^2 \sum_{m=1}^{M} v_m^2 (k_m - k_m^0)^2,$$

$$\tag{4.4.4}$$

where k_m^0 and v_m $(m = 1, 2, \ldots, M)$ are components of $\mathbf{k}_0$ and weighting coefficients, respectively.

The regularized *OLS* problem involves finding a $\mathbf{k}_\alpha \in K_{\text{ad}}$ such that

$$E_\alpha(\mathbf{k}_\alpha) = \min E_\alpha(\mathbf{k}), \quad \mathbf{k} \in K_{\text{ad}}. \tag{4.4.5}$$

Note that the regularized problem (4.4.5) differs from the original *OLS* problem in two aspects. First, the domain of minimization is restricted in a sphere around $\mathbf{k}_0$ when α is large enough. Second, the shape of objective function is changed by the regularization term. It is more like a quadratic function when α is larger. In other words, both the solution domain and objective function are more regular than that of the original *OLS* problem. As a result, all numerical optimization methods given before will be more effective.

The solution k_α of regularization problem (4.4.5) depends on the regularization coefficient α. When α tends to zero, the regularization term vanishes and k_α tends to the solution of the original *OLS* problem. On the other hand, when α tends to infinity, the regularization term will dominate the *OLS* term and thus solution k_α tends to its prior estimation k_0. In the former case, the solution completely depends on the observation data, but may be unstable when observation and/or model errors exist. In the latter case, the solution is absolutely stable, but no consideration is given to the observation data.

Taking $0 < \alpha < \infty$ is a compromise between the stability and accuracy of the inverse solution, or a compromise between the confidence of observations and the confidence of prior information. If α is taken appropriately, the regularized problem will be well-posed and its solution will be a reasonable approximation to the original *OLS* problem.

In the field of groundwater modeling, objective function (4.4.1) has been used for a long time (Neuman, 1973; Gavalas *et al.*, 1976; Cooley, 1977; Shah *et al.*, 1978; Neuman and Yakowitz, 1979; Cooley, 1982; Carrera and Neuman, 1986a among others). It was explained by different authors from different points of view. For example, from the point of view of multiobjective optimization, objective function (4.4.1) can be seen as a weighted combination of two objectives $E_{LS}(\mathbf{k})$ in (4.4.2) and $E_R(\mathbf{k})$ in (4.4.3). From the point of view of constrained optimization, $E_R(\mathbf{k})$ can be seen as a penalty function.

In a similar manner, we can incorporate other kinds of prior information into the optimization problem. For example, if the Darcy's velocity of groundwater is measured at some points in the flow region, we can add another regularization terms to objective function (4.4.1):

$$E_{\alpha,\beta}(\mathbf{k}) = E_{LS}(\mathbf{k}) + \alpha^2 E_R(\mathbf{k}) + \beta^2 E_Q(\mathbf{k}), \qquad (4.4.6)$$

where

$$E_Q(\mathbf{k}) = \|\mathbf{q}_D(\mathbf{k}) - \mathbf{q}_D^{\text{obs}}\|_{Q_D}^2 \qquad (4.4.7)$$

measures the "distance" between observed velocities $\mathbf{q}_D^{\text{obs}}$ and computed velocities $q_D(\mathbf{k})$ in the velocity observation space Q_D, and β is the regularization coefficient for the velocity term.

4.4.2 THE REGULARIZATION COEFFICIENT

Kravaris and Seinfeld (1985) proved that under certain conditions, the regularized *OLS* problem (4.1.5) is well-posed, provided that the regularization coefficient α is appropriately selected.

An appropriate regularization coefficient must depend on the observation error. In fact, if the observation error is small, a small regularization coefficient should be used to get a good match. Otherwise, we have to increase the value of the regularization coefficient to improve the stability of the solution. Let η be the norm of observation error, i.e.,

$$\|\mathbf{u}_D(\mathbf{k}^*) - \mathbf{u}_D^{\text{obs}}\|_{U_D} = \eta, \tag{4.4.8}$$

where $\mathbf{u}_D(\mathbf{k}^*)$ is the model output corresponding to the true model parameter $\mathbf{k}^*$. Thus, any parameter $\mathbf{k}$ that causes the norm of least squares residuals to be less or equal to η, i.e.,

$$\|\mathbf{u}_D(\mathbf{k}) - \mathbf{u}_D^{\text{obs}}\|_{U_D} \leq \eta \tag{4.4.9}$$

must not be rejected. In other words, any further improvement of this match is meaningless. If there are two parameters $\mathbf{k}_1$ and $\mathbf{k}_2$ and both of them satisfy

$$\|\mathbf{u}_D(\mathbf{k}_1) - \mathbf{u}_D^{\text{obs}}\|_{U_D} < \eta \quad \text{and} \quad \|\mathbf{u}_D(\mathbf{k}_2) - \mathbf{u}_D^{\text{obs}}\|_{U_D} < \eta, \tag{4.4.10}$$

then we can not determine which one is better, unless we have other information.

Therefore, Tikhonov and Arsenin (1977) suggested that the regularization coefficient α can be so selected that the following condition is satisfied:

$$\|\mathbf{u}_D(\mathbf{k}_\alpha) - \mathbf{u}_D^{\text{obs}}\|_{U_D} = \eta, \tag{4.4.11}$$

where $\mathbf{k}_\alpha$ is the solution of problem (4.4.5). Kravaris and Seinfeld (1985) examined the existence of such a coefficient α. They found that this method is mathematically well defined and that $\eta \to 0$ implies $\mathbf{k}_\alpha \to \mathbf{k}^*$.

To find a coefficient α satisfying condition (4.4.11), we must use a trial-and-error process. First, guess a coefficient α and solve problem (4.4.5) to obtain a solution $\mathbf{k}_\alpha$, then calculate the right-hand side of (4.4.11), $R_\alpha = \|\mathbf{u}_D(\mathbf{k}_\alpha) - \mathbf{u}_D^{\text{obs}}\|_{U_D}$. If $R_\alpha < \eta$, then increase α and repeat the above procedure; if $R_\alpha > \eta$, then decrease α and repeat the procedure, until $R_\alpha \approx \eta$ is achieved. This approach, of course, is time consuming.

If an upper bound of $E_R(\mathbf{k})$ can be estimated, for example, $\|\mathbf{k} - \mathbf{k}_0\|_{U_K} \leq \Delta$ is known, Miller (1970) suggested that α may be selected as

$$\alpha = \frac{\eta}{\Delta}. \tag{4.4.12}$$

When α is determined by (4.4.12), the stability requirement of the regularized *OLS* solution will be satisfied (Kravaris and Seinfeld, 1985). Furthermore, if $\tilde{\mathbf{k}}_\alpha$ is a minimizer of the following objective function

$$E_\alpha(\mathbf{k}) = E_{LS}(\mathbf{k}) + \left(\frac{\eta}{\Delta}\right)^2 E_R(\mathbf{k}), \tag{4.4.13}$$

then

$$\|\mathbf{u}_D(\tilde{\mathbf{k}}_\alpha) - \mathbf{u}_D^{obs}\|^2 + \left(\frac{\eta}{\Delta}\right)^2 \|\tilde{\mathbf{k}} - \mathbf{k}_0\|^2$$

$$\leq E_\alpha(\mathbf{k}^*) = \|\mathbf{u}_D(\mathbf{k}^*) - \mathbf{u}_D^{obs}\|^2 + \left(\frac{\eta}{\Delta}\right)^2 \|\mathbf{k}^* - \mathbf{k}_0\|^2 \leq 2\eta^2. \tag{4.4.14}$$

Thus,

$$\|u_D(\tilde{\mathbf{k}}_\alpha) - \mathbf{u}_D^{obs}\|_{U_D} \leq \sqrt{2}\,\eta. \tag{4.4.15}$$

Equation (4.4.15) means that $\tilde{\mathbf{k}}_\alpha$ gives a satisfactory match for values up to a factor $\sqrt{2}$.

In Chapter 7, we will determine the regularization coefficient from the point of view of statistics.

Once the regularization coefficient α is selected, the regularized *OLS* problem (4.4.5) can be solved by any numerical optimization algorithm mentioned in Sections 4.2 and 4.3. Since the objective function (4.4.1) is in the form of the sum of square functions, the Gauss–Newton method and its modified forms can still be used. Examples of using regularization method in aquifer identification can be found in Gavalas *et al.* (1976), Neuman and Yakowitz (1979), Neuman *et al.* (1980), Cooley (1982), Chen (1985), and Lee and Seinfeld (1987) among others.

Exercise 4.4.1.   What are the advantages of the regularized *OLS* method in comparison with the original *OLS* method?

Exercise 4.4.2.   Derive the Gauss–Newton method for the regularized *OLS* problem (4.4.5) and give the expression of matrix $[\mathbf{A}_n]$ in (4.3.12).

Exercise 4.4.3.   Incorporate a regularization term into the inverse solution program given in Appendix C.

4.5   The L$_p$-Norm Criteria

In Chapter 3, the general criterion for parameter identification is given by

$$\min_{\mathbf{k} \in K_{\mathrm{ad}}} E(\mathbf{k}), \quad \text{where } E(\mathbf{k}) = \|u_D(\mathbf{k}) - u_D^{\mathrm{obs}}\|_{U_D}. \tag{4.5.1}$$

Using different norms in the observation space U_D to measure the "distance" between model output and observations, (4.5.1) gives different criteria. So far, we only considered the use of L_2-norm. When the probability distribution of observation error is normal, L_2-norm is the best selection. But, when outliers are suspected in observations, L_2-norm is not a good selection.

For the general L_p-norm defined in (A.23) of Appendix A, the objective function in (4.5.1) becomes

$$E(\mathbf{k}) = \left(\sum_{l=1}^{L} w_{D,l}^p |u_{D,l}(\mathbf{k}) - u_{D,l}^{\mathrm{obs}}|^p \right)^{1/p} \tag{4.5.2}$$

for a pointwise observation system D, where $w_{D,l}$ $(l = 1, 2, \ldots, L)$ are weighting coefficients, and

$$E(\mathbf{k}) = \left(\int_{(D)} w_D^p |u_D(\mathbf{k}) - u_D^{\mathrm{obs}}|^p \, \mathrm{d}D \right)^{1/p} \tag{4.5.3}$$

for a distributed observation system D, where w_D is a weighting function. Equation (4.5.2) and (4.5.3) are called L_p-norm criteria for parameter identification. In practice, however, only L_1-norm and L_∞-norm criteria are used sometimes.

When $p = 1$, (4.5.2) reduces to

$$E(\mathbf{k}) = \sum_{l=1}^{L} w_{D,l} |u_{D,l}(\mathbf{k}) - u_{D,l}^{\mathrm{obs}}|. \tag{4.5.4}$$

Because only the absolute values of residuals other than their square values are included in (4.5.4), the negative effect of outliers in L_1-norm criterion is not as much as that in L_2-norm criterion.

Let $p \to \infty$ in (4.5.2), we obtain

$$E(\mathbf{k}) = \max_{1 \leq l \leq L} w_{D,l} |u_{D,l}(\mathbf{k}) - u_{D,l}^{\mathrm{obs}}|. \tag{4.5.5}$$

Equation (4.5.5) means that the L_∞-norm criterion requires to minimize the maximum residual. It is only used in the case that the observation errors are small or there is no any outlier.

To minimize the L_1-norm objective function (4.5.4), we define

$$u_{D,l}(\mathbf{k}) - u_{D,l}^{\text{obs}} = V_l - U_l \quad (l = 1, 2, \ldots L), \tag{4.5.6}$$

where $V_l \geq 0$ and $U_l \geq 0$. When $u_{D,l}(\mathbf{k}) - u_{D,l}^{\text{obs}} \geq 0$ let $U_l = 0$; when $u_{D,l}(\mathbf{k}) - u_{D,l}^{\text{obs}} \leq 0$, let $V_l = 0$. Thus, objective function $E(\mathbf{k})$ can be rewritten as

$$E(\mathbf{k}) = \sum_{l=1}^{L} w_{D,l}(V_l + U_l). \tag{4.5.7}$$

Note that $u_{D,l}(\mathbf{k}) - u_{D,l}^{\text{obs}}$ may be positive or negative, but U_l and V_l are always nonnegative.

Let $\mathbf{k}_n$ be a known point in the search sequence (4.2.4). The linearization of (4.5.6) yields:

$$\sum_{m=1}^{M} a_{lm} k_m - b_l - V_l + U_l = 0, \quad (l = 1, 2, \ldots, L), \tag{4.5.8}$$

where

$$a_{lm} = \frac{\partial u_{D,l}}{\partial k_m}, \qquad b_l = \frac{\partial u_{D,l}}{\partial k_m} k_m^{(n)} - u_{D,l}(\mathbf{k}_n) + u_{D,l}^{\text{obs}}. \tag{4.5.9}$$

In (4.5.9), partial derivatives are evaluated at $\mathbf{k}_n$, $k_m^{(n)}$ is the mth component of $\mathbf{k}_n$. The matrix form of (4.5.8) is

$$\mathbf{J}_D \mathbf{k} - \mathbf{b} - \mathbf{V} + \mathbf{U} = \mathbf{0}, \tag{4.5.10}$$

where $\mathbf{J}_D$ is the Jacobian defined in (4.3.18).

If we consider $(\mathbf{k}, \mathbf{V}, \mathbf{U})$ as a combination variable and set constraints

$$\mathbf{k}_{\min} \leq \mathbf{k} \leq \mathbf{k}_{\max}, \qquad \mathbf{V} \geq \mathbf{0}, \qquad \mathbf{U} \geq \mathbf{0}, \tag{4.5.11}$$

then objective function (4.5.7) in combination with constraints (4.5.10) and (4.5.11) forms a *linear programming* (LP) problem (Luenberger, 1973). It can be solved by the classical simplex method. Thus, the next search point $\mathbf{k}_{n+1}$ can be obtained. A subroutine of simplex method can be found in *Numerical Recipes* (Press *et al.*, 1992).

A detailed procedure of using L_1-norm criterion in the solution of inverse problems for groundwater modeling was given by Xiang *et al.* (1993).

The regularized L_1-norm and L_∞-norm criteria are associated with the following objective funtions

$$E_\alpha(\mathbf{k}) = \sum_{l=1}^{L} w_{D,l} |u_{D,l}(\mathbf{k}) - u_{D,l}^{\text{obs}}| + \alpha \sum_{m=1}^{M} v_m |k_m - k_m^0| \tag{4.5.12}$$

and

$$E(\mathbf{k}) = \max_{1 \le l \le L} w_{D,l} |u_{D,l}(\mathbf{k}) - u_{D,l}^{obs}| + \alpha \max_{1 \le m \le M} v_m |k_m - k_m^0|, \qquad (4.5.13)$$

respectively.

Exercise 4.5.1. When the steepest descent method is used to solve the minimization problem (4.4.5) with the objective function given in (4.5.12), how do we calculate the gradient vector $\nabla E_\alpha(\mathbf{k})$?

Exercise 4.5.2. Solve the same problem as given in Exercise 4.5.1, but using (4.5.13) as the objective function.

4.6 Coupled Inverse Problems

In recent years, researchers working in the field of groundwater modeling are paying more and more attention to coupled problems, such as flow and mass transport in saturated and unsaturated zones, flow and mass transport in deformable porous media, saltwater intrusion, geothermal reservoir evaluation, and biological transformations in groundwater systems. Coupled problems are also encountered in the oil reservoir simulation and nuclear waste repositories.

The coupled inverse problem of flow and mass transport was considered by Carrera (1987) and Wagner and Gorelick (1987), where the unknown transmissivities and dispersivities are determined simultaneously. Ewing *et al.* (1987) described a coupled inverse problem associated the simulation of secondary recovery processes in oil reservoirs. Lee and Seinfeld (1987), Ewing and Lin (1989), and Tang *et al.* (1989) considered the parameter identification problem of two-phase flow. Woodbury *et al.* (1987) and Woodbury and Smith (1988) used temperature measurement to improve the estimation of hydraulic conductivities, in which a coupled inverse problem of flow and heat transport was solved. The same problem was also considered by Wang and Beck (1989) and Wang *et al.* (1989) with the analysis of uncertainty. Kool and Parker (1987) and Mishra and Parker (1989) considered the inverse problem for coupled unsaturated flow and mass transportation. A complete description of coupled inverse problems in groundwater modeling was given by Sun and Yeh (1990a,b).

Basic concepts and methods for *coupled inverse problems* (with multi-state variables) are extremely similar to those with only one state variable. There are several differences between them, however, which make coupled inverse problems more complex and challenging. First, state variables of a coupled problem often have different dimensions and accuracies because they are measured by different instruments. Second, there are crossover effects between state variables and parameters. For example, in a leaky aquifer the head of the unconfined layer often depends upon the transmissivity of the confined layer, and the solute concentration may depend upon the hydraulic conductivity indirectly through the velocity. Third, there are many choices in the formulation of a performance criterion. Finally, there are many options in designing an experiment for the identification of unknown parameters which may be determined by only oberving a part of states.

In this section, the coupled inverse problem is generally defined in the framework of vector optimization (Sun and Yeh, 1990a). Some basic solution methods will be presented briefly.

4.6.1   A GENERAL DEFINITION OF COUPLED INVERSE PROBLEMS

In Section 1.1.3, we have given a general form of mathematical models that can describe any coupled, distributed parameter system through a vector operator $\mathbf{L}$ as follows:

$$\mathbf{L}(\mathbf{u}, \mathbf{p}; \mathbf{x}, t) = \mathbf{0}, \quad \mathbf{x} \in (\Omega), \quad 0 \le t \le t_f, \tag{4.6.1}$$

where $\mathbf{L} = (L_1, L_2, \ldots, L_n)^T$ is a set of n *PDE* operators; $\mathbf{u} = (u_1, u_2, \ldots, u_n)^T$ a set of n state variables; $\mathbf{p} = (p_1, p_2, \ldots, p_m)^T$ a set of m parameters. In (4.6.1), $\mathbf{x}$ are space variables, t the time, (Ω) a space region bounded by surface (Γ), and $[0, t_f]$ a time interval. Equation (4.6.1) must be supplemented by initial conditions

$$\mathbf{u} = \mathbf{f}_0, \quad \text{when } t = 0 \tag{4.6.2a}$$

and boundary conditions

$$\mathbf{u} = \mathbf{f}_1 \quad \text{on } (\Gamma_1), \tag{4.6.2b}$$

$$L_B(\mathbf{u}, \mathbf{p}; \mathbf{x}, t) = \mathbf{f}_2 \quad \text{on } (\Gamma_2), \tag{4.6.2c}$$

where $\mathbf{L}_B = (L_{B,1}, L_{B,2}, \ldots, L_{B,n})^T$ is a vector operator describing flux boundary conditions; $\mathbf{f}_1$ and $\mathbf{f}_2$ are known vector functions; $(\Gamma_1) \cup (\Gamma_2) = (\Gamma)$.

The inverse problem of (4.6.1) and (4.6.2) is defined as finding a $\mathbf{p}^* \in P_{\text{ad}}$ such that

$$\mathbf{E}(\mathbf{p}^*) = \min \mathbf{E}(\mathbf{p}), \quad \mathbf{p} \in P_{\text{ad}}, \tag{4.6.3}$$

where

$$\mathbf{E}(\mathbf{p}) = [E_1(\mathbf{p}), E_2(\mathbf{p}), \ldots, E_n(\mathbf{p})]^T \tag{4.6.4}$$

and $E_i(\mathbf{p})$ $(i = 1, 2, \ldots, n)$ is a performance criterion that measures the error between the model output and observations of state u_i when $\mathbf{p}$ is used as the model parameter. Usually, $E_i(\mathbf{p})$ is defined by the least squares criterion as

$$E_i(\mathbf{p}) = \|u_i(\mathbf{p}) - \tilde{u}_i\|^2_{U_{D,i}}, \quad (i = 1, 2, \ldots, n), \tag{4.6.5}$$

where $\|\cdot\|_{U_{D,i}}$ is the L_2-norm defined in the observation space $U_{D,i}$ of state u_i.

Equation (4.6.3) is a *vector optimization problem* (*VOP*) or *multiobjective optimization problem* (Chankong and Haimes, 1983). A solution $\hat{\mathbf{p}}$ is named a *noninferior solution* or Pareto optimum of problem (4.6.3), if there exists no other $\mathbf{p} \in P_{\text{ad}}$, such that $\mathbf{E}(\mathbf{p}) \leq \mathbf{E}(\hat{\mathbf{p}})$. The vector inequality means that $E_i(\mathbf{p}) \leq E_i(\hat{\mathbf{p}})$ for all $i = 1, 2, \ldots, n$, and with strict inequality for at least one i. To solve *VOP* (4.6.3), first we have to find a set of noninferior solutions P_{nf}. If P_{nf} has only one element $\hat{\mathbf{p}}$, then take it as the optimal solution (or the preference solution of the *VOP*), i.e., let $\mathbf{p}^* = \hat{\mathbf{p}}$; otherwise, $\mathbf{p}^*$ must be selected from P_{nf} using additional criteria.

In order to incorporate prior information of the estimated parameters to alleviate the ill-posedness of inverse solution, objective functions (4.6.4) can be regularized to

$$\mathbf{E}_R(\mathbf{p}) = [E_1(\mathbf{p}), \ldots, E_n(\mathbf{p}), E_{n+1}(\mathbf{p}), \ldots, E_{n+m}(\mathbf{p})]^T, \tag{4.6.6}$$

where $E_{n+j}(\mathbf{p})$ $(j = 1, 2, \ldots, m)$ is a performance criterion that measures the difference between parameter p_j and its prior estimation p_j°. When L_2-norm is used, we define

$$E_{n+j}(\mathbf{p}) = \|p_j - p_j^\circ\|^2_{L_2}. \tag{4.6.7}$$

The solution of the coupled inverse problem is then defined as finding a $\mathbf{p}^* \in P_{\text{ad}}$ such that

$$\mathbf{E}_R(\mathbf{p}^*) = \min \mathbf{E}_R(\mathbf{p}), \quad \mathbf{p} \in P_{\text{ad}}. \tag{4.6.8}$$

4.6.2 THE SOLUTION OF VECTOR OPTIMIZATION PROBLEMS

There are several approaches for finding the noninferior solutions of *VOP* (4.6.8) through solving appropriate scalar optimization problems (*SOP*). One common approach is the weighting method. We define a scalar objective as follows:

$$E(\mathbf{p}) = \sum_{i=1}^{n+m} w_i E_i(\mathbf{p}), \tag{4.6.9}$$

where $\mathbf{w} = (w_1, w_2, \ldots, w_{n+m})^T$ is a weighting coefficient vector, and all its components are positive. A noninferior solution can be found by solving the following *SOP*

$$E(\hat{\mathbf{p}}) = \min E(\mathbf{p}), \quad \mathbf{p} \in P_{\text{ad}}. \tag{4.6.10}$$

Systematically changing the weighting coefficient vector, a set of noninferior solutions P_{nf} can be obtained. The optimal solution $\mathbf{p}^*$ is then determined by designating an optimal weighting coefficient vector $\mathbf{w}^*$.

Another approach of finding noninferior solutions for the *VOP* (4.6.8) is the *k*th-objective-ε-constraint method. The corresponding *SOP* is defined by

$$\min E_k(\mathbf{p}), \quad \mathbf{p} \in P_{ad} \tag{4.6.11}$$

subject to

$$E_l(\mathbf{p}) \leq \varepsilon_l, \quad l = 1, 2, \ldots, n + m, \text{ but } l \neq k,$$

where k is selected from $1 \leq k \leq n+m$. $\varepsilon = (\varepsilon_1, \varepsilon_2, \ldots, \varepsilon_{k-1}, \varepsilon_{k+1}, \ldots, \varepsilon_{n+m})^T$ is a constraint vector. The preference solution of the *VOP* (4.6.8) can be obtained by designating an optimal constraint vector ε^*.

To designate an optimal weighting coefficient vector $\mathbf{w}^*$ for *SOP* (4.6.9) is not a trivial task. Neuman and Yakowitz (1979) presented several methods for determining $\mathbf{w}^*$ for the case of $n = 1$, $m = 1$; Carrera and Neuman (1986a) described an iterative process for the case of $n = 1$, $m > 1$; Wagner and Gorelick (1987) considered the case of $n = 2$, $m = 0$; and Woodbury and Smith (1988) extended the iterative process used by Carrera and Neuman (1986a) to the case of $n = 2, m = 1$.

In fact, from the point of view of vector optimization, the selection of $\mathbf{w}^*$ depends upon the preference of a modeler. Generally speaking, the determination of optimal weights should be an interactive procedure. However, it is also possible to design a noninteractive procedure based on the statistical frame (Chapter 7). If the observation errors of u_i $(i = 1, 2, \ldots, n)$ and the estimation errors of p_j $(j = 1, 2, \ldots, m)$ are independent, normally distributed random variables with zero mean, then we can simply take

$$\mathbf{w}^* = (\sigma_{u_1}^{-2}, \sigma_{u_2}^{-2}, \ldots, \sigma_{u_n}^{-2}, \sigma_{p_1}^{-2}, \sigma_{p_2}^{-2}, \ldots, \sigma_{p_m}^{-2})^T,$$

where $\sigma_{u_i}^2$ and $\sigma_{p_j}^2$ are variances of observation error of u_i and estimation error of p_j, respectively. They may be determined by an iterative process as described by Carrera and Neuman (1986a). In each iteration, let

$$\sigma_{u_i}^2 = \frac{E_i}{N_i}, \quad \sigma_{p_j}^2 = \frac{E_{n+j}}{N_j}, \tag{4.6.12}$$

where N_i and N_j are numbers of observations of state u_i and estimations of parameter p_j, respectively.

4.6.3   INVERSE SOLUTION METHODS

All numerical optimization methods mentioned in Section 4.2 and Section 4.3 can be used to solve the *SOP* (4.6.9). If a gradient method is used, it is necessary to calculate the total gradient vector in each iteration:

$$\mathbf{g}_E = \left(\left[\frac{\partial E}{\partial p_1} \right], \left[\frac{\partial E}{\partial p_2} \right], \dots, \left[\frac{\partial E}{\partial p_m} \right] \right)^T, \tag{4.6.13}$$

where $\partial E / \partial p_j$ is the gradient vector of $E(\mathbf{p})$ with respect to the discretized unknown parameter p_j.

If the Gauss–Newton method is used, it is necessary to calculate the total sensitivity matrix in each iteration:

$$[\mathbf{J}_D] = \begin{bmatrix} \left[\dfrac{\partial u_{1,D}}{\partial p_1} \right] & \left[\dfrac{\partial u_{2,D}}{\partial p_1} \right] & \cdots & \left[\dfrac{\partial u_{n,D}}{\partial p_1} \right] \\[2ex] \left[\dfrac{\partial u_{1,D}}{\partial p_2} \right] & \left[\dfrac{\partial u_{2,D}}{\partial p_2} \right] & \cdots & \left[\dfrac{\partial u_{n,D}}{\partial p_2} \right] \\[2ex] \vdots & \vdots & \ddots & \vdots \\[2ex] \left[\dfrac{\partial u_{1,D}}{\partial p_m} \right] & \left[\dfrac{\partial u_{2,D}}{\partial p_m} \right] & \cdots & \left[\dfrac{\partial u_{n,D}}{\partial p_m} \right] \end{bmatrix}, \tag{4.6.14}$$

where $[\partial u_{i,D} / \partial p_j]$ is the sensitivity matrix of the observations of state u_i with respect to the discretized unknown parameter p_j.

Example 4.6.1.   *(Sun and Yeh, 1990a).* Let us consider the mass transport in a two-dimensional confined aquifer. The flow region (Ω) is a rectangle (Figure 4.6.1) with given head (100 m) and given concentration (zero) boundary (Γ_1) = $\overline{AB}$ and no-flow boundary (Γ_2) = $\overline{BCDA}$. The lengths of $\overline{AB}$ and $\overline{BC}$ are 480 m and 640 m, respectively. The thickness m of the aquifer is assumed to be 50 m. The flow region consists of two zones, with the upper part being zone 1 and lower part zone 2 (See Figure 4.6.1). In order to identify hydraulic conductivities and dispersivities of the two zones, $K_1, K_2, \alpha_{L,1}, \alpha_{T,1}, \alpha_{L,2}$, and $\alpha_{T,2}$, the following tracer test is designed:

(1) a pumping well is located at node F (See Figure 4.6.1) with pumping rate $W = 10^4$ m^3/day;

(2) a recharge well is located at node E with recharge rate $Q = 10^3$ m^3/day, and the concentration of the recharged water is $C' = 10^3$ g/m^3;

(3) four observation wells are located at nodes E, G_1, G_2 and G_3, and three observation times are $t_1 = 1, t_2 = 2$ and $t_3 = 3$ days.

The performance criterion of this problem is defined as

$$E = w_c \sum_{l=1}^{12} \frac{(c_l - c_l{}^\circ)}{\sigma_c^2} + w_h \sum_{l=1}^{12} \frac{(h_l - h_l{}^\circ)^2}{\sigma_h^2}, \tag{4.6.15}$$

where subscript l runs through the four observation wells and three observation times; c_l and h_l are model outputs; $c_l{}^\circ$ and $h_l{}^\circ$ are corresponding observations; σ_c^2 and σ_h^2 are variances of observation errors for c and h, respectively; w_c and w_h are weighting coefficients.

The assumed true parameter values are given in Table 4.6.1. Using these parameters, we can generate a set of noise-free "observation data" by solving the coupled groundwater flow mass transport equations. By adding different noise to these data, field observations of $c_l{}^\circ$ and $h_l{}^\circ$ can be simulated.

Let us consider three cases in the identification of K_1 and K_2:

(1) only using head observations, i.e., let $w_c = 0$ and $w_h = 1$.

(2) only using concentration observation, i.e., let $w_c = 1$ and $w_h = 0$.

(3) using both head and concentration observations and taking the same weight, i.e., let $w_c = 1$ and $w_h = 1$.

The initial estimations of K_1 and K_2 are both 30 m/day. The Gauss–Newton method is used to minimize the performance criterion E in (4.6.15). After four iterations, results of different cases are presented in Table 4.6.2. The table shows that when $\sigma_c < 1$ g/m^3 and $\sigma_h < 0.05$ m, the solution of the inverse problem almost converge to the true parameters for any case. However, when $\sigma_c > 5$ g/m^3 or $\sigma_h > 0.1$ m, the solutions of the inverse problem are significantly different from the true parameters.

When $\sigma_c < 1$ g/m^3 and $\sigma_h < 0.05$ m, six parameters $K_1, K_2, \alpha_{L,1}, \alpha_{T,1}$, $\alpha_{L,2}$ and $\alpha_{T,2}$ can be identified simultaneously using both head and concentration observations. With the Gauss–Newton method, these parameters will almost converge to their true values after only four iterations as shown in Table 4.6.3.

Exercise 4.6.1. Present an approach of calculating the gradient g_E defined in (4.6.13) for the performance E given in Example 4.6.1.

Exercise 4.6.2. Present an approach of calculating the Gauss–Newtion matrix defined in (4.3.7) for minimizing the performance E given in Example 4.6.1.

Exercise 4.6.3. Explain why is it possible to identify the hydraulic conductivity by only observing the concentration.

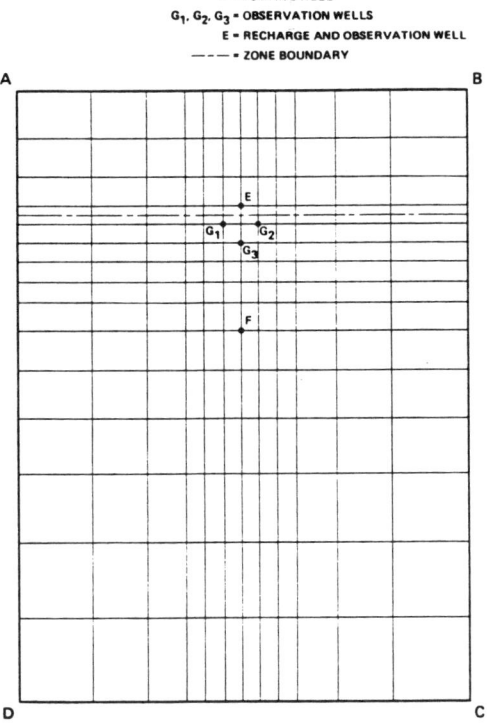

Fig. 4.6.1. Flow region, discretization, and locations of pumping, recharge, and observation wells.

TABLE 4.6.1
Model Parameters

Zone 1	Zone 2
$K_1 = 40$ m/d	$K_2 = 20$ m/d
$S_1 = 0.002$	$S_2 = 0.001$
$\theta_1 = 0.1$	$\theta_2 = 0.05$
$\alpha_{L,1} = 90$ m	$\alpha_{L,2} = 60$ m
$\alpha_{T,1} = 30$ m	$\alpha_{T,2} = 20$ m

TABLE 4.6.2
Identified Values of Hydraulic Conductivities in Different Cases

Noise Levels		Weighting Coefficients		Identified Parameters	
σ_C	σ_h	W_c	W_h	K_1, m/d	K_2, m/d
0.001	0.001	1	1	39.99	19.99
		1	0	39.54	19.35
		1	0	39.94	20.00
0.5	0.005	1	1	39.96	20.01
		1	0	40.23	20.49
		0	1	39.96	19.97
1.0	0.01	1	1	39.66	20.11
		1	0	44.19	19.91
		0	1	39.66	20.14
5.0	0.05	1	1	38.95	20.86
		1	0	55.83	32.92
		0	1	38.96	20.88
1.0	0.1	1	1	39.14	22.09
		1	0	44.19	19.91
		0	1	40.62	14.93

TABLE 4.6.3
True Values, Initial Estimates, and Identifed Values of Hydraulic Conductivities and Dispersivities

Parameters	K_1, m/d	K_2, m/d	$\alpha_{L,1}$, m	$\alpha_{T,1}$, m	$\alpha_{L,2}$, m	$\alpha_{T,2}$, m
True Values	40	20	90	30	60	20
Initial Estimates	30	30	70	50	50	30
Identified Values	39.97	19.28	90.95	29.56	61.31	20.26

Direct Methods for the Solution of Inverse Problems

5.1 Equation Error Criteria and the Matrix Method

In this section, a general form of equation error criteria is introduced. The inverse solution is then obtained by solving a system of superdeterministic equations. The ill-posedness of inverse problems, however, may strongly manifest itself in this case through the ill-condition of coefficient matrix. Data processing, adding constraints and parameterization are always necessary in this case for improving the stability of inverse solutions.

5.1.1 EQUATION ERROR CRITERIA

We have noticed that the general form of groundwater models can be represented by an operator equation

$$L(u; p; \mathbf{x}; t) = 0 \qquad (5.1.1)$$

with appropriate subsidiary conditions, where L is a *PDE* operator, u the state variable and p the model parameter.

When a numerical method is used to discretize the distributed parameter model, we can generate a system of algebraic equations

$$\begin{cases} L_1(u_1, u_2, \ldots, u_N; p_1, p_2, \ldots, p_M) = 0, \\ L_2(u_1, u_2, \ldots, u_N; p_1, p_2, \ldots, p_M) = 0, \\ \quad \vdots \\ L_N(u_1, u_2, \ldots, u_N; p_1, p_2, \ldots, p_M) = 0 \end{cases} \qquad (5.1.2)$$

or in compact form

$$\mathbf{L}(\mathbf{u}, \mathbf{p}) = \mathbf{0}, \tag{5.1.3}$$

where L is an N-dimensional vector operator with components $L_1, L_2, \ldots, L_N$; $N = N_S \times N_T$, N_S is the number of nodes, N_T the number of time steps. In Equation (5.1.3),

$$\mathbf{u} = (u_1, u_2, \ldots, u_N)^T, \qquad \mathbf{p} = (p_1, p_2, \ldots, p_M)^T \tag{5.1.4}$$

are discretized state variable and model parameter, respectively.

The forward problem involves solving $\mathbf{u}$ from Equations (5.1.2) when $\mathbf{p}$ is given, while the inverse problem solving $\mathbf{p}$ from the same system of equations when $\mathbf{u}$ is given. In Section 3.4.2, we knew that a necessary condition for the identifiability of parameter $\mathbf{p}$ is $M < N$. In other words, Equations (5.1.2) must be a superdeterministic system, and thus, there is no any $\mathbf{p} \in R^M$ exactly satisfying all these equations. Let vector

$$\mathbf{r}(\mathbf{p}) = (r_1, r_2, \ldots, r_N) \tag{5.1.5}$$

represent the *equation errors* or *residual* when $\mathbf{u}$ is replaced by observations $\tilde{\mathbf{u}}$ and $\mathbf{p}$ is used as model parameter in (5.1.3), i.e.,

$$\mathbf{L}(\tilde{\mathbf{u}}, \mathbf{p}) = \mathbf{r}(\mathbf{p}). \tag{5.1.6}$$

We define $\hat{\mathbf{p}}$ as a solution of the superdeterministic system (5.1.3), if it is a minimizer of the following optimization problem

$$D(\hat{\mathbf{p}}) = \min_{\mathbf{p} \in P_{ad}} D(\mathbf{p}), \tag{5.1.7}$$

where

$$D(\mathbf{p}) = \|\mathbf{r}(\mathbf{p})\|_{R^N} \tag{5.1.8}$$

and $\|\cdot\|$ is a norm defined in space R^N. When L_1-norm is used, objective function $D(\mathbf{p})$ in (5.1.8) may be replaced by

$$D_1(\mathbf{p}) = \sum_{n=1}^{N} w_n |r_n| = \sum_{n=1}^{N} w_n |L_n(\tilde{\mathbf{u}}, \mathbf{p})| \tag{5.1.9}$$

which is called the L_1-*norm equation error criterion*. In (5.1.9), $w_n \geq 0$ ($n = 1, 2, \ldots, N$) is a set of weighting coefficients.

Similarly, when L_2-norm is used, objective function $D(\mathbf{p})$ may be replaced by

$$D_2(\mathbf{p}) = \sum_{n=1}^{N} w_n^2 r_n^2 = \sum_{n=1}^{N} w_n^2 L_n^2(\tilde{\mathbf{u}}, \mathbf{p}), \tag{5.1.10}$$

which is known as the L_2-*norm equation error criterion*. The weighting coefficients $\{w_n\}$ in (5.1.10) are not necessary the same as those in (5.1.9).

Thus, with the help of equation error criteria, the inverse problem is again transformed into an optimization problem.

5.1.2 THE MATRIX METHOD

Suppose that the L_2-norm equation error criterion $D_2(\mathbf{p})$ is used in the optimization problem (5.1.7). In this case, the necessary condition of optimum yields

$$\sum_{n=1}^{N} w_n^2 L_n \frac{\partial L_n}{\partial p_i} = 0, \quad (i = 1, 2, \ldots, M). \tag{5.1.11}$$

In groundwater modeling, L_n $(n = 1, 2, \ldots, N)$ are usually linear functions with respect to $\mathbf{p}$. Let

$$L_n = a_{n,1} p_1 + a_{n,2} p_2 + \cdots + a_{n,M} p_M - b_n, \quad (n = 1, 2, \ldots, M). \tag{5.1.12}$$

Equations (5.1.11) can be rewritten in matrix form

$$\mathbf{A}^T \mathbf{W}(\mathbf{Ap} - \mathbf{b}) = 0, \tag{5.1.13}$$

where $\mathbf{A}$ is an $N \times M$ matrix with elements a_{nm} $(n = 1, 2, \ldots, N; m = 1, 2, \ldots, M)$, $\mathbf{b}$ is a vector with components b_n $(n = 1, 2, \ldots, N)$. Both $\mathbf{A}$ and $\mathbf{b}$ depend on observations $\tilde{\mathbf{u}}$ and the numerical method used in solving the forward problem. In (5.1.13), $\mathbf{W}$ is an $N \times N$ diagonal matrix with w_n^2 $(n = 1, 2, \ldots, N)$ as its elements. The solution of (5.1.13) is

$$\hat{\mathbf{p}} = (\mathbf{A}^T \mathbf{W} \mathbf{A})^{-1} \mathbf{A}^T \mathbf{W} \mathbf{b}. \tag{5.1.14}$$

When all $w_n = 1$ $(n = 1, 2, \ldots, N)$, it reduces to

$$\hat{\mathbf{p}} = (\mathbf{A}^T \mathbf{A})^{-1} \mathbf{A}^T \mathbf{b}. \tag{5.1.15}$$

The above procedure is known as *the matrix method for parameter identification*. Let us consider the two-dimensional groundwater flow problem presented in Example 1.1.4. If the *MCB* method is used for solving the forward problem, for node i, we have mass balance equation (Sun and Yeh, 1983)

$$\sum_{e_i} (A_{ii}^e \phi_i + A_{ij}^e \phi_j + A_{ik}^e \phi_k) +$$

$$+ \sum_{e_i} \left(B_{ii}^e \frac{d\phi_i}{dt} + B_{ij}^e \frac{d\phi_j}{dt} + B_{ik}^e \frac{d\phi_k}{dt} \right) + R_i = 0, \tag{5.1.16}$$

where $\sum_{e_i}$ represents the summation for all elements around node i;

$$A_{il}^e = -\frac{1}{12\Delta}(T_i + T_j + T_k)(b_i b_l + c_i c_l), \quad (l = i, j, k); \tag{5.1.17}$$

$$B_{ii}^e = \frac{S_i \Delta}{3} \frac{22}{36}, \qquad B_{ij}^e = B_{ik}^e = \frac{S_i \Delta}{3} \frac{7}{36}; \tag{5.1.18}$$

$$R_i = Q_i P_i. \tag{5.1.19}$$

In Equations (5.1.17)–(5.1.19), T_i, T_j, T_k are nodal values of transmissivity at node i, j, k, respectively; S_i the storage coefficient at node i; Δ the area of element Δijk; P_i the area of the exclusive subdomain of node i; b_i, b_j, b_k and c_i, c_j, c_k are defined by nodal coordinates (x_i, y_i), (x_j, y_j) and (x_k, y_k) of nodes i, j and k as follows

$$b_i = y_j - y_k, \quad c_i = x_k - x_j,$$

$$b_j = y_k - y_i, \quad c_j = x_i - x_k, \tag{5.1.20}$$

$$b_k = y_i - y_j, \quad c_k = x_j - x_i.$$

When the finite element method (FEM) with linear basis functions is used, we need only to change B_{ii}^e, B_{ij}^e, B_{ik}^e in (5.1.18) to $S_i\Delta/6$, $S_i\Delta/12$ and $S_i\Delta/12$, respectively.

Substituting ϕ_l $(l = i, j, k)$ by their observation values $\tilde{\phi}_l$ and using

$$\frac{\mathrm{d}\phi_l}{\mathrm{d}t} \approx \frac{\tilde{\phi}_l(t + \Delta t) - \tilde{\phi}_l(t)}{\Delta t}, \tag{5.1.21}$$

Equation (5.1.16) can be rewritten as

$$\sum_{e_i} [F_i^e(T_i + T_j + T_k) + E_i^e S_i] + R_i = 0, \tag{5.1.22}$$

where

$$F_i^e = -\frac{1}{12\Delta}[(b_i b_i + c_i c_i)\tilde{\phi}_i + (b_i b_j + c_i c_j)\tilde{\phi}_j + (b_i b_k + c_i c_k)\tilde{\phi}_k], \tag{5.1.23}$$

$$E_i^e = \frac{\Delta}{3}\left[\frac{22}{36}\frac{\mathrm{d}\tilde{\phi}_i}{\mathrm{d}t} + \frac{7}{36}\frac{\mathrm{d}\tilde{\phi}_j}{\mathrm{d}t} + \frac{7}{36}\frac{\mathrm{d}\tilde{\phi}_k}{\mathrm{d}t}\right]. \tag{5.1.24}$$

If distributed head observations are available for a series of time steps, we can build Equation (5.1.22) for each node and each time step. Matrix $\mathbf{A}$ and vector $\mathbf{b}$ in (5.1.15) are then formed by coefficients of these equations. The inverse solution thus can be obtained directly by matrix inversion.

5.1.3   ILL-POSEDNESS IN THE MATRIX METHOD

From Equation (5.1.15), the existence and uniqueness of the inverse solution seem no problem, provided that coefficient matrix $\mathbf{A}^T\mathbf{A}$ is not singular. However, there is no guarantee concerning the stability. In Example 2.2.6, We have seen that small change in head observations may cause large change in the identified parameters.

In practice, the observed state vector $\tilde{\mathbf{u}}$ must include an error $\delta\mathbf{u}$, that is

$$\tilde{\mathbf{u}} = \mathbf{u}^* + \delta\mathbf{u}, \tag{5.1.25}$$

where $\mathbf{u}^*$ is the true state vector. When matrix method (5.1.15) is used, error $\delta\mathbf{u}$ will cause changes in both matrix $\mathbf{A}$ and vector $\mathbf{b}$. Let

$$\tilde{\mathbf{A}} = \mathbf{A} + \delta\mathbf{A}, \qquad \tilde{\mathbf{b}} = \mathbf{b} + \delta\mathbf{b}, \tag{5.1.26}$$

where $\delta\mathbf{A}$ and $\delta\mathbf{b}$ are errors of $\mathbf{A}$ and $\mathbf{b}$ caused by $\delta\mathbf{u}$, respectively. The true parameter $\mathbf{p}^*$ now can not exactly satisfy the discretization equations. Instead, we have

$$\tilde{\mathbf{A}}\mathbf{p}^* - \tilde{\mathbf{b}} = \tilde{\mathbf{r}}, \tag{5.1.27}$$

where $\tilde{\mathbf{r}}$ is the residual vector. Thus

$$\mathbf{p}^* = (\tilde{\mathbf{A}}^T\tilde{\mathbf{A}})^{-1}\tilde{\mathbf{A}}^T(\tilde{\mathbf{b}} + \tilde{\mathbf{r}}). \tag{5.1.28}$$

On the other hand, the identified parameter obtained from (5.1.15) is

$$\hat{\mathbf{p}} = (\tilde{\mathbf{A}}^T\tilde{\mathbf{A}})^{-1}\tilde{\mathbf{A}}^T\tilde{\mathbf{b}}. \tag{5.1.29}$$

From (5.1.28) and (5.1.29), we obtain

$$\delta\mathbf{p} = \mathbf{p}^* - \hat{\mathbf{p}} = (\tilde{\mathbf{A}}^T\tilde{\mathbf{A}})^{-1}\tilde{\mathbf{A}}^T\tilde{\mathbf{r}}. \tag{5.1.30}$$

Unfortunately, the square matrix $\tilde{\mathbf{A}}^T\tilde{\mathbf{A}}$ in the above is often nearly singular. As a result, small residual $\tilde{\mathbf{r}}$ may cause a large error $\delta\mathbf{p}$.

In groundwater modeling, the nearly singular behavior of matrix $\tilde{\mathbf{A}}^T\tilde{\mathbf{A}}$ may occur when head differences between a node and its neighboring nodes are small for all time steps. In fact, (5.1.23) can be rewritten as

$$F_i^e = -\frac{1}{12\Delta}[(b_ib_j + c_ic_j)(\tilde{\phi}_j - \tilde{\phi}_i) + (b_ib_k + c_ic_k)(\tilde{\phi}_k - \tilde{\phi}_i)]. \tag{5.1.31}$$

When $\tilde{\phi}_j - \tilde{\phi}_i \approx 0$, $\tilde{\phi}_k - \tilde{\phi}_i \approx 0$ for all elements around node i, we will have $\sum_{e_i} F_i^e \approx 0$, and the coefficients of T_i in all discretization equations will either be zero or near zero. Consequently, matrix $\tilde{\mathbf{A}}^T\tilde{\mathbf{A}}$ is nearly singular and that may cause significant error on estimated parameter in the whole region (Weir, 1989).

5.1.4 DATA PROCESSING, CONSTRAINTS AND PARAMETERIZATION

The main difficulty in using the matrix method is its excessive data requirement. In practice, we can never have observations that cover all nodes and all time steps. Alternatively, there are usually only a few observation wells in the interested region for measuring the state variables. As a result, interpolation techniques are always needed to generate distributed observations to meet the data requirement of the matrix method. Various data processing techniques, such as polynomial interpolation, cubic spline, Kriging and so on, can serve this purpose.

Due to the following reasons, we can never expect the matrix method to give satisfactory results.

- · For any field problem, the distribution of observation wells is always very sparse.

- · The elements of matrix **A** in (5.1.15) actually depend on space and time derivatives of the state variable. Thus, it is impossible to obtain their accurate estimates from sparse observations.

- · The inverse solution is very sensitive to errors in the elements of matrix **A**.

When model error, observation error and data processing error are introduced into the input data, unreasonable parameters may be obtained from (5.1.15). For example, some nodal values of hydraulic conductivity are negative and some nodal values of storage coefficient are larger than one. When these cases occur, we have to impose constraints on the estimated parameters based on prior information. The following steps are thus suggested.

- · *Step* 1. If S components of **p** in (5.1.4) can be estimated from the measurements or prior information, then substitute these values into Equations (5.1.13) and solve for other components. The number of unknown components is thus decreased to $M - S$.

- · *Step* 2. Check the components of the direct solution $\hat{p}$ one by one. If any one is larger than its upper bound or less than its lower bound, replace it by the corresponding bound and again solve Equations (5.1.13). When R bounds are used, the number of unknown components is further decreased to $M - S - R$. This procedure may be repeated until all components solved have reasonable values.

- · *Step* 3. Try to relax part of constraints and solve Equations (5.1.13) again, while keeping those reasonable solutions. The number of unknown components will be less than R. Repeat this procedure until there is no constraint that can be relaxed.

If the observations and prior information are insufficient, we have to decrease the dimension of the unknown parameter through a kind of parameterization technique (see Section 3.4.1). Assume that parameter **p** is linearly represented by a lower dimensional vector **k**, i.e.,

$$p = Gk, \tag{5.1.32}$$

where G is a $M \times K$ matrix, k is a K-dimensional vector $(K \ll M)$. Substituting (5.1.32) into (5.1.12), we will have a system of superdeterministic equations

$$AGk = b \tag{5.1.33}$$

and its solution will be

$$k = (G^T A^T A G)^{-1} G^T A^T b. \tag{5.1.34}$$

Generally, the stability of k increases as the dimension of parameterization decreases.

Sagar *et al.* (1975) used the Spline interpolation method to obtain head distributions. The unknown distributed parameters were then identified using the direct method to each node individually. Frind and Pinder (1973) used the matrix method in combination with the finite element method to identify aquifer parameters. Transmissivity was parameterized by lower order basis functions, while the head was represented by higher order basis functions. In order to increase the stability of inverse solutions, they assumed that some Cauchy data of transmissivity (see Section 2.3.1) were given. Emsellem and de Marsily (1971) imposed "smoothness" as a criterion to decrease the dimension of parameterization.

Yeh *et al.* (1983) presented a numerical example of using the matrix method to identify the transmissivity of a two-dimensional aquifer, where a finite difference method was applied to obtain the discretization system. Head distributions were reconstructed by Kriging interpolation based on head observations at 30 wells. The dimension of parameterization was changed from one to sixteen. Yeh *et al.* (1983) found that when Gaussian noises with standard deviation $\delta = 1.0$ (ft) were added to the head observations, the optimal dimension of parameterization is $K = 4$. A further increase in the value of K will cause a large increase in parameter uncertainty. In other words, we have to assume that the identified transmissivity has a very simple structure.

Exercise 5.1.1. Express the equation error criterion when L_∞ norm is used.

Exercise 5.1.2. When the finite difference method is used in the simulation model, find the elements of matrix A in (5.1.15) for the groundwater flow problem discussed in Section 5.1.2.

Exercise 5.1.3. Write a program based on the matrix method that can estimate the hydraulic conductivity T and storage coefficient S of a two-dimensional confined aquifer (using either the *MCB* method or the *FDM* to generate the discretization system).

Exercise 5.1.4. In Exercise 5.1.3, suppose that the aquifer is homogeneous, find the values of T and S by the matrix method and test its stability.

5.2 Mathematical Programming Methods

In this section, the *linear programming (LP) and quadratic programming (QP)* methods will be used to solve optimization problem (5.1.7) derived from different equation error criteria. These methods were clearly described by Hefez *et al.* (1975). The *quasi-linearization* method presented by Yeh and Tauxe (1971), Chang and Yeh (1976) for parameter identification is also considered.

5.2.1 L_1-NORM CRITERION AND LINEAR PROGRAMMING

Let us turn back to the L_1-norm criterion $D_1(\mathbf{p})$ expressed in (5.1.9) and assume that equation errors $L_n(\tilde{u}, \mathbf{p}) = r_n$ $(n = 1, 2, \ldots, N)$ are linear functions of $\mathbf{p}$ as given in (5.1.12). Because r_n may be positive or negative, we define

$$r_n = U_n - V_n. \tag{5.2.1}$$

When $r_n \geq 0$, let $V_n = 0$; when $r_n < 0$, let $U_n = 0$. Thus, we always have $U_n \geq 0, V_n \geq 0$, and $|r_n| = U_n + V_n$. Optimization problem (5.1.7) with criterion (5.1.9) now becomes

$$\min D_1(\mathbf{p}) = \sum_{n=1}^{N} w_n(U_n + V_n) \tag{5.2.2}$$

subject to

$$\sum_{m=1}^{M} a_{n,m} p_m - U_n + V_n = b_n, \ (n = 1, 2, \ldots, N),$$

$$p_m \geq 0, \ (m = 1, 2, \ldots, M),$$

$$U_n \geq 0, V_n \geq 0, \ (n = 1, 2, \ldots, N).$$

If we define a new $(M + 2N)$-dimensional vector

$$\mathbf{x} = (p_1, p_2, \ldots, p_m; U_1, U_2, \ldots, U_N; V_1, V_2, \ldots, V_N)^T$$

or

$$\mathbf{x} = (\mathbf{p}, \mathbf{U}, \mathbf{V})^T, \tag{5.2.3}$$

then problem (5.2.2) can be rewritten as

$$\min \mathbf{cx} \tag{5.2.4}$$

subject to

$$\overline{A}x = b,$$

$$x \geq 0,$$

(5.2.5)

where

$$b = (b_1, b_2, \ldots, b_N),$$
$$c = (0, 0, \ldots, 0; w_1, w_2, \ldots, w_N; w_1, w_2, \ldots, w_N),$$
$$\overline{A} = [A, I, -I].$$

(5.2.6)

In the above, c is a $(M + 2N)$-dimensional vector, $\overline{A}$ is an $N \times (M + 2N)$ matrix consisting of matrix A and two $N \times N$ unit matrices I and $-I$.

Problem (5.2.4) is a canonical linear programming problem. We have mentioned in Section 4.5 that this problem can be solved by the classical simplex method. An advantage of using *LP* to solve inverse problems is that it can incorporate constraints of unknown parameters directly into the optimization problem. When parameter p is restricted by lower bounds p_{min} and upper bounds p_{max}, constraints $p \geq 0$ in (5.2.2) should be changed to $p_{min} \leq p \leq p_{max}$ and the resulting problem becomes a *bounded LP* problem (Luenberger, 1973). Introducing slack variables $y \geq 0$ and $z \geq 0$, such that

$$p - y = p_{min}, \qquad p + z = p_{max}$$

(5.2.7)

and defining a new $(3M + 2N)$-dimensional vector

$$\overline{x} = (x, y, z)^T,$$

(5.2.8)

this bounded *LP* problem can be transformed into a canonical linear programming problem

$$\min \overline{c}\,\overline{x}$$

(5.2.9)

subject to

$$\overline{\overline{A}}\overline{x} = \overline{b},$$

$$\overline{x} \geq 0,$$

where $\overline{c} = (c, 0, 0)$ is a $(3M + 2N)$-dimensional vector, and $(N \times 2M) \times (3M + 2N)$ matrix

$$\overline{\overline{A}} = \begin{bmatrix} A & I_N & -I_N & 0 & 0 \\ I_M & 0 & 0 & -I_M & 0 \\ I_M & 0 & 0 & 0 & I_M \end{bmatrix}.$$

(5.2.10)

In the above equation, I_N and I_M are $N \times N$ and $M \times M$ unit matrices, respectively.

The *LP* method was used by Kleinecke (1971) for the identification of aquifer parameters. He found that inverse solutions are very sensitive to errors in head observations. In order to increase the stability of *LP* solutions, Neuman (1973) used another error criterion relating a prior estimate p^o of the unknown parameter,

$$E_p = \sum_{m=1}^{M} |p_m - p_m^o|, \tag{5.2.11}$$

which can be incorporated into the *LP* problem as constraints.

5.2.2 L_2-NORM CRITERION AND QUADRATIC PROGRAMMING

Substituting (5.1.12) into the equation error criterion (5.1.10), objective function $D_2(\mathbf{p})$ becomes a quadratic function

$$D_2(\mathbf{p}) = \sum_{n=1}^{N} w_n^2 (a_{n,1}p_1 + a_{n,2}p_2 + \ldots + a_{n,m}p_M - b_n)^2. \tag{5.2.12}$$

Using matrix notation, it can be rewritten as

$$D_2(\mathbf{p}) = \mathbf{c}^T \mathbf{p} + \frac{1}{2} \mathbf{p}^T \mathbf{Q} \mathbf{p}, \tag{5.2.13}$$

where

$$\mathbf{c} = -\mathbf{b}^T \mathbf{W} \mathbf{A}, \qquad \mathbf{Q}^T = \mathbf{A}^T \mathbf{W}^T \mathbf{W} \mathbf{A}.$$

In this equation, $\mathbf{W}$ is a diagonal matrix with elements w_n $(n = 1, 2, \ldots, N)$. Now, parameter $\mathbf{p}$ can be identified by solving minimization problem

$$\min D_2(\mathbf{p}) \tag{5.2.14}$$

subject to

$$\mathbf{p}_{min} \leq \mathbf{p} \leq \mathbf{p}_{max}.$$

As in Section 5.2.1, constraints $\mathbf{p}_{min} \leq \mathbf{p} \leq \mathbf{p}_{max}$ can be transformed into a system of linear equations by introducing some slack variables. Thus, (5.2.14) is a *quadratic programming* (*QP*) problem. Using Wolfe's method (Luenberger, 1973; Fletcher, 1987), it can be transformed into a series of *LP* problems.

Hefez *et al.* (1975) presented a synthetic example using mathematical programming methods to identify the transmissivity and the storage coefficient. When water level observations are made without noise, both the *LP* method and the *QP* method give accurate answers to the inverse problem. Significant deviations of the identified parameters are observed, however, when only ± 5 cm noises are added to those observations.

5.2.3 THE QUASI-LINEARIZATION METHOD

In fact, the *quasi-linearization method* belongs to indirect methods. We introduce it here because it can also transfer the inverse problem into a *QP* problem. Let us go back to the regularized *LS* criterion

$$E(\mathbf{p}) = \sum_{l=1}^{L} w_l^2 [u_l(\mathbf{p}) - u_l^{\text{obs}}]^2 + \alpha \sum_{m=1}^{M} (p_m - p_m^o)^2. \tag{5.2.15}$$

Since $u_l(\mathbf{p})$ $(l = 1, 2, \ldots, L)$ are not quadratic functions, $E(\mathbf{p})$ is not quadratic either. However, if we consider both vectors $\mathbf{u}$ and $\mathbf{p}$ in (5.1.4) to be unknown variables and introduce a composed $(N + M)$-dimensional vector

$$\mathbf{x} = (\mathbf{u}, \mathbf{p})^T, \tag{5.2.16}$$

then objective function (5.2.15) can be seen as a quadratic function with respect to $\mathbf{x}$, and the inverse problem becomes

$$\min E(\mathbf{u}, \mathbf{p}) \tag{5.2.17}$$

subject to

$$\mathbf{L}(\mathbf{u}, \mathbf{p}) = \mathbf{0},$$
$$\mathbf{p}_{\min} \leq \mathbf{p} \leq \mathbf{p}_{\max},$$

where constraints $\mathbf{L}(\mathbf{u}, \mathbf{p}) = \mathbf{0}$ show that $\mathbf{u}$ and $\mathbf{p}$ must satisfy the simulation equations (5.1.3). Since functions $\mathbf{L}(\mathbf{u}, \mathbf{p})$ are usually nonlinear, (5.2.17) is a nonlinear optimization problem. The basic idea of the quasi-linearization method is to replace $\mathbf{L}(\mathbf{u}, \mathbf{p})$ approximately by linear functions and reduce (5.2.17) to a series of *QP* problems.

Suppose that $\mathbf{k}^0$ is an initial guess of the unknown parameter. Its corresponding state $\mathbf{u}^0$ can be found by solving the simulation equations

$$\mathbf{L}(\mathbf{u}^0, \mathbf{p}^0) = \mathbf{0}. \tag{5.2.18}$$

Taking first terms in the Taylor's expansion of $\mathbf{L}(\mathbf{u}, \mathbf{p}) = \mathbf{0}$ at point $(\mathbf{u}^0, \mathbf{p}^0)$ yields

$$\mathbf{L}(\mathbf{u}, \mathbf{p}) \approx \frac{\partial \mathbf{L}}{\partial \mathbf{u}}(\mathbf{u} - \mathbf{u}^0) + \frac{\partial \mathbf{L}}{\partial \mathbf{p}}(\mathbf{p} - \mathbf{p}^0), \tag{5.2.19}$$

where $\partial \mathbf{L}/\partial \mathbf{u}$ and $\partial \mathbf{L}/\partial \mathbf{p}$ are Jacobian matrices. All partial derivatives are evaluated at point $(\mathbf{u}^0, \mathbf{k}^0)$. Thus, nonlinear constraints $\mathbf{L}(\mathbf{u}, \mathbf{p}) = \mathbf{0}$ can be replaced approximately by linear constraints and the nonlinear optimization problem (5.2.17) becomes a *QP* problem

$$\min E(\mathbf{u}, \mathbf{p}) \tag{5.2.20}$$

subject to

$$\frac{\partial \mathbf{L}}{\partial \mathbf{u}}\mathbf{u} + \frac{\partial \mathbf{L}}{\partial \mathbf{p}}\mathbf{p} = \frac{\partial \mathbf{L}}{\partial \mathbf{u}}\mathbf{u}^0 + \frac{\partial \mathbf{L}}{\partial \mathbf{p}}\mathbf{p}^0.$$

Solving (5.1.20), we can obtain a updated parameter $\mathbf{p}^1$. Then we replace $\mathbf{p}^0$ by $\mathbf{p}^1$ and repeat this procedure until a given convergent criterion is satisfied.

Let us consider the linearization of the simulation equations discussed in Section 5.1.2. Assume that all nodal values of transmissivity T are unknown, and the simulation equation $L_n = 0$ is associated with node i and time step r. From (5.1.16), we have

$$L_n \equiv \sum_{e_i}(A_{ii}^e\phi_i^{r+1} + A_{ij}^e\phi_j^{r+1} + A_{ik}^e\phi_k^{r+1}) +$$

$$+ \sum_{e_i}\left(B_{ii}^e\frac{\phi_i^{r+1} - \phi_i^r}{\Delta t} + B_{ij}^e\frac{\phi_j^{r+1} - \phi_j^r}{\Delta t} + B_{ik}^e\frac{\phi_k^{r+1} - \phi_k^r}{\Delta t}\right) + R_i = 0,$$

$$\tag{5.2.21}$$

where $A_{ii}^e, A_{ij}^e, \ldots, B_{ik}^e, R_i$ are given by (5.1.17)–(5.1.19). Differentiating (5.2.21) yields

$$\frac{\partial L_n}{\partial \phi_l^{r+1}} = \sum_{e_i}\left(A_{il}^e + \frac{B_{il}^e}{\Delta t}\right),$$

$$\frac{\partial L_n}{\partial \phi_l^r} = -\sum_{e_i}\frac{B_{il}^e}{\Delta t},$$

$$\frac{\partial L_n}{\partial T_l} = -\frac{1}{4\Delta}\sum_{e_i}[(b_i^2 + c_i^2)\phi_i^{r+1} + (b_ib_j + c_ic_j)\phi_j^{r+1} + (b_ib_k + c_ic_k)\phi_k^{r+1}].$$

where $l = i, j, k$. Thus, we have the following first-order approximation:

$$L_n(\phi, \mathbf{T}) \approx \sum_{e_i}[(T_e^0\alpha_i + \beta_i)(\phi_i^{r+1} - \phi_i^{r+1,0})$$

$$+ (T_e^0\alpha_j + \beta_j)(\phi_j^{r+1} - \phi_j^{r+1,0})$$

$$+ (T_e^0\alpha_k + \beta_k)(\phi_k^{r+1} - \phi_k^{r+1,0})$$

$$+ \beta_i(\phi_i^r - \phi_i^{r,0}) + \beta_j(\phi_j^r - \phi_j^{r,0}) + \beta_k(\phi_k^r - \phi_k^{r,0})$$

$$+ \gamma(T_i - T_i^0) + \gamma(T_j - T_j^0) + \gamma(T_k - T_k^0)], \tag{5.2.22}$$

where

$$T_e^0 = T_i^0 + T_j^0 + T_k^0,$$

$$\alpha_l = -\frac{1}{12\Delta}(b_ib_l + c_ic_l), \qquad \beta_l = -\frac{B_{il}^e}{\Delta t} \quad (l = i, j, k),$$

$$\gamma = \alpha_i\phi_i^{r+1,0} + \alpha_j\phi_j^{r+1,0} + \alpha_k\phi_k^{r+1,0}.$$

Superscript 0 in the above equations shows that the variable is evaluated at the initial estimate ($\mathbf{T}^0$) of the unknown parameter. Since the right-hand side of (5.2.22) is a linear function with respect to ϕ and $\mathbf{T}$, the linearization of the simulation equations is completed.

Chang and Yeh (1976) presented a numerical example in which the quasi-linearization method was combined with the finite difference method to identify 84 components of the transmissivity of a confined aquifer. When the head observations were made without noise, the identified values of transmissivity almost return to their true values after 6 iterations. They also found that the so obtained inverse solution is relatively stable if the maximum percentage error introduced to the observations is less than 1%.

Exercise 5.2.1. When the unknown parameter is parameterized by a low dimensional vector **k**, transform the inverse problem into a *LP* problem.

Exercise 5.2.2. Present a direct method for parameter identification when the L_∞-norm criterion is used.

Exercise 5.2.3. Why do we say that the quasi-linear method is an indirect method of inverse solution? Can we combine it with a parameterization method?

Exercise 5.2.4. Compare direct methods with indirect methods of inverse solutions from the following aspects:

(1) Basic idea.

(2) Performance criteria.

(3) The selection of different norms.

(4) Data requirement.

(5) The use of parameterization.

(6) Solution procedure.

(7) Stability.

(8) Applicability.

The Adjoint State Method

6.1 The Adjoint State Method for Groundwater Flow Problems

The *adjoint state method based* on the *variational theory* has been used in ground-water modeling for more than two decades (Carter *et al.*, 1974; Chavent *et al.*, 1975; Seinfeld and Chen 1978; Neuman *et al.*, 1980; Sun and Yeh, 1985; Townley and Wilson, 1985; Ahlferd *et al.*, 1988; Sun and Yeh, 1990a; and etc.). Its applications include not only parameter identification, but also sensitivity analysis, reliability estimates, observation design, and so forth. In this section, we will show how to derive the adjoint equations for groundwater flow problems and how to solve them.

6.1.1 GREEN'S THEOREMS

In multivariable calculus or in vector analysis, we have learned *Gauss' divergence theorem*

$$\int_{(R)} \nabla \cdot \mathbf{F} \, dR = \int_{(S)} \mathbf{F} \cdot \mathbf{n} \, dS, \tag{6.1.1}$$

where $\mathbf{F}$ is a continuously differentiable vector function defined on the space domain (R), ∇ the gradient operator, (S) the bounding surface of (R), $\mathbf{n}$ the outward normal direction of (S). $\nabla \cdot \mathbf{F}$ is known as the divergence of $\mathbf{F}$, and $\mathbf{F} \cdot \mathbf{n}$ is the flux. Divergence theorem (6.1.1) has a clear physical meaning that the total divergence in a domain should be equal to the total flux through its bounding surface.

In a cartesian coordinate system, we have

$$\nabla \cdot \mathbf{F} = \frac{\partial F_x}{\partial x} + \frac{\partial F_y}{\partial y} + \frac{\partial F_z}{\partial z} \tag{6.1.2}$$

and

$$\mathbf{F}\cdot\mathbf{n} = F_x n_x + F_y n_y + F_z n_z, \tag{6.1.3}$$

where (F_x, F_y, F_z) and $(n_x, , n_y, n_z)$ are components of $\mathbf{F}$ and $\mathbf{n}$, respectively. *Two-dimensional* divergence theorem has the same form:

$$\int_{(\Omega)} \nabla\cdot\mathbf{F}\,d\Omega = \int_{(\Gamma)} \mathbf{F}\cdot\mathbf{n}\,d\Gamma, \tag{6.1.4}$$

where (Γ) is the boundary curve of the two-dimensional domain (Ω).

When $\mathbf{F}$ is defined by two scalar functions ψ and ϕ that $\mathbf{F} = \psi\nabla\phi$, Equation (6.1.1) can be rewritten as

$$\int_{(R)} (\nabla\psi\cdot\nabla\phi + \psi\nabla^2\phi)\,dV = \int_{(S)} \psi\nabla\phi\cdot\mathbf{n}\,dS. \tag{6.1.5}$$

Equation (6.1.5) is called *Green's first theorem*. Similarly, let $\mathbf{F} = \psi\nabla\phi - \phi\nabla\psi$, we obtain *Green's second theorem*

$$\int_{(R)} (\psi\nabla^2\phi - \phi\nabla^2\psi)\,dD = \int_{(S)} (\psi\nabla\phi - \phi\nabla\psi)\cdot\mathbf{n}\,dS. \tag{6.1.6}$$

In what follows, we will use some extended forms of Green's theorems. Let u and K be scalar functions and $\mathbf{v}$ a vector function. The following identities can then be derived directly from (6.1.5) and (6.1.6):

$$\int_{(R)} \phi\nabla\cdot(u\nabla\psi)\,dR = -\int_{(R)} u(\nabla\phi\cdot\nabla\psi)\,dR + \int_{(S)} u\phi\nabla\psi\cdot\mathbf{n}\,dS, \tag{6.1.7}$$

$$\int_{(R)} \phi\nabla\cdot(K\nabla\psi)\,dR = \int_{(R)} \psi\nabla\cdot(K\nabla\phi)\,dR - \int_{(S)} K(\phi\nabla\psi - \psi\nabla\phi)\cdot\mathbf{n}\,dS, \tag{6.1.8}$$

$$\int_{(R)} \phi\nabla\cdot(\mathbf{v}\psi)\,dR = \int_{(R)} \psi\mathbf{v}\cdot\nabla\phi\,d\Omega + \int_{(S)} \phi\psi\mathbf{v}\cdot\mathbf{n}\,dS. \tag{6.1.9}$$

Two-dimensional forms of these identities are identical with the above, except that the dimension of all integrals is decreased by one.

6.1.2   THE ADJOINT PROBLEM FOR GROUNDWATER FLOW IN CONFINED AQUIFERS

Two-dimensional groundwater flow in an isotropic and confined aquifer is governed by (see Example 1.1.4)

$$S\frac{\partial\phi}{\partial t} - \nabla\cdot(T\nabla\phi) - Q = 0 \quad (x, y) \in (\Omega),\ 0 \le t \le t_f \tag{6.1.10a}$$

subject to initial and boundary conditions

$$\phi(x, y, 0) = f_0, \quad (x, y) \in (\Omega), \tag{6.1.10b}$$

$$\phi|_{(\Gamma_1)} = f_1, \quad T\nabla\phi\cdot\mathbf{n}|_{(\Gamma_2)} = f_2, \tag{6.1.10c}$$

where flow region (Ω) is bounded by $(\Gamma) = (\Gamma_1) \cup (\Gamma_2)$; f_0, f_1, and f_2 are known functions, t_f is the final time.

Let transmissivity T be the unknown parameter. Any variation δT of T must cause a variation $\delta\phi$ of ϕ, because ϕ is dependent on T through the governing equation and subsidiary conditions. Taking the variation of (6.1.10), we have

$$S\frac{\partial \delta\phi}{\partial t} - \nabla\cdot(\delta T \nabla\phi) - \nabla\cdot(T\nabla\delta\phi) - Q = 0, \tag{6.1.11a}$$

and

$$\delta\phi|_{t=0} = 0, \tag{6.1.11b}$$

$$\delta\phi|_{(\Gamma_1)} = 0, \qquad (\delta T\nabla\phi + T\nabla\delta\phi)\cdot\mathbf{n}|_{(\Gamma_2)} = 0. \tag{6.1.11c}$$

Problem (6.1.11) is named the *variational problem* of the *primary problem* (6.1.10). It relates variations δT and $\delta\phi$.

Let $\psi(x, y, t)$ be an arbitrary function having continuous second-order space derivatives on (Ω) and first-order time derivative in $[0, t_f]$. Multiplying (6.1.11a) by ψ and integrating the result over the time-space region $[0, t_f] \times (\Omega)$ yields

$$\int_{(\Omega)} (S\psi\delta\phi)|_0^{t_f} \, d\Omega -$$
$$- \int_0^{t_f} \int_{(\Omega)} \left[S\delta\phi\frac{\partial\psi}{\partial t} - \psi\nabla\cdot(T\nabla\delta\phi) - \psi\nabla\cdot(\delta T\nabla\phi) \right] d\Omega \, dt = 0. \tag{6.1.12}$$

If we define

$$\psi(x, y, t_f) = 0, \quad (x, y) \in (\Omega), \tag{6.1.13}$$

then the first term of (6.1.12) vanishes because of condition (6.1.11b).

Using the two-dimensional forms of identities (6.1.7) and (6.1.8), Equation (6.1.12) can be rewritten as

$$\int_0^{t_f} \left\{ \int_{(\Omega)} \left[S\frac{\partial\psi}{\partial t} + \nabla\cdot(T\nabla\psi) \right] \delta\phi \, d\Omega - \int_{(\Omega)} (\nabla\psi\cdot\nabla\phi)\delta T \, d\Omega \right.$$
$$- \int_{(\Gamma)} \delta\phi T\nabla\psi\cdot\mathbf{n} \, d\Gamma + \int_{(\Gamma)} \psi T\nabla\delta\phi\cdot\mathbf{n} \, d\Gamma$$
$$\left. + \int_{(\Gamma)} \delta T\psi\nabla\phi\cdot\mathbf{n} \, d\Gamma \right\} dt = 0, \tag{6.1.14}$$

Using conditions (6.1.11c) and defining

$$\psi|_{(\Gamma_1)} = 0 \text{ and } T\nabla\psi\cdot\mathbf{n}|_{(\Gamma_2)} = 0, \tag{6.1.15}$$

all integrals along (Γ_1) in (6.1.14) vanish. Thus, we have

$$\int_0^{t_f} \int_{(\Omega)} \left[S\frac{\partial \psi}{\partial t} + \nabla \cdot (T\nabla \psi) \right] \delta\phi \, d\Omega \, dt - \int_0^{t_f} \int_{(\Omega)} (\nabla\psi \cdot \nabla\phi)\delta T \, d\Omega \, dt = 0.$$

$$(6.1.16)$$

In order to identify transmissivity $T(x,y)$ by fitting head observations, we have to define a performance or a criterion to measure the "goodness of fit". A general form of the performance function may be written as

$$E(\phi, T) = \int_0^{t_f} \int_{(\Omega)} f(\phi, T; x, y, t) \, d\Omega \, dt.$$

$$(6.1.17)$$

Corresponding to different selections of function $f(\phi, T; x, y, t)$, all L_p-norm criteria with either distributed or pointwise observations are special cases of (6.1.17). For instance, if we define

$$f(\phi, T; x, y, t) = \sum_{i=1}^{I} \sum_{j=1}^{J} w_{i,j}^2 [u_{i,j}(T) - u_{i,j}^{\text{obs}}]^2 \delta(x - x_i)\delta(y - y_i)\delta(t - t_j),$$

$$(6.1.18)$$

then (6.1.17) reduces to the *OLS* criterion used in Chapter 4. In (6.1.18), $\delta(\cdot)$ is the Dirac δ function, (x_i, y_i) $(i = 1, 2, \ldots, I)$ are locations of observations wells, t_j $(j = 1, 2, \ldots, J)$ the observation times.

Taking the variation of $E(\phi, T)$ yields

$$\delta E = \int_0^{t_f} \int_{(\Omega)} \left(\frac{\partial f}{\partial T}\delta T + \frac{\partial f}{\partial \phi}\delta\phi \right) d\Omega \, dt.$$

$$(6.1.19)$$

The summation of (6.1.19) and (6.1.16) gives

$$\delta E = \int_0^{t_f} \int_{(\Omega)} \left[\frac{\partial f}{\partial \phi} - S\frac{\partial \psi}{\partial t} - \nabla \cdot (T\nabla\psi) \right] \delta\phi \, d\Omega \, dt$$

$$+ \int_0^{t_f} \int_{(\Omega)} \left[\frac{\partial f}{\partial T} + \nabla\psi \cdot \nabla\phi \right] \delta T \, d\Omega \, dt.$$

$$(6.1.20)$$

If $\psi(x, y, t)$ is selected to satisfy the following *PDE*

$$S\frac{\partial \psi}{\partial t} + \nabla \cdot (T\nabla\psi) - \frac{\partial f}{\partial \phi} = 0$$

$$(6.1.21a)$$

subject to conditions (6.1.13) and (6.1.15), namely,

Final condition: $\psi(x, y, t_f) = 0,$

$$(6.1.21b)$$

Boundary conditions: $\psi|_{(\Gamma_1)} = 0, \qquad T\frac{\partial \psi}{\partial n}\bigg|_{(\Gamma_2)} = 0,$

$$(6.1.21c)$$

then the first integral on the right-hand side of (6.1.20) related to unknown $\delta\phi$ will vanish, and we obtain

$$\delta E = \int_0^{t_f} \int_{(\Omega)} \left[\frac{\partial f}{\partial T} + \nabla\psi\cdot\nabla\phi \right] \delta T \, d\Omega \, dt. \tag{6.1.22a}$$

From (6.1.22a), we can find the *functional partial derivative*

$$\frac{\delta E}{\delta T} = \int_0^{t_f} \int_{(\Omega)} \left[\frac{\partial f}{\partial T} + \nabla\psi\cdot\nabla\phi \right] d\Omega \, dt. \tag{6.1.22b}$$

Now, let us explain how the above result can be used. Assume that $T(x,y)$ is parameterized by

$$T(x,y) \approx \sum_{m=1}^{M} k_m N_m(x,y), \tag{6.1.23}$$

then $\mathbf{k} = (k_1, k_2, \ldots, k_M)$ becomes the vector to be identified. For any partial increment Δk_m, (6.1.23) gives $\delta T \approx N_m \Delta k_m$. The partial increment δE then can be determined by (6.1.22b), and we have

$$\frac{\partial E}{\partial k_m} \approx \int_0^{t_f} \int_{(\Omega)} \left[\frac{\partial f}{\partial T} + \nabla\psi\cdot\nabla\phi \right] N_m \, d\Omega \, dt, \quad (m = 1, 2, \ldots, M). \tag{6.1.24}$$

Problem (6.1.21) is called the *adjoint problem* of the primary problem (6.1.10) and ψ is called the *adjoint state* of ϕ. Solving each equation once, all components of gradient $\nabla E(\mathbf{k})$ can be obtained by (6.1.24). Remember that the gradient $\nabla E(\mathbf{k})$ was calculated in Section 4.2.3 by finite difference approximation (4.2.14), where the primary problem (6.1.10) had to be solved $M + 1$ times!

Introducing the following time transformation

$$\tau = t_f - t, \tag{6.1.25}$$

adjoint problem (6.1.21) can be transformed into

$$S\frac{\partial\psi}{\partial\tau} - \nabla\cdot(T\nabla\psi) + \frac{\partial f}{\partial\phi} = 0 \tag{6.1.26a}$$

subject to

$$\psi|_{\tau=0} = 0, \quad (x,y) \in (\Omega), \tag{6.1.26b}$$

$$\psi|_{(\Gamma_1)} = 0, \quad T\nabla\psi\cdot\mathbf{n}|_{(\Gamma_2)} = 0, \tag{6.1.26c}$$

where $\psi = \psi(x, y, t_f - \tau)$. Note that the form of (6.1.26) is exactly identical to the primary problem (6.1.10), only the sink/source term, initial and bounday conditions of them are different. Therefore, we can use the same subroutine to solve both primary and adjoint problems. This means that gradient ∇E can be obtained with only two simulation runs when the adjoint state method is used.

If storage coefficient S is unknown, parallel results can be derived. We may obtain functional partial derivative

$$\frac{\partial E}{\partial S} = \int_0^{t_f} \int_{(\Omega)} \left[\frac{\partial f}{\partial S} + \psi \frac{\partial \phi}{\partial t} \right] d\Omega \, dt, \tag{6.1.27}$$

which can be used to calculate the gradient of E with respect to S.

6.1.3 THE ADJOINT PROBLEM FOR GROUNDWATER FLOW IN UNCONFINED AQUIFERS

When Dupuit's assumptions are applicable, groundwater flow in a unconfined aquifer is governed by

$$S_y \frac{\partial h}{\partial t} - \nabla \cdot [K(h - b)\nabla h] - Q = 0 \tag{6.1.28a}$$

subject to initial and boundary conditions

$$h(x, y, 0) = g_0, \quad (x, y) \in (\Omega), \tag{6.1.28b}$$

$$h|_{(\Gamma_1)} = g_1, \qquad K(h - b)\nabla h \cdot \mathbf{n}|_{(\Gamma_2)} = g_2. \tag{6.1.28c}$$

All notations in the above equations were explained in Example 1.1.5.

If both K and S_y are unknown parameters, we will have the following *variational problem* for (6.1.28):

$$\delta S_y \frac{\partial h}{\partial t} + S_y \frac{\partial \delta h}{\partial t} - \nabla \cdot [\delta K(h - b)\nabla h]$$

$$- \nabla \cdot [K\delta h \nabla h] - \nabla \cdot [K(h - b)\nabla \delta h] = 0 \tag{6.1.29a}$$

subject to initial and boundary conditions

$$\delta h|_{t=0} = 0, \quad (x, y) \in (\Omega), \tag{6.1.29b}$$

$$\delta h|_{(\Gamma_1)} = 0, \qquad [\delta K(h - b)\nabla h + K\delta h \nabla h + K(h - b)\nabla \delta h] \cdot \mathbf{n}|_{(\Gamma_2)} = 0. \tag{6.1.29c}$$

Multiplying (6.1.29a) by function $\psi(x, y, t)$ and integrating the result over the time-space region $[0, t_f] \times (\Omega)$, and then using Green's theorems to exchange the positions of ψ and h, we can obtain

$$\int_0^{t_f} \int_{(\Omega)} \left\{ S_y \frac{\partial \psi}{\partial t} + \nabla \cdot [K(h - b)\nabla \psi] - K\nabla h \cdot \nabla \psi \right\} \delta h \, d\Omega \, dt$$

$$- \int_0^{t_f} \int_{(\Omega)} [(h-b)\nabla h \cdot \nabla \psi] \delta K \, d\Omega \, dt$$

$$- \int_0^{t_f} \int_{(\Omega)} \psi \frac{\partial h}{\partial t} \delta S_y \, d\Omega \, dt = 0, \tag{6.1.30}$$

where the following conditions are assumed:

$$\psi(x, y, t_f) = 0, \quad (x, y) \in (\Omega), \tag{6.1.31a}$$

$$\psi|_{(\Gamma_1)} = 0, \quad K(h-b)\nabla\psi \cdot \mathbf{n}|_{(\Gamma_2)} = 0, \tag{6.1.31b}$$

Taking the variation of the general performance function

$$E(h, S, K) = \int_0^{t_f} \int_{(\Omega)} f(h, S, K; x, y, t) \, d\Omega \, dt \tag{6.1.32}$$

yields

$$\delta E = \int_0^{t_f} \int_{(\Omega)} \left(\frac{\partial f}{\partial h} \delta h + \frac{\partial f}{\partial S} \delta S + \frac{\partial f}{\partial K} \delta K \right) d\Omega \, dt. \tag{6.1.33}$$

Using (6.1.30), the first term on the right-hand side of (6.1.33), which is related to the unknown variation δh, can be eliminated. Hence, we obtain

$$\delta E = \int_0^{t_f} \int_{(\Omega)} \left[\frac{\partial f}{\partial K} + (h-b)\nabla h \cdot \nabla \psi \right] \delta K \, d\Omega dt$$

$$+ \int_0^{t_f} \int_{(\Omega)} \left[\frac{\partial f}{\partial S} + \psi \frac{\partial h}{\partial t} \right] \delta S \, d\Omega \, dt, \tag{6.1.34}$$

provided that ψ satisfies the following *PDE*:

$$S_y \frac{\partial \psi}{\partial t} + \nabla \cdot [K(h-b)\nabla\psi] - K\nabla h \cdot \nabla\psi - \frac{\partial f}{\partial h} = 0. \tag{6.1.35}$$

Equation (6.1.35) with final condition (6.1.31a) and boundary conditions (6.1.31b) form the adjoint problem of the primary problem (6.1.28). Its solution ψ is called the adjoint state of water level h. Solving the primary problem to get the distribution of h, and then substituting it into the adjoint problem, ψ can be solved. Once h and ψ are obtained, both $\partial E/\partial K$ and $\partial E/\partial S$ can be determined from (6.1.34).

It is interesting to note that Equation (6.1.35) is in the form of a "convection-diffusion" equation. The third term on the left-hand side of (6.1.35) is a "convective term" with "convective velocity" $v = -K\nabla h$. Therefore, any subroutine which is designed for solving mass transport problems can also be used to solve the adjoint equation (6.1.35).

6.1.4   THE ADJOINT PROBLEM FOR LEAKY AQUIFER SYSTEMS

Yeh and Sun (1990) extended the adjoint state method to leaky aquifer systems. Using equations presented in Example 1.1.5 and replacing the notation $1/\sigma$ by K_z, the *variational problem for a leaky system* with respect to δK_z may be written as

$$S_y\frac{\partial \delta h}{\partial t} = \nabla\cdot[K(h-b)\nabla\delta h] + \nabla\cdot[K\delta h\nabla h]$$

$$-K_z(\delta h - \delta\phi) - \delta K_z(h-\phi), \tag{6.1.36a}$$

$$S\frac{\partial \delta\phi}{\partial t} = \nabla\cdot[T\nabla\delta\phi] + K_z(\delta h - \delta\phi) + \delta K_z(h-\phi) \tag{6.1.36b}$$

subject to initial and boundary conditions

$$\delta h|_{t=0} = 0, \qquad \delta\phi|_{t=0} = 0, \tag{6.1.36c}$$

$$\delta h|_{(\Gamma_1)} = 0, \qquad \delta\phi|_{(\Gamma_3)} = 0, \tag{6.1.36d}$$

$$[K(h-b)\nabla\delta h + K\delta h\nabla h]\cdot\mathbf{n}|_{(\Gamma_2)} = 0,$$

$$T\nabla\delta\phi\cdot\mathbf{n}|_{(\Gamma_4)} = 0. \tag{6.1.36e}$$

Let $\psi_1(x,y,t)$ and $\psi_2(x,y,t)$ be functions chosen below. Multiplying (6.1.36a) by ψ_1 and (6.1.36b) by ψ_2, integrating their summation over $(\Omega) \times [0,t_f]$, using Green's theorems to exchange the positions of h, and ψ_1 as well as the position of ϕ and ψ_2, and assuming

$$\psi_1|_{t=t_f} = 0, \qquad \psi_2|_{t=t_f} = 0, \tag{6.1.37a}$$

$$\psi_1|_{(\Gamma_1)} = 0, \qquad \psi_2|_{(\Gamma_3)} = 0, \tag{6.1.37b}$$

$$K(h-b)\nabla\psi_1\cdot\mathbf{n}|_{(\Gamma_2)} = 0, \qquad T\nabla\psi_2\cdot\mathbf{n}|_{(\Gamma_4)} = 0, \tag{6.1.37c}$$

we can obtain

$$\int_0^{t_f}\int_{(\Omega)}\left\{S_y\frac{\partial\psi_1}{\partial t} + \nabla\cdot[K(h-b)\nabla\psi_1] - K\nabla h\cdot\nabla\psi_1\right.$$

$$\left.- K_z(\psi_1 - \psi_2)\right\}\delta h\,d\Omega\,dt$$

$$-\int_0^{t_f}\int_{(\Omega)}\left\{S\frac{\partial\psi_2}{\partial t} + \nabla\cdot[T\nabla\psi_2] + K_z(\psi_1 - \psi_2)\right\}\delta\phi\,d\Omega\,dt$$

$$-\int_0^{t_f}\int_{(\Omega)}(\psi_1 - \psi_2)(h-\phi)\delta K_z\,d\Omega\,dt = 0. \tag{6.1.38}$$

Suppose that the performance function is

$$E(h,\phi,\mathbf{p}) = \int_0^{t_f}\int_{(\Omega)}f(h,\phi,\mathbf{p};x,t)\,d\Omega\,dt, \tag{6.1.39}$$

where $\mathbf{p} = (K, T, K_z, S_y, S)$, and ψ_1 and ψ_2 satisfy the *PDEs*

$$S_y \frac{\partial \psi_1}{\partial t} + \nabla \cdot [K(h - b)\nabla \psi_1] - K\nabla h \cdot \nabla \psi_1 - K_z(\psi_1 - \psi_2) = \frac{\partial f}{\partial h}, \quad (6.1.40a)$$

$$S \frac{\partial \psi_2}{\partial t} + \nabla \cdot [T\nabla \psi_2] + K_z(\psi_1 - \psi_2) = \frac{\partial f}{\partial \phi}, \quad (6.1.40b)$$

then functional partial derivative $\partial E / \partial K_z$ can be represented as

$$\frac{\partial E}{\partial K_z} = \int_0^{t_f} \int_{(\Omega)} \left[\frac{\partial f}{\partial K_z} + (h - \phi)(\psi_1 - \psi_2) \right] d\Omega \, dt. \quad (6.1.41)$$

Similarly, we can find other functional partial derivatives:

$$\frac{\partial E}{\partial K} = \int_0^{t_f} \int_{(\Omega)} \left[\frac{\partial f}{\partial K} + (h - b)(\nabla h \cdot \nabla \psi_1) \right] d\Omega \, dt, \quad (6.1.42)$$

$$\frac{\partial E}{\partial T} = \int_0^{t_f} \int_{(\Omega)} \left[\frac{\partial f}{\partial T} + \nabla \phi \cdot \nabla \psi_2 \right] d\Omega \, dt, \quad (6.1.43)$$

$$\frac{\partial E}{\partial S_y} = \int_0^{t_f} \int_{(\Omega)} \left[\frac{\partial f}{\partial S_y} + \psi_1 \frac{\partial h}{\partial t} \right] d\Omega \, dt, \quad (6.1.44)$$

$$\frac{\partial E}{\partial S} = \int_0^{t_f} \int_{(\Omega)} \left[\frac{\partial f}{\partial S} + \psi_2 \frac{\partial \phi}{\partial t} \right] d\Omega \, dt. \quad (6.1.45)$$

Equations (6.1.40) with subsidiary conditions (6.1.37) form the *adjoint problem of leaky aquifer systems*. ψ_1 and ψ_2 are the adjoint states. Since the form of (6.1.40) is identical to that of (1.1.6), we can use the same subroutine to solve both of them. In Yeh and Sun (1990), a numerical example is given where both primary and adjoint systems are solved by the *MCB* method.

Exercise 6.1.1. Provide a proof for the two-dimensional forms of Equations (6.1.7)–(6.1.9).

Exercise 6.1.2. Give the expression of function $f(\phi, T; x, y, t)$ in (6.1.17), such that $E(\phi, T)$ can reduce to the regularized *OLS* criterion (4.4.4).

Exercise 6.1.3. Suppose that transmissivity $T(x, y)$ is parameterized by the zonation method. Find the expression of $\partial E / \partial T_i$ from (6.1.24), where T_i is transmissivity in the i-th zone.

Exercise 6.1.4. Extend the results obtained in Section 6.1.2 to three-dimensional groundwater flow problems.

Exercise 6.1.5. Provide details in the derivation of (6.1.38).

6.2 The Adjoint State Method for Mass Transport Problems

In this section, the adjoint state method is extended to mass transport problems and coupled groundwater flow and mass transport problems. The derivation procedure is essentially the same as before. However, we should keep in mind that the solute concentration may depend on the hydraulic conductivity through the velocity and that the dispersion coefficient is always a tensor.

6.2.1 THE ADJOINT PROBLEM FOR ADVECTION-DISPERSION EQUATIONS

Solute transport in groundwater is governed by

$$\frac{\partial(\theta C)}{\partial t} - \nabla\cdot(\theta \mathbf{D} \nabla C) + \nabla\cdot(\theta \mathbf{V} C) + M(C) = 0$$
$$(x, y) \in (\Omega),\ 0 \le t \le t_f \tag{6.2.1a}$$

with appropriate initial and boundary conditions

$$C(x, y, 0) = g_0, \quad (x, y) \in (\Omega), \tag{6.2.1b}$$
$$C|_{(\Gamma_1)} = g_1, \qquad \theta(\mathbf{D}\nabla C - \mathbf{V}C)\cdot\mathbf{n}|_{(\Gamma_2)} = g_2. \tag{6.2.1c}$$

All notations in (6.2.1) were explained in Example 1.1.6.

Let the performance function be

$$E(C, \mathbf{p}) = \int_0^{t_f} \int_{(\Omega)} f(C, \mathbf{p}; x, y, t)\, d\Omega\, dt, \tag{6.2.2}$$

where $\mathbf{p}$ represents all unknown parameters. First, assume that only the longitudinal dispersivity α_L is unknown. A variation of α_L, $\delta\alpha_L$, will cause a variation $\delta\mathbf{D}$, and the latter will cause a variation δC. The relationship between them is determined by the following variational problem of (6.2.1):

$$\frac{\partial(\theta\delta C)}{\partial t} - \nabla\cdot(\theta\mathbf{D}\nabla\delta C) - \nabla\cdot\left(\theta\frac{\partial\mathbf{D}}{\partial\alpha_L}\delta\alpha_L\nabla C\right)$$
$$+\nabla\cdot(\theta\mathbf{V}\delta C) + M'(C)\delta C = 0, \tag{6.2.3a}$$
$$\delta C|_{t=0} = 0, \tag{6.2.3b}$$

$$\delta C|_{(\Gamma_1)} = 0, \qquad \theta\left(\mathbf{D}\nabla\delta C + \frac{\partial\mathbf{D}}{\partial\alpha_L}\delta\alpha_L\nabla C - \mathbf{V}\delta C\right)\cdot\mathbf{n}\bigg|_{(\Gamma_2)} = 0,$$
$$\tag{6.2.3c}$$

where

$$\frac{\partial \mathbf{D}}{\partial \alpha_L} = \frac{1}{V^2} \begin{bmatrix} V_x^2 & V_x V_y \\ V_x V_y & V_y^2 \end{bmatrix}. \tag{6.2.4}$$

Following the steps used in Section 6.1.2, we can find functional partial derivative

$$\frac{\partial E}{\partial \alpha_L} = \int_0^{t_f} \int_{(\Omega)} \left[\frac{\partial f}{\partial \alpha_L} + \theta \frac{\partial \mathbf{D}}{\partial \alpha_L} \nabla \psi \cdot \nabla C \right] d\Omega \, dt, \tag{6.2.5}$$

where function $\psi(x, y, t)$ satisfies

$$\theta \frac{\partial \psi}{\partial t} + \nabla \cdot (\theta \mathbf{D} \nabla \psi) + \theta \mathbf{V} \cdot \nabla \psi + M'(C) \psi - \frac{\partial f}{\partial C} = 0, \tag{6.2.6a}$$

$$\psi|_{t=t_f} = 0, \tag{6.2.6b}$$

$$\psi|_{(\Gamma_1)} = 0, \qquad \theta \mathbf{D} \nabla \psi \cdot \mathbf{n}|_{(\Gamma_2)} = 0, \tag{6.2.6c}$$

Equation (6.2.5) is known as the adjoint problem of the primary problem (6.2.1) and ψ the adjoint sate.

Next, let us consider hydraulic conductivity K as an unknown parameter. A variation δK causes a velocity variation $\delta \mathbf{V}$ which further causes $\delta \mathbf{D}$ and δC. The chain rule gives

$$\frac{\partial \mathbf{D}}{\partial K} = \frac{\partial \mathbf{D}}{\partial V_x} \frac{\partial V_x}{\partial K} + \frac{\partial \mathbf{D}}{\partial V_y} \frac{\partial V_y}{\partial K}, \tag{6.2.7}$$

where $\partial \mathbf{D}/\partial V_x$ and $\partial \mathbf{D}/\partial V_y$ can be found from (1.1.10), $\partial V_x/\partial K$ and $\partial V_y/\partial K$ are determined by the Darcy's Law

$$\frac{\partial V_x}{\partial K} = -\frac{1}{\theta} \frac{\partial \phi}{\partial x}, \qquad \frac{\partial V_y}{\partial K} = -\frac{1}{\theta} \frac{\partial \phi}{\partial y}. \tag{6.2.8}$$

In (6.2.8), we assume the head distribution $\phi(x, y, t)$ to be known, and consequently, $\partial \mathbf{V}/\partial K = -(1/\theta) \nabla \phi$ is also known.

The variational problem of (6.2.1) with respect to δK is

$$\frac{\partial (\theta \delta C)}{\partial t} - \nabla \cdot (\theta \mathbf{D} \nabla \delta C) - \nabla \cdot \left(\theta \frac{\partial \mathbf{D}}{\partial K} \delta K \nabla C \right) +$$
$$+ \nabla \cdot (\theta \mathbf{V} \delta C) - \nabla \cdot (C \nabla \phi \delta K) + M'(C) \delta C = 0, \tag{6.2.9a}$$

$$\delta C|_{t=0} = 0, \tag{6.2.9b}$$

$$\delta C|_{(\Gamma_1)} = 0,$$

$$\theta \left(\mathbf{D} \nabla \delta C + \frac{\partial \mathbf{D}}{\partial K} \delta K \nabla C - \mathbf{V} \delta C + \frac{1}{\theta} C \nabla \phi \delta K \right) \cdot \mathbf{n}|_{(\Gamma_2)} = 0. \tag{6.2.9c}$$

Using the procedure that we have used several times, the adjoint problem (6.2.6) can again be obtained, and

$$\frac{\partial E}{\partial K} = \int_0^{t_f} \int_{(\Omega)} \left[\frac{\partial f}{\partial K} - \theta \frac{\partial \mathbf{D}}{\partial K} \nabla C \cdot \nabla \psi + C \nabla \phi \cdot \nabla \psi \right] d\Omega \, dt. \tag{6.2.10}$$

Using time transformation (6.1.25), the form of adjoint problem (6.2.6) will becomes exactly the same as that of the primary problem (6.2.1). Therefore, the two problems can be solved by the same subroutine.

6.2.2 THE ADJOINT PROBLEM FOR COUPLED GROUNDWATER FLOW AND MASS TRANSPORT PROBLEMS

In Section 6.2.1, we assumed that the head distribution ϕ is known. In practice, however, head ϕ is also unknown and dependent on hydraudic conductivity K and other parameters. Both head ϕ and concentration C must be considered as unknown state variables.

The governing equations and subsidiary conditions of coupled groundwater flow and mass transport problems have been given in Example 1.1.6. Taking the variation of these equations with respect to the variation of hydraulic conductivity, δK, we can obtain

$$S\frac{\partial \delta\phi}{\partial t} - \nabla\cdot(m\nabla\phi\delta K) - \nabla\cdot(Km\nabla\delta\phi) = 0, \tag{6.2.11a}$$

$$\frac{\partial(\theta\delta C)}{\partial t} - \nabla\cdot(\theta\mathbf{D}\nabla\delta C) - \nabla\cdot\left(\theta\frac{\partial\mathbf{D}}{\partial\mathbf{V}}\delta\mathbf{V}\nabla C\right)$$
$$-\nabla\cdot(\theta C\delta\mathbf{V}) - \nabla\cdot(\theta\mathbf{V}\delta C) - M'(C)\delta C = 0, \tag{6.2.11b}$$
$$(x,y) \in (\Omega), \ 0 \le t \le t_f,$$

subject to initial and boundary conditions

$$\delta\phi|_{t=0} = 0, \qquad \delta\phi|_{(\Gamma_1)} = 0,$$

$$-(m\nabla\phi\delta K + Km\nabla\delta\phi)\cdot\mathbf{n}|_{(\Gamma_2)} = 0, \tag{6.2.12a}$$

$$\delta C|_{t=0} = 0, \qquad \delta C|_{(\Gamma_3)} = 0,$$

$$-\left(\theta\frac{\partial\mathbf{D}}{\partial\mathbf{V}}\delta\mathbf{V}\nabla C + \theta\mathbf{D}\nabla\delta C - \theta C\delta\mathbf{V} - \theta\mathbf{V}\delta C\right)\cdot\mathbf{n}|_{(\Gamma_4)} = 0. \tag{6.2.12b}$$

Let the performance function be

$$E(\phi, C, \mathbf{p}; x, y, t) = \int_0^{t_f} \int_{(\Omega)} f(\phi, c, \mathbf{p}; x, y, t) \, d\Omega \, dt \tag{6.2.13}$$

Its variation with respect to δK is

$$\delta E = \int_0^{t_f} \int_{(\Omega)} \left(\frac{\partial f}{\partial \phi} \delta \phi + \frac{\partial f}{\partial C} \delta C + \frac{\partial f}{\partial K} \delta K \right) d\Omega \, dt. \tag{6.2.14}$$

In order to eliminate the terms associated with unknown variations $\delta \phi$ and δC in the above equation, let us introduce two adjoint states ψ_1 and ψ_2 that have continuous second order space derivatives on (Ω) and continuous first order time derivatives in $[0, t_f]$. Multiplying (6.2.11a) by ψ_1, integrating the result over $(\Omega) \times [0, t_f]$, and then using Green's theorems, we can obtain

$$\int_0^{t_f} \int_{(\Omega)} \left\{ \left[s\frac{\partial \psi_1}{\partial t} + \nabla \cdot (Km\nabla\psi_1) \right] \delta\phi - m\nabla\phi \cdot \nabla\psi_1 \delta K \right\} d\Omega \, dt$$

$$- \int_0^{t_f} \int_{(\Gamma_2)} (Km\nabla\psi_1) \cdot \mathbf{n} \delta\phi \, d\Gamma \, dt = 0, \tag{6.2.15}$$

where conditions

$$\psi_1|_{t=t_f} = 0 \quad \text{and} \quad \psi_1|_{(\Gamma_1)} = 0 \tag{6.2.16}$$

have been used.

Similarly, multiplying (6.2.11b) by ψ_2, we can obtain

$$\int_0^{t_f} \int_{(\Omega)} \left[-\theta\frac{\partial \psi_2}{\partial t} - \nabla \cdot (\theta \mathbf{D} \nabla \psi_2) - \theta \mathbf{V} \nabla \psi_2 + M'(C)\psi_2 \right] \delta C \, d\Omega \, dt$$

$$+ \int_0^{t_f} \int_{(\Omega)} \left[\theta\frac{\partial \mathbf{D}}{\partial \mathbf{V}} \delta \mathbf{V} \nabla C \cdot \nabla \psi_2 - \theta C \nabla \psi_2 \cdot \delta \mathbf{V} \right] d\Omega \, dt = 0, \tag{6.2.17}$$

where we have assumed that

$$\psi_2|_{t=t_f} = 0, \qquad \psi_2|_{(\Gamma_3)} = 0, \qquad \theta \mathbf{D} \nabla \psi_2 \cdot \mathbf{n}|_{(\Gamma_4)} = 0. \tag{6.2.18}$$

From Darcy's law, $\delta \mathbf{V}$ can be represented as

$$\delta \mathbf{V} = -\frac{K}{\theta} \nabla \delta\phi - \frac{1}{\theta} \nabla\phi \delta K. \tag{6.2.19}$$

As a result, the second integral on the left-hand side of (6.2.17) can be rewritten as

$$\int_0^{t_f} \int_{(\Omega)} \left[\theta\frac{\partial \mathbf{D}}{\partial \mathbf{V}} \delta \mathbf{V} \nabla C \nabla \psi_2 - \theta C \nabla \psi_2 \cdot \delta \mathbf{V} \right] d\Omega \, dt$$

$$= \int_0^{t_f} \int_{(\Omega)} \left[\nabla \cdot (K\mathbf{E}\nabla\psi_2) - \nabla \cdot (KC\nabla\psi_2) \right] \delta h \, d\Omega \, dt$$

$$+ \int_0^{t_f} \int_{(\Omega)} [\mathbf{F}\nabla C \cdot \nabla\psi_2 - C\psi_2 \cdot \nabla\phi] \delta K \, d\Omega \, dt$$

$$- \int_0^{t_f} \int_{(\Gamma)} [K\mathbf{E}\nabla\psi_2 - KC\nabla\psi_2] \cdot \mathbf{n} \delta h \, d\Gamma \, dt, \tag{6.2.20}$$

where **E** and **F** are tensors with elements

$$E_{ij} = \frac{\partial D_{il}}{\partial V_j} \frac{\partial C}{\partial x_l}, \qquad F_{ij} = \frac{\partial D_{ij}}{\partial V_l} \frac{\partial h}{\partial x_l}. \tag{6.2.21}$$

In (6.2.21), $i = 1, 2$; $j = 1, 2$; $D_{11} = D_{xx}$, $D_{12} = D_{xy}$, $D_{21} = D_{yx}$, $D_{22} = D_{yy}$; $x_1 = x$, $x_2 = y$; and the summation for $l = 1, 2$ is implied.

Substituting (6.2.20) into (6.2.17) and adding the result to (6.2.15), we have

$$\int_0^{t_f} \int_{(\Omega)} \left\{ \left[\theta \frac{\partial \psi_2}{\partial t} + \nabla \cdot (\theta \mathbf{D} \nabla \psi_2) + \theta \mathbf{V} \nabla \psi_2 - M'(C)\psi_2 \right] \delta C \right.$$

$$+ \left[S \frac{\partial \psi_1}{\partial t} + \nabla \cdot (Km \nabla \psi_1) - \nabla \cdot (K\mathbf{E} \nabla \psi_2) + \nabla \cdot (KC \nabla \psi_2) \right] \delta h$$

$$\left. + [\mathbf{F} \nabla C \cdot \nabla \psi_2 - C \nabla \phi \cdot \nabla \psi_2 - m \nabla \phi \cdot \nabla \psi_1] \delta K \right\} \, d\Omega \, dt = 0,$$

$$\tag{6.2.22}$$

where the following condition

$$Km \nabla \psi_1 \cdot \mathbf{n}|_{(\Gamma_2)} = K(\mathbf{E} \nabla \psi_2 - C \nabla \psi_2) \cdot \mathbf{n}|_{(\Gamma_2)} \tag{6.2.23}$$

has been used. Adding (6.2.22) to (6.2.14), we can find that all terms associated with δh and δc are eliminated, provided that ψ_1 and ψ_2 satisfy

$$S \frac{\partial \psi_1}{\partial t} + \nabla \cdot (Km \nabla \psi_1) = \frac{\partial f}{\partial \phi} + \nabla \cdot (K\mathbf{E} \nabla \psi_2) - \nabla \cdot (KC \nabla \psi_2), \tag{6.2.24a}$$

$$\theta \frac{\partial \psi_2}{\partial t} + \nabla \cdot (\theta \mathbf{D} \nabla \psi_2) + \theta \mathbf{V} \nabla \psi_2 - M'(C)\psi_2 = \frac{\partial f}{\partial C}. \tag{6.2.24b}$$

Finally, we obtain the following *functional partial derivative*:

$$\frac{\partial E}{\partial K} = \int_0^{t_f} \int_{(\Omega)} \left[\frac{\partial f}{\partial K} - \mathbf{F} \nabla \psi_2 \cdot \nabla C + m \nabla \phi \cdot \nabla \psi_1 + C \nabla \phi \cdot \nabla \psi_2 \right] \, d\Omega \, dt. \tag{6.2.25}$$

To calculate $\partial E / \partial K$, both the primary problem governed by (1.1.8) and its adjoint problem governed by (6.2.24) have to be solved. The primary state variables are ϕ and C, while the adjoint state variables are ψ_1 and ψ_2. When ϕ and C satisfy the subsidiary conditions (1.1.9), ψ_1 and ψ_2 must satisfy (6.2.16), (6.2.23) and (6.2.18).

Using a similar procedure, we can find other functional partial derivatives as follows:

$$\frac{\partial E}{\partial S} = \int_0^{t_f} \int_{(\Omega)} \frac{\partial \phi}{\partial t} \psi_1 \, d\Omega \, dt, \tag{6.2.26a}$$

$$\frac{\partial E}{\partial \theta} = \int_0^{t_f} \int_{(\Omega)} \left(\frac{\partial C}{\partial t} \psi_2 + \mathbf{D} \nabla C \cdot \nabla \psi_2 - C \mathbf{V} \cdot \nabla \phi_2 \right) d\Omega \, dt, \qquad (6.2.26b)$$

$$\frac{\partial E}{\partial \alpha_L} = \int_0^{t_f} \int_{(\Omega)} \frac{\partial \mathbf{D}}{\partial \alpha_L} \nabla C \cdot \nabla \psi_2 \, d\Omega \, dt, \qquad (6.2.26c)$$

$$\frac{\partial E}{\partial \alpha_T} = \int_0^{t_f} \int_{(\Gamma)} \frac{\partial \mathbf{D}}{\partial \alpha_T} \nabla C \cdot \nabla \psi_2 \, d\Omega \, dt. \qquad (6.2.26d)$$

6.2.3 THE SOLUTION OF ADJOINT PROBLEMS

The coupled groundwater flow-mass transport problem and its adjoint problem can be solved sequentially by the following procedure:

Step 1. Solve Equation (1.1.8a) with subsidiary condition (1.1.9a) to obtain the head distribution $\phi(x, y, t)$;

Step 2. Compute velocity $\mathbf{V}$ and dispersion coefficient $\mathbf{D}$ by Darcy's law (1.1.11) and (1.1.10);

Step 3. Solve Equation (1.1.8b) with subsidiary condition (1.1.9b) to obtain the concentration distribution $c(x, y, t)$;

Step 4. Solve Equation (6.2.24b) with subsidiary condition (6.2.18) to obtain the adjoint state $\psi_2(x, y, t)$;

Step 5. Calculate the sink/source term $\nabla \cdot (K \mathbf{E} \nabla \psi_2) - \nabla \cdot (K C \nabla \psi_2)$ of Equation (6.2.24a) and boundary condition (6.2.23) using the calculated h, C and ψ_2;

Step 6. Solve Equation (6.2.24a) with subsidiary conditions (6.2.16) and (6.2.23) to obtain the adjoint state $\psi_1(x, y, t)$.

All these equations can be solved by a numerical method. Since the form of adjoint equations is the same as that of primary equations, the same subroutine can be used for solving all of them.

The numerical solutions of ϕ, C and ψ_2 are generally accurate enough. The accuracy of ψ_1, however, is often poor since it depends upon ∇h, ∇C, and $\nabla \phi_2$. It is well known that the gradient of a unknown function can not be accurately calculated when linear basis functions are used. Moreover, overshooting and numerical dispersion of solutions C and ψ_2 may cause serious errors in ∇C and $\nabla \psi_2$, and consequently, ψ_1 will also be in error. Use of a high-order finite element method, therefore, is strongly sugguested. Numerical examples of solving for ϕ, C, ψ_1 and ψ_2 can be found in Sun and Yeh (1990a).

Exercise 6.2.1. Give the expression of $\partial \mathbf{D} / \partial K$ in (6.2.10).

Exercise 6.2.2. Derive the expression of $\partial E / \partial \theta$ in (6.2.26).

Exercise 6.2.3. Derive the adjoint problem for one-dimensional coupled ground-water flow and mass transport problems.

6.3 The Adjoint State Method for General Coupled Problems

Applying the adjoint state method to study coupled problems has been considered by many authors. Seinfeld and Chen (1978) and Watson *et al.* (1980) used this method in the analysis of oil-water two-phase flow problems. Oblow (1978) and Cacuci *et al.* (1980) described a general procedure of deriving the adjoint equations for any type of coupled problems. Sun and Yeh (1990a) presented some adjoint operation rules that simplify the derivation of adjoint problems.

6.3.1 THE DERIVATION OF ADJOINT STATE EQUATIONS

For coupled problems, a general performance criterion is defined as

$$E = \int_0^{t_f} \int_{(\Omega)} f(\mathbf{u}, \mathbf{p}; \mathbf{x}, t) \, d\Omega \, dt, \tag{6.3.1}$$

where $f(\mathbf{u}, \mathbf{p}; \mathbf{x}, t)$ is a user-chosen function. The first-order variation of E is

$$\delta E = \int_0^{t_f} \int_{(\Omega)} \left(\frac{\partial f}{\partial \mathbf{u}} \delta \mathbf{u} + \frac{\partial f}{\partial \mathbf{p}} \delta \mathbf{p} \right) d\Omega \, dt. \tag{6.3.2}$$

In order to find the functional derivative of E with respect to the unknown parameter $\mathbf{p}$, we have to eliminate the unknown terms associated with $\delta \mathbf{u}$ in (6.3.2).

The first-order variations of governing equations (4.6.1) with subsidiary conditions (4.6.2) for a coupled problem are

$$\nabla_{\mathbf{u}} \mathbf{L} \delta \mathbf{u} + \nabla_{\mathbf{p}} \mathbf{L} \delta \mathbf{p} = \mathbf{0}, \tag{6.3.3}$$

$$\delta \mathbf{u} = \mathbf{0} \quad \text{when } t = 0, \tag{6.3.4a}$$

$$\delta \mathbf{u} = \mathbf{0} \quad \text{on } (\Gamma_1), \tag{6.3.4b}$$

$$\nabla_{\mathbf{u}} \mathbf{L}_B \delta \mathbf{u} + \nabla_{\mathbf{p}} \mathbf{L}_B \delta \mathbf{p} = \mathbf{0} \quad \text{on } (\Gamma_2). \tag{6.3.4c}$$

In Equation (6.3.3), matrices $\nabla_{\mathbf{u}} \mathbf{L}$ and $\nabla_{\mathbf{p}} \mathbf{L}$, the gradient operators of $\mathbf{L}$ with repsect to $\mathbf{u}$ and $\mathbf{p}$, respectively, are defined as

$$
\nabla_{\mathbf{u}}\mathbf{L} = \begin{bmatrix}
\dfrac{\partial L_1}{\partial u_1} & \dfrac{\partial L_1}{\partial u_2} & \cdots & \dfrac{\partial L_1}{\partial u_n} \\[2ex]
\dfrac{\partial L_2}{\partial u_1} & \dfrac{\partial L_2}{\partial u_2} & \cdots & \dfrac{\partial L_2}{\partial u_n} \\[2ex]
\vdots & \vdots & \ddots & \vdots \\[2ex]
\dfrac{\partial L_n}{\partial u_1} & \dfrac{\partial L_n}{\partial u_2} & \cdots & \dfrac{\partial L_n}{\partial u_n}
\end{bmatrix}, \tag{6.3.5}
$$

$$
\nabla_{\mathbf{p}}\mathbf{L} = \begin{bmatrix}
\dfrac{\partial L_1}{\partial p_1} & \dfrac{\partial L_1}{\partial p_2} & \cdots & \dfrac{\partial L_1}{\partial p_m} \\[2ex]
\dfrac{\partial L_2}{\partial p_1} & \dfrac{\partial L_2}{\partial p_2} & \cdots & \dfrac{\partial L_2}{\partial p_m} \\[2ex]
\vdots & \vdots & \ddots & \vdots \\[2ex]
\dfrac{\partial L_n}{\partial p_1} & \dfrac{\partial L_n}{\partial p_2} & \cdots & \dfrac{\partial L_n}{\partial p_m}
\end{bmatrix}. \tag{6.3.6}
$$

Matrix operators $\nabla_{\mathbf{u}}\mathbf{L}_B$ and $\nabla_{\mathbf{p}}\mathbf{L}_B$ have the same forms as $\nabla_{\mathbf{u}}\mathbf{L}$ and $\nabla_{\mathbf{p}}\mathbf{L}$, except changing $\mathbf{L}$ to $\mathbf{L}_B$.

The scalar product of any two vector functions $\mathbf{u}$ and $\mathbf{v}$ on space domain (Ω) and time interval $(T) = [0, t_f]$ is defined as

$$
(\mathbf{u}, \mathbf{v})_{\Omega,T} = \int_0^{t_f} \int_{(\Omega)} u_i v_i \, d\Omega \, dt, \tag{6.3.7}
$$

where the summation for all components of $\mathbf{u}$ and $\mathbf{v}$ is implied. Similarly, we can also define the scalar product of $\mathbf{u}$ and $\mathbf{v}$ on boundary (Γ) of (Ω) and time interval (T) as

$$
(\mathbf{u}, \mathbf{v})_{\Gamma,T} = \int_0^{t_f} \int_{(\Gamma)} u_i v_i \, d\Gamma \, dt. \tag{6.3.8}
$$

For any *vector function* ψ having continuous second-order derivatives, we have the following identities:

$$
(\psi, \nabla_{\mathbf{u}}\mathbf{L}\delta\mathbf{u})_{\Omega,T} = (\delta\mathbf{u}, \nabla_{\mathbf{u}}^+\mathbf{L}\psi)_{\Omega,T} + \text{BIT}, \tag{6.3.9}
$$

$$
(\psi, \nabla_{\mathbf{p}}\mathbf{L}\delta\mathbf{p})_{\Omega,T} = (\delta\mathbf{p}, \nabla_{\mathbf{p}}^+\mathbf{L}\psi)_{\Omega,T} + \text{BIT}, \tag{6.3.10}
$$

where BIT denotes all boundary integral terms, and $\nabla_{\mathbf{u}}^+\mathbf{L}$ and $\nabla_{\mathbf{p}}^+\mathbf{L}$ are the *transposed adjoint operators* of $\nabla_{\mathbf{u}}\mathbf{L}$ and $\nabla_{\mathbf{p}}\mathbf{L}$, respectively. Identities (6.3.9) and

(6.3.10) are obtained by using Green's theorems several times to transfer the differential operation from δu (or δp) to ψ. Each time, one or two boundary integral terms are generated. In what follows, some rules by which $\nabla_u^+ L$ and $\nabla_p^+ L$ can be found from $\nabla_u L$ and $\nabla_p L$ will be given. Adding (6.3.9) and (6.3.10) and using (6.3.3), we obtain

$$(\delta u, \nabla_u^+ L\psi)_{\Omega,T} + (\delta p, \nabla_p^+ L\psi)_{\Omega,T} + BIT = 0. \tag{6.3.11}$$

The term *BIT* in the above equation may be represented as

$$BIT = (\delta p, \nabla_p^+ L_B \psi)_{\Gamma,T} + (\delta u, \nabla_u^+ L_B \psi)_{\Gamma,T} + (\delta u, \psi)_{\Omega,t_f}$$
$$-(\psi, \delta u)_{\Omega,0} - (\psi, \nabla_u L_B \delta u + \nabla_p L_B \delta p)_{\Gamma,T}. \tag{6.3.12}$$

The initial and boundary conditions for δu have been given in (6.3.6). If we further define the final and boundary conditions for ψ such that

$$\psi = 0 \quad \text{when } t = t_f, \tag{6.3.13a}$$
$$\psi = 0 \quad \text{on } (\Gamma_1), \tag{6.3.13b}$$
$$\nabla_u^+ L_B \psi = 0 \quad \text{on } (\Gamma_2), \tag{6.3.13c}$$

then all terms on the right-hand side of (6.3.12), except the first one, will vanish. Consequently,

$$BIT = (\delta p, \nabla_p^+ L_B \psi)_{\Gamma,T}, \tag{6.3.14}$$

where $\nabla_p^+ L_B$ is often equal to a zero operator. If there is no unknown boundary parameter, then BIT = 0.

Now, let us return to (6.3.2). It can be rewritten as

$$\delta E = \left(\delta u, \frac{\partial f}{\partial u}\right)_{\Omega,T} + \left(\delta p, \frac{\partial f}{\partial p}\right)_{\Omega,T}. \tag{6.3.15}$$

Adding (6.3.11) to (6.3.15), we obtain

$$\delta E = \left(\delta u, \nabla_u^+ L\psi + \frac{\partial f}{\partial u}\right)_{\Omega,T} + \left(\delta p, \nabla_p^+ L\psi + \frac{\partial f}{\partial p}\right)_{\Omega,T}$$
$$+(\delta p, \nabla_p^+ L_B \psi)_{\Gamma,T}. \tag{6.3.16}$$

The first scalar product on the right-hand side of the above equation can be eliminated if ψ is selected to satisfy

$$\nabla_u^+ L\psi + \frac{\partial f}{\partial u} = 0. \tag{6.3.17}$$

Equation (6.3.17) with subsidiary conditions (6.3.13) form the (vector) *adjoint problem* of the primary coupled problem defined by (4.6.1) and (4.6.2). Its solution

ψ is named the *adjoint state* (vector) of the primary state vector **u**. With the adjoint state ψ, (6.3.15) reduces to

$$\delta E = \left(\delta \mathbf{p}, \nabla_{\mathbf{p}}^{+} \mathbf{L} \psi + \frac{\partial f}{\partial \mathbf{p}} \right)_{\Omega,T} + (\delta \mathbf{p}, \nabla_{\mathbf{p}}^{+} \mathbf{L}_B \psi)_{\Gamma,T}. \tag{6.3.18}$$

Therefore, for any unknown parameter p_j we can obtain the partial functional derivative

$$\frac{\partial E}{\partial p_j} = \int_0^{t_f} \int_{(\Omega)} \left[\nabla_{\mathbf{p}}^{+} \mathbf{L} \psi + \frac{\partial f}{\partial \mathbf{p}} \right]_j \, d\Omega \, dt, \tag{6.3.19}$$

where $[\cdot]_j$ represents the j-th component of $[\cdot]$. In (6.3.19), we have assumed that the boundary integral term in (6.3.18) is zero.

The derivation of *adjoint equations for coupled problems* is now completed. The reader should be aware that all results given in Sections 6.1 and 6.2 are special cases of the results given in this section.

6.3.2 ADJOINT OPERATION RULES

Let us consider how to find the transposed adjoint operator matrices $\nabla_{\mathbf{u}}^{+} \mathbf{L}$, $\nabla_{\mathbf{p}}^{+} \mathbf{L}$, $\nabla_{\mathbf{u}}^{+} \mathbf{L}_B$ and $\nabla_{\mathbf{p}}^{+} \mathbf{L}_B$. For any operator matrix $\mathbf{R}$, we define $\mathbf{R}^{+} = (\mathbf{R}^{*})^{T}$, where the asterisk indicates the adjoint operation and superscript T represents the transpose operation.

Table 6.3.1 lists some operation rules for elements of $\nabla \mathbf{L}$ and $\nabla \mathbf{L}_B$, respectively, which are derived from Green's theorem's and often encountered in groundwater modeling.

In Table 6.3.1, f and g are scalar functions, $\mathbf{V}$ is a vector function, $\mathbf{K}$ is a symmetric tensor function, and $\mathbf{n}$ is the normal direction of boundary (Γ). Using rules given in Table 6.3.1, the derivation of adjoint problems becomes simple and straightforward. Steps of deriving the adjoint problem for a coupled problem can be summarized as follows:

Step 1. Find gradient operator matrices $\nabla_{\mathbf{u}} \mathbf{L}, \nabla_{\mathbf{p}} \mathbf{L}, \nabla_{\mathbf{u}} \mathbf{L}_B$ and $\nabla_{\mathbf{p}} \mathbf{L}_B$ by differentiation.

Step 2. Find $\nabla_{\mathbf{u}}^{*} \mathbf{L}, \nabla_{\mathbf{p}}^{*} \mathbf{L}, \nabla_{\mathbf{u}}^{*} \mathbf{L}_B$ and $\nabla_{\mathbf{p}}^{*} \mathbf{L}_B$ using the adjoint operation rules, and then obtain $\nabla_{\mathbf{u}}^{+} \mathbf{L}, \nabla_{\mathbf{p}}^{+} \mathbf{L}, \nabla_{\mathbf{u}}^{+} \mathbf{L}_B$ and $\nabla_{\mathbf{p}}^{+} \mathbf{L}_B$ by transposition.

Step 3. Form the vector adjoint equation using (6.3.17).

Step 4. Form the final and boundary conditions of the adjoint problem using (6.3.13).

Step 5. Obtain all partial functional derivatives of performance E using (6.3.19).

TABLE 6.3.1
Adjoint operation rules for operators ∇L and ∇L_B

Rules for ∇L		Rules for ∇L_B	
Elements of ∇L	Elements of $(\nabla L)^*$	Elements of ∇L_B	Elements of $(\nabla L_B)^*$
$f \bullet$	$f \bullet$	$\mathbf{V} \bullet \cdot n$	0
$f \dfrac{\partial}{\partial t} \bullet$	$-\dfrac{\partial}{\partial t}(f \bullet)$	$f \nabla \bullet \cdot n$	$f \nabla \bullet \cdot n$
$\dfrac{\partial}{\partial t}(f \bullet)$	$-f \dfrac{\partial}{\partial t} \bullet$	$\mathbf{K} \nabla \bullet \cdot n$	$\mathbf{K} \nabla \bullet \cdot n$
$\nabla \cdot (f \nabla \bullet)$	$\nabla \cdot (f \nabla \bullet)$	$\mathbf{K} \nabla (f \bullet) \cdot n$	$f \mathbf{K} \nabla \bullet \cdot n$
$\nabla \cdot (\mathbf{K} \nabla \bullet)$	$\nabla \cdot (\mathbf{K} \nabla \bullet)$		
$\mathbf{V} \cdot (\nabla \bullet)$	$-\nabla \cdot (\mathbf{V} \bullet)$		
$\nabla \cdot (\mathbf{V} \bullet)$	$-\mathbf{V} \cdot (\nabla \bullet)$		
$\nabla \cdot (f \nabla g \bullet)$	$-f \nabla g \cdot \nabla \bullet$		
$\nabla \cdot (\mathbf{K} \nabla g \bullet)$	$-\mathbf{K} \nabla g \cdot \nabla \bullet$		
$\nabla \cdot [\mathbf{K} \nabla (f \bullet)]$	$f \nabla \cdot (\mathbf{K} \nabla \bullet)$		

Example 6.3.1. The governing equation of temperature $T(x, z, t)$ in a fuel rod is

$$\rho C_p(T) \frac{\partial T}{\partial t} - \frac{1}{r} \frac{\partial}{\partial r} \left[k(T) r \frac{\partial T}{\partial r} \right] = Q(r, z, t), \tag{6.3.20}$$

where ρ = density, C_p = heat capacity, T = temperature, k = thermal conductivity, Q = nuclear heat source density, (r, z) = cylinder coordintates, t = time.

Let us consider a more general form of (6.3.20):

$$L \equiv \rho C_p(T) \frac{\partial T}{\partial t} - \nabla \cdot [k(T) \nabla T] - Q = 0. \tag{6.3.21}$$

Differentiating L with respect to T gives

$$\begin{aligned} \nabla_T L = {} & \rho C_p'(T) \frac{\partial T}{\partial t} \bullet + \rho C_p(T) \frac{\partial}{\partial t} \bullet \\ & - \nabla \cdot [k'(T) \nabla T \bullet] - \nabla \cdot [k(T) \nabla \bullet]. \end{aligned} \tag{6.3.22}$$

Using the rules given in Table 6.3.1, we can find

$$\begin{aligned} (\nabla_T L)^* = {} & \rho C_p'(T) \frac{\partial T}{\partial t} \bullet - \frac{\partial}{\partial t} [\rho C_p(T) \bullet] \\ & + [k'(T) \nabla T \cdot \nabla \bullet] - \nabla \cdot [k(T) \nabla \bullet]. \end{aligned} \tag{6.3.23}$$

Thus, the adjoint state equation of (6.3.21) is

$$\rho C_p(T)\frac{\partial \psi}{\partial t} + \nabla \cdot [k(T)\nabla \psi] - k'(T)\nabla T \cdot \nabla \psi = \frac{\partial f}{\partial T}, \tag{6.3.24}$$

where ψ is the adjoint state of temperature T. Returning to the cylinder coordinate system (r, z), (6.3.24) can be rewritten as

$$\rho C_p(T)\frac{\partial \psi}{\partial t} + \frac{k}{r}\frac{\partial}{\partial r}\left(r\frac{\partial \psi}{\partial r}\right) = \frac{\partial f}{\partial T}. \tag{6.3.25}$$

6.3.3   ADJOINT STATE EQUATIONS FOR OIL-WATER TWO PHASE FLOW PROBLEMS

Using the adjoint rules given above, it is now easy to obtain the adjoint state equations for oil-water two phase flow problems.

The oil-water two phase flow in reservoirs is governed by the following equations (Huyakorn and Pinder, 1983):

$$L_1 \equiv \frac{\partial}{\partial t}(\theta S_w) - \nabla \cdot \left[\mathbf{k}\frac{k_{rw}}{\mu_w}(\nabla P_w + \rho_w g \nabla z) \right] + Q_w = 0, \tag{6.3.26a}$$

$$L_2 \equiv \frac{\partial}{\partial t}(\theta S_o) - \nabla \cdot \left[\mathbf{k}\frac{k_{ro}}{\mu_o}(\nabla P_o + \rho_o g \nabla z) \right] + Q_o = 0 \tag{6.3.26b}$$

subject to initial and boundary conditions

$$P_w = f_0, \qquad P_o = g_0 \quad \text{when } t = t_0, \tag{6.3.27a}$$

$$P_w = f_1, \qquad P_o = g_1 \quad \text{on } (\Gamma_1), \tag{6.3.27b}$$

$$-\left[\mathbf{k}\frac{k_{rw}}{\mu_w}(\nabla P_w + \rho_w g \nabla z) \right] \cdot \mathbf{n} = f_2, \tag{6.3.27c}$$

$$-\left[\mathbf{k}\frac{k_{ro}}{\mu_o}(\nabla P_o + \rho_o g \nabla z) \right] \cdot \mathbf{n} = g_2 \quad \text{on } (\Gamma_2). \tag{6.3.27d}$$

In (6.3.26) and (6.3.27), S is the saturation, P the pressure, ρ the density, μ the dynamic viscosity, Q the sink/source term, $\mathbf{k}$ the intrinsic permeability tensor and k_r the relative permeability. Subscripts w and o are associated with water phase and oil phase, respectively. Since there are two supplementary equations

$$S_o + S_w = 1, \tag{6.3.28a}$$

$$P_o - P_w = P_{o/w}, \tag{6.3.28b}$$

where $P_{o/w}$ is the capillary pressure, we can select any two variables from S_o, S_w, P_o, P_w as the unknown states. Usually, pressure P_w and saturation S_w of the water phase are selected. Unknown parameters may include porosity θ, components k_{ij} of $\mathbf{k}$, relative permeabilities k_{rw} and k_{ro}. Thus, we have

$$\mathbf{u} = (P_w, S_w)^T, \qquad \mathbf{p} = (\theta, k_{ij}, k_{ro}, k_{rw})^T. \tag{6.3.29}$$

Since k_{rw} and k_{ro} depend on S_w and S_o, both Equation (6.3.26a) and (6.3.26b) are nonlinear. Now assume that $k_{rw} = k_{rw}(S_w)$, $k_{ro} = k_{ro}(S_w)$, and $P_{o/w} = P_{o/w}(S_w)$. From (6.3.26), we can obtain

$$\frac{\partial L_1}{\partial P_w} = -\nabla\cdot\left(\frac{kk_{rw}}{\mu_w}\nabla\cdot\right),$$

$$\frac{\partial L_1}{\partial S_w} = \frac{\partial}{\partial t}(\theta\cdot) - \nabla\cdot\left[\frac{kk'_{rw}}{\mu_w}(\nabla P_w + \rho_w g\nabla z)\cdot\right],$$

$$\frac{\partial L_2}{\partial P_w} = -\nabla\cdot\left(\frac{kk_{ro}}{\mu_o}\nabla\cdot\right),$$

$$\frac{\partial L_2}{\partial S_w} = -\frac{\partial}{\partial t}(\theta\cdot) - \nabla\cdot\left[\frac{kk'_{ro}}{\mu_o}(\nabla P_o + \rho_o g\nabla z)\cdot\right]$$
$$- \nabla\cdot\left[\frac{kk_{ro}}{\mu_o}\nabla(P'_{o/w}\cdot)\right].$$

Adjoint operations yield

$$\left(\frac{\partial L_1}{\partial P_w}\right)^* = -\nabla\cdot\left(\frac{kk_{rw}}{\mu_w}\nabla\cdot\right),$$

$$\left(\frac{\partial L_1}{\partial S_w}\right)^* = -\theta\frac{\partial}{\partial t}\cdot + \frac{kk'_{rw}}{\mu_w}(\nabla P_w + \rho_w g\nabla z)\cdot\nabla\cdot,$$

$$\left(\frac{\partial L_2}{\partial P_w}\right)^* = -\nabla\cdot\left(\frac{kk_{ro}}{\mu_o}\nabla\cdot\right),$$

$$\left(\frac{\partial L_2}{\partial S_w}\right)^* = \theta\frac{\partial}{\partial t}\cdot + \frac{kk'_{ro}}{\mu_o}(\nabla P_o + \rho_o g\nabla z)\cdot\nabla\cdot + P'_{o/w}\nabla\cdot\left(\frac{kk_{ro}}{\mu_o}\nabla\cdot\right).$$

Therefore, the adjoint equations of this problem are

$$\frac{\partial f}{\partial P_w} - \nabla\cdot\left(\frac{kk_{rw}}{\mu_w}\nabla\phi_1\right) - \nabla\cdot\left(\frac{kk_{ro}}{\mu_o}\nabla\phi_2\right) = 0, \qquad (6.3.30a)$$

$$\frac{\partial f}{\partial S_w} - \theta\frac{\partial\phi_1}{\partial t} + \frac{kk'_{rw}}{\mu_w}(\nabla P_w + \rho_w g\nabla z)\cdot\nabla\phi_1 + \theta\frac{\partial\phi_2}{\partial t}$$
$$+ \frac{kk'_{ro}}{\mu_o}(\nabla P_o + \rho_o g\nabla z)\cdot\nabla\phi_2 + P'_{o/w}\nabla\cdot\left(\frac{kk_{ro}}{\mu_o}\nabla\phi_2\right) = 0. \qquad (6.3.30b)$$

From (6.3.27), we have

$$\frac{\partial L_{B1}}{\partial P_w} = -\left(\frac{kk_{rw}}{\mu_w}\nabla\cdot\right)\cdot\mathbf{n}, \qquad \frac{\partial L_{B2}}{\partial P_w} = \left(\frac{kk_{ro}}{\mu_o}\nabla\cdot\right)\cdot\mathbf{n},$$

$$\frac{\partial L_{B1}}{\partial S_w} = -\left[\frac{kk'_{rw}}{\mu_w}(\nabla P_w + \rho_w g\nabla z)\cdot\right]\cdot\mathbf{n},$$

$$\frac{\partial L_{B2}}{\partial S_w} = -\left[\frac{kk'_{ro}}{\mu_o}(\nabla P_o + \rho_o g\nabla z)\cdot + \frac{kk_{ro}}{\mu_o}\nabla(P'_{o/w}\cdot)\right]\cdot\mathbf{n}.$$

Adjoint operations yield

$$\left(\frac{\partial L_{B1}}{\partial P_w}\right)^* = -\left(\frac{kk_{rw}}{\mu_w}\nabla\cdot\right)\cdot\mathbf{n}, \qquad \left(\frac{\partial L_{B2}}{\partial P_w}\right)^* = -\left(\frac{kk_{ro}}{\mu_o}\nabla\cdot\right)\cdot\mathbf{n},$$

$$\left(\frac{\partial L_{B1}}{\partial S_w}\right)^* = 0, \qquad \left(\frac{\partial L_{B2}}{\partial S_w}\right)^* = -\left(P'_{o/w}\frac{kk_{ro}}{\mu_o}\nabla\cdot\right)\cdot\mathbf{n},$$

Therefore, the final and boundary conditions for (6.3.30) are

$$\phi_1 = 0, \qquad \phi_2 = 0, \quad \text{when } t = t_f, \tag{6.3.31a}$$
$$\phi_1 = 0, \qquad \phi_2 = 0, \quad \text{on } (\Gamma_1), \tag{6.3.31b}$$
$$-\left(\frac{kk_{rw}}{\mu_w}\nabla\phi_1 + \frac{kk_{ro}}{\mu_o}\nabla\phi_2\right)\cdot\mathbf{n} = 0,$$
$$-\left(P'_{o/w}\frac{kk_{ro}}{\mu_o}\nabla\phi_2\right)\cdot\mathbf{n} = 0, \quad \text{on } (\Gamma_2). \tag{6.3.31c}$$

Differentiating Equation (6.3.26) with respect to the unknown parameters, we obtain

$$\frac{\partial L_1}{\partial\theta} = \frac{\partial}{\partial t}(S_w\cdot), \qquad \frac{\partial L_2}{\partial\theta} = \frac{\partial}{\partial t}(S_o\cdot),$$

$$\frac{\partial L_1}{\partial k_{ij}} = -\nabla\cdot\left[\frac{k_{rw}}{\mu_w}\frac{\partial\mathbf{k}}{\partial k_{ij}}(\nabla P_w + \rho_w g\nabla z)\cdot\right],$$

$$\frac{\partial L_2}{\partial k_{ij}} = -\nabla\cdot\left[\frac{k_{ro}}{\mu_o}\frac{\partial\mathbf{k}}{\partial k_{ij}}(\nabla P_o + \rho_o g\nabla z)\cdot\right],$$

$$\frac{\partial L_1}{\partial k_{rw}} = -\nabla\cdot\left[\frac{\mathbf{k}}{\mu_w}(\nabla P_w + \rho_w g\nabla z)\cdot\right], \qquad \frac{\partial L_2}{\partial k_{rw}} = 0\cdot,$$

$$\frac{\partial L_1}{\partial k_{ro}} = 0\cdot, \qquad \frac{\partial L_2}{\partial k_{ro}} = -\nabla\cdot\left[\frac{\mathbf{k}}{\mu_o}(\nabla P_o + \rho_o g\nabla z)\cdot\right].$$

Their adjoint operators are

$$\left(\frac{\partial L_1}{\partial\theta}\right)^* = -S_w\frac{\partial}{\partial t}\cdot, \qquad \left(\frac{\partial L_2}{\partial\theta}\right)^* = -S_o\frac{\partial}{\partial t}\cdot,$$

$$\left(\frac{\partial L_1}{\partial k_{ij}}\right)^* = \frac{k_{rw}}{\mu_w}\frac{\partial\mathbf{k}}{\partial k_{ij}}(\nabla P_w + \rho_w g\nabla z)\cdot\nabla\cdot,$$

$$\left(\frac{\partial L_2}{\partial k_{ij}}\right)^* = \frac{k_{ro}}{\mu_o}\frac{\partial\mathbf{k}}{\partial k_{ij}}(\nabla P_o + \rho_o g\nabla z)\cdot\nabla\cdot,$$

$$\left(\frac{\partial L_1}{\partial k_{rw}}\right)^* = \frac{\mathbf{k}}{\mu_w}(\nabla P_w + \rho_w g\nabla z)\cdot\nabla\cdot, \qquad \left(\frac{\partial L_2}{\partial k_{rw}}\right)^* = 0\cdot,$$

$$\left(\frac{\partial L_1}{\partial k_{ro}}\right)^* = 0\cdot, \qquad \left(\frac{\partial L_2}{\partial k_{ro}}\right)^* = \frac{\mathbf{k}}{\mu_o}(\nabla P_o + \rho_o g\nabla z)\cdot\nabla\cdot.$$

Therefore, the partial functional derivatives are

$$\frac{\partial E}{\partial \theta} = \int_{t_0}^{t_f} \int_{(\Omega)} \left[\frac{\partial f}{\partial \theta} - S_w \frac{\partial \phi_1}{\partial t} - S_o \frac{\partial \phi_2}{\partial t} \right] d\Omega \, dt, \tag{6.3.32a}$$

$$\frac{\partial E}{\partial k_{ij}} = \int_{t_0}^{t_f} \int_{(\Omega)} \left[\frac{\partial f}{\partial k_{ij}} + \frac{k_{rw}}{\mu_w} \frac{\partial k}{\partial k_{ij}} (\nabla P_w + \rho_w g \nabla z) \nabla \phi_1 \right.$$

$$\left. + \frac{k_{ro}}{\mu_o} \frac{\partial k}{\partial k_{ij}} (\nabla P_o + \rho_o g \nabla z) \nabla \phi_2 \right] d\Omega \, dt, \tag{6.3.32b}$$

$$\frac{\partial E}{\partial k_{rw}} = \int_{t_0}^{t_f} \int_{(\Omega)} \left[\frac{\partial f}{\partial k_{rw}} + \frac{k}{\mu_w} (\nabla P_w + \rho_w g \nabla z) \nabla \phi_1 \right] d\Omega \, dt, \tag{6.3.32c}$$

$$\frac{\partial E}{\partial k_{ro}} = \int_{t_0}^{t_f} \int_{(\Omega)} \left[\frac{\partial f}{\partial k_{ro}} + \frac{k}{\mu_o} (\nabla P_o + \rho_o g \nabla z) \nabla \phi_2 \right] d\Omega \, dt. \tag{6.3.32d}$$

Exercise 6.3.1. Use the adjoint operation rules to reproduce the adjoint problem and partial functional derivatives for leaky aquifer systems given in Section 6.1.4.

Exercise 6.3.2. Use the adjoint operation rules to reproduce the adjoint problem and partial functional derivatives for advection-dispersion equations given in Section 6.2.1.

6.4 The Discrete Approach for Deriving Adjoint State Equations

There are two approaches for obtaining the discretized adjoint system for a given problem. In our previous discussions we have only considered one of them, i.e., the *continuous approach*. It includes two principal steps: first, to derive the continuous adjoint system represented by a set of *PDEs* with the variational method, then discretize these *PDEs* into a system of algebraic equations by a numerical method. Another way of obtaining the discretized adjoint system is the discrete approach. It also includes two principal steps: first to discretize the primary problem into a system of algebraic equations with a numerical method, then to find the adjoint system of these algebraic equations by the variational method.

Let us return to the groundwater flow problem discussed in Section 6.1.2. Using a numerical method, problem (6.1.10) can be discretized into a system of algebraic equations for each time step. Suppose that the time period $[0, t_f]$ is divided into K time steps, we then have

$$\mathbf{G}\phi_k = \mathbf{H}\phi_{k-1} + \mathbf{q}_k \quad (k = 1, 2, \dots, K), \tag{6.4.1}$$

where $\phi_k = (\phi_{k,1}, \phi_{k,2}, \dots, \phi_{k,N})^T$ is the head distribution of the k-th time step defined by its nodal values; $\mathbf{q}_k$ the sink/source term; In (6.4.1), coefficient matrices

G and H depend on transmissivity T and storage coefficient S, respectively. However, if we use vector $\mathbf{p}$ to represent all parameters, then both G and H are functions of $\mathbf{p}$.

The discrete form of performance function (6.1.17) can then be written as

$$E = \sum_{l=1}^{L} w_l f_l(\phi_1, \phi_2, \ldots, \phi_k, \mathbf{p}), \tag{6.4.2}$$

where $(w_1, w_2, \ldots, w_L)$ is a set of weighting coefficients. Taking the variation of E, we obtain

$$\delta E = \sum_{l=1}^{L} w_l \left[\sum_{k=1}^{K} \frac{\partial f_l}{\partial \phi_k} \delta \phi_k + \frac{\partial f_l}{\partial \mathbf{p}} \delta \mathbf{p} \right]. \tag{6.4.3}$$

All terms associated with unknown variations $\delta \phi_k$ $(k = 1, 2, \ldots, K)$ can be eliminated by the following procedure.

First, taking the variation of (6.4.1) to obtain

$$\delta \mathbf{G} \phi_k + \mathbf{G} \delta \phi_k - \delta \mathbf{H} \phi_{k-1} - \mathbf{H} \delta \phi_{k-1} = 0 \quad (k = 1, 2, \ldots, K), \tag{6.4.4}$$

then, multiplying (6.4.4) by an arbitrary vector ψ_k^T for each k, adding them together, and rearranging the order of summation, the final result becomes:

$$\sum_{k=1}^{K-1} \{\psi_k^T \mathbf{G} - \psi_{k+1}^T \mathbf{H}\} \delta \phi_k + \psi_K^T \mathbf{G} \delta \phi_K$$

$$+ \sum_{k=1}^{K} \psi_k^T \{\delta \mathbf{G} \phi_k - \delta \mathbf{H} \phi_{k-1}\} = 0. \tag{6.4.5}$$

Adding (6.4.5) to (6.4.3) yields

$$\delta E = \sum_{k=1}^{K-1} \left\{ \psi_k^T \mathbf{G} - \psi_{k+1}^T \mathbf{H} + \sum_{l=1}^{L} w_l \frac{\partial f_l}{\partial \phi_k} \right\} \delta \phi_k$$

$$+ \left\{ \psi_K^T \mathbf{G} + \sum_{l=1}^{L} w_l \frac{\partial f_l}{\partial \phi_K} \right\} \delta \phi_K$$

$$+ \left\{ \sum_{k=1}^{K} \psi_k^T \left(\frac{\partial \mathbf{G}}{\partial \mathbf{p}} \phi_{k+1} - \frac{\partial \mathbf{H}}{\partial \mathbf{p}} \phi_k \right) + \sum_{l=1}^{L} w_l \frac{\partial f_l}{\partial \mathbf{p}} \right\} \delta \mathbf{p}. \tag{6.4.6}$$

Now, it is obvious that if ψ_k $(k = 1, 2, \ldots, K)$ satisfy

$$\mathbf{G} \psi_k = \mathbf{H} \psi_{k+1} - \sum_{l=1}^{L} w_l \frac{\partial f_l}{\partial \phi_k} \quad (k = 1, 2, \ldots, K - 1) \tag{6.4.7a}$$

and

$$\mathbf{G}\psi_K = -\sum_{l=1}^{L} w_l \frac{\partial f_l}{\partial \phi_K}, \qquad (6.4.7b)$$

then all terms associated with $\delta\phi_k$ $(k = 1, 2, \ldots, K)$ in (6.4.6) will be eliminated. Thus functional derivative $\delta E/\delta\mathbf{p}$ can be represented as

$$\frac{\delta E}{\delta \mathbf{p}} = \sum_{k=1}^{K} \psi_k^T \left(\frac{\partial \mathbf{G}}{\partial \mathbf{p}} \phi_{k+1} - \frac{\partial \mathbf{H}}{\partial \mathbf{p}} \phi_k \right) + \sum_{l=1}^{L} w_l \frac{\partial f_l}{\partial \mathbf{p}}. \qquad (6.4.8)$$

Equation (6.4.7) is the adjoint system of the discretized primary system (6.4.1). Since they have the same coefficient matrices, we can form and solve them with the same routine.

The discrete approach can be easily applied to any coupled problem. We can also develop some adjoint operation rules for the discrete approach. It has been shown by Sun and Yeh (1992b) that both approaches, continuous and discrete, generate consistent results.

Exercise 6.4.1. If (6.4.1) is the finite difference system of groundwater flow problem (6.1.10), derive equations (6.4.7) and (6.4.8) and then compare them with the corresponding results derived by the continuous approach.

6.5 Applications of the Adjoint State Method

The adjoint state method was first used in the petroleum industry for identifying reservoir parameters (Jacquard and Jain, 1965; Chavent *et al.*, 1975). Neuman (1980) used this method for identification of log-transmissivities. Thereafter, more applications of this method were found in groundwater modeling. We may list the following aspects: *parameter identification* (Carrera and Neuman, 1986b; Sun and Yeh, 1990a); *sensitivity analysis* (Sykes *et al.*, 1985; Wilson and Metcalfe, 1985); *reliability evaluation* (Townley and Wilson, 1985); *parameter structure identification* (Sun and Yeh, 1985); *remediation design* (Ahlfeld *et al.*, 1988); *experimental design* (Sun and Yeh, 1990b; Sun and Yeh, 1992a); *observation network design* (Mckinney and Loucks, 1992).

In this section, however, only two applications of the adjoint method will be introduced: parameter estimation and sensitivity analysis. We plan to present other applications of this method in the following chapters.

6.5.1 PARAMETER IDENTIFICATION

The parameter identification problem has been transformed into a minimization problem in Chapter 4. When a gradient method is used to solve the problem, in each iteration, we must calculate the gradient of objective function

$$E(\mathbf{k}) = \sum_{l=1}^{L} w_l^2 [u_l(\mathbf{k}) - \tilde{u}_l]^2. \tag{6.5.1}$$

We have known that $M+1$ simulation runs are needed for obtaining all components $\partial E/\partial k_1$, $\partial E/\partial k_2$, ..., $\partial E/\partial k_M$ by the finite difference approximation, where M is the dimension of parameterization. Using the adjoint state method, however, these components can be obtained with only two simulation runs, no matter what value is assigned to M. In fact, if we select

$$f(\mathbf{u}, \mathbf{k}; \mathbf{x}, t) = \sum_{l=1}^{L} w^2(\mathbf{x}, t)[u(\mathbf{k}; \mathbf{x}, t) - \tilde{u}(\mathbf{x}, t)]^2 \delta(\mathbf{x} - \mathbf{x}_l)\delta(t - t_l),$$

$$\tag{6.5.2}$$

objective function (6.5.1) can be expressed in a general form

$$E(\mathbf{k}) = \int_0^{t_f} \int_{(\Omega)} f(\mathbf{u}, \mathbf{k}; \mathbf{x}, t) \, d\Omega \, dt. \tag{6.5.3}$$

In (6.5.2), the l-th observation is taken at point $\mathbf{x}_l$ and time t_l. All gradient components of E can thus be obtained from the functional derivative of E. A special case has been shown in (6.1.24), where the unknown parameter is the transmissivity and the observed state is the head. Similar expressions can be derived for other cases.

It is obvious that the adjoint method can also handle regularized *OLS* problems, where extra regularization terms are added to the objective function.

In order to calculate the total gradient vector $\mathbf{g}_E$ defined in (4.6.13) for solving coupled inverse problems, we can select

$$f_i(\mathbf{u}, \mathbf{p}; \mathbf{x}, t) = \sum_{l=1}^{L_i} w_i^2(\mathbf{x}, t)[u_i(\mathbf{p}; \mathbf{x}, t) - \tilde{u}_i(\mathbf{x}, t)]^2 \delta(\mathbf{x} - \mathbf{x}_l)\delta(t - t_l)$$

$$\tag{6.5.4}$$

as function $f(\mathbf{u}, \mathbf{p}; \mathbf{x}, t)$ in criterion (6.3.1), where L_i is the number of observations associated with state u_i. As a result, the general performance criterion reduces to the least squares criterion E_i defined in (4.6.5). Hence, using (6.3.19), we have

$$\frac{\partial E_i}{\partial p_j} = \int_0^{t_f} \int_{(\Omega)} \left[\frac{\partial f_i}{\partial \mathbf{p}} + \nabla_{\mathbf{p}}^+ \mathbf{L} \psi_i \right]_j d\Omega \, dt. \tag{6.5.5}$$

In the above equation, ψ_i is the adjoint state vector defined in the adjoint equation (6.3.17), where f is replaced by f_i. Using the adjoint method to calculate the gradient vector and the Fletcher–Reeves algorithm to update the search sequence, we can form a highly efficient numerical scheme for solving inverse problems (Neuman, 1980; Carrera and Neuman, 1986b).

With the adjoint state method, we can increase the number of identified parameters without significantly increasing the computation effort. The number of identified parameters, of course, is mainly dependent on the quantity and quality of available data. As previously denoted, over-parameterization may cause the inverse problem to be ill-posed.

Chen and his co-workers developed a so called generalized pluse-spectrum technique (GPST) which is similar to the adjoint state method in certain sense (Chen and Liu, 1984; Chen, 1985; Tang *et al.*, 1989). Combination of this technique with a regularization method and a quasi-Newton method is also a highly efficient numerical technique for inverse solution.

6.5.2   THE COMPUTATION OF SENSITIVITY MATRIX

Let $(u_1, u_2, \ldots, u_L)$ be a set of values of a distributed state $u(\mathbf{x}, t)$, and $(p_1, p_2, \ldots, p_K)$ be a set of values of a distributed parameter $p(\mathbf{x})$. The *Jacobian sensitivity matrix* of $(u_1, u_2, \ldots, u_L)$ with respect to $(p_1, p_2, \ldots, p_K)$ is defined as

$$
\mathbf{J}_D = \begin{bmatrix}
\dfrac{\partial u_1}{\partial p_1} & \dfrac{\partial u_2}{\partial p_1} & \cdots & \dfrac{\partial u_L}{\partial p_1} \\[2ex]
\dfrac{\partial u_1}{\partial p_2} & \dfrac{\partial u_2}{\partial p_2} & \cdots & \dfrac{\partial u_L}{\partial p_2} \\[2ex]
\vdots & \vdots & \ddots & \vdots \\[2ex]
\dfrac{\partial u_1}{\partial p_K} & \dfrac{\partial u_2}{\partial p_K} & \cdots & \dfrac{\partial u_L}{\partial p_K}
\end{bmatrix}, \tag{6.5.6}
$$

where the subscript D is used to denote a set of locations and/or times. The following are some examples:

Case 1. $(u_1, u_2, \ldots, u_L)$ is a set of observations associated with L observation wells and/or observation times, and $p_1, p_2, \ldots, p_K$ are parameter values associated with K zones.

Case 2. $(u_1, u_2, \ldots, u_L)$ is a set of predictions associated with L wells and/or times of interest, and $p_1, p_2, \ldots, p_K$ are distributed values associated with all nodes.

Case 3. $(u_1, u_2, \ldots, u_L)$ is a set of values associated with L designed locations and/or times, and $p_1, p_2, \ldots, p_K$ are parameter values associated with K designed locations.

Case 1 is often met in the solution of inverse problems when the Gauss–Newton method is used to identify K values of the unknown parameter. If we want to estimate the reliability of a calibrated model for predicition, Case 2 is encountered. Case 3 may be met in experimental design problems.

To calculate the sensitivity matrix (6.5.6), we have given two approaches in Section 4.3.3: the finite difference method and the sensitivity equation method. Both of them need to run the simulation model for $K + 1$ times. Now, let us consider using the adjoint state method.

Let $u_l = u(x_l, t_l)$, $\quad l = 1, 2, \ldots, L$. If we set

$$f_l(u, p; x, t) = u(x, t)\delta(x - x_l)\delta(t - t_l), \tag{6.5.7}$$

where $\delta(\cdot)$ is the Dirac-δ function, then

$$u_l = \int_0^{t_f} \int_{(\Omega)} f_l(u, p, x, t)\, d\Omega\, dt. \tag{6.5.8}$$

Equation (6.5.8) means that u_l can be seen as a performance function, and thus its variational can be expressed by

$$\delta u_l = \int_0^{t_f} \int_{(\Omega)} \nabla_p^+ L \psi_l \delta p\, d\Omega\, dt, \tag{6.5.9}$$

where ψ_l is the adjoint state determined by equation

$$\nabla_u^+ L \psi_l + \frac{\partial f_l}{\partial u} = 0 \tag{6.5.10}$$

subject to appropriate subsidiary conditions. In (6.5.9) and (6.5.10), $L(u, p) = 0$ is the primary *PDE* operator, $\nabla_p^+ L$ and $\nabla_u^+ p$ are transposed adjoint operators of gradient operators $\nabla_p L$ and $\nabla_u L$, respectively.

Furthermore, if parameter $p(x)$ is parameterized by the zonation method, Equation (6.5.9) gives

$$\frac{\partial u_l}{\partial p_k} = \int_0^{t_f} \int_{(\Omega_k)} \nabla_p^+ L \psi_l\, d\Omega\, dt, \tag{6.5.11}$$

where (Ω_k) is the zone associated with p_k.

Using (6.5.11) to calculate all elements for sensitivity matrix (6.5.6), we have to solve the primary problem once and the adjoint problem (6.5.10) L times. The total computation effort is equivalent to running the simulation model $L + 1$ times.

Therefore, for the problems with $L < K$, we should select the adjoint state method to calculate the sensitivity matrix, otherwise, the sensitivity equation method should be selected. Li *et al.* (1986) shown that the two approaches produce exactly the same results when the same simulation model is used.

Example 6.5.1. (*Yeh and Sun, 1990*). Consider the leaky system described in Example 1.1.5. Figures 6.5.1 and 6.5.2 show its cross-sectional view and plane view, respectively. A river cuts off the system and defines boundary conditions for both the confined and unconfined aquifers with a constant head 100 m. Remaining boundaries are all impervious. The initial head is 100 m in the whole system. The flow region of each aquifer is divided into 162 triangle elements with 100 nodes as depicted in Figure 6.5.2. The leakage parameter K_z has different values at different locations as defined in Table 6.5.1. Other parameters of the system are given in Table 6.5.2. There are four pumping wells in the confined aquifer. Their locations can be found in Figure 6.5.2. Pumping rates are all 4000 m^3/day.

Adjoint equations and algorithms for calculating the functional derivatives for a leaky system have been given in Section 6.1.4. Now, assume an observation well to be located at node $l = $#85, both h and ϕ are observed at time = 7.5 days, i.e., we have two state observations ($L = 2$). Solving the primary system once and the adjoint system twice with

$$f = h(x, y, t)\delta(x - x_l)\delta(y - y_l)\delta(t - 7.5) \qquad (6.5.12a)$$

and

$$f = \phi(x, y, t)\delta(x - x_l)\delta(y - y_l)\delta(t - 7.5) \qquad (6.5.12b)$$

as the user-chosen functions, respectively, then all sensitivity coefficients $\partial h/\partial K_z$, $\partial h/\partial K_1$, $\partial h/\partial K_2$, $\partial \phi/\partial K_z$, $\partial \phi/\partial K_1$, and $\partial \phi/\partial K_2$ with respect to all nodal parameters can be obtained. Table 6.5.3 to Table 6.5.8 list their values associated with four nodes only (#54, #67, #72 and #86). Corresponding results obtained by the finite difference approximation are also listed in these Tables for comparison. The closest values between them are denoted by superscript *.

From these Tables, it is easy to find that the sensitivity coefficients produced by the finite difference approximation are dependent on the size of perturbation increments. Generally, there is no rule to determine the best perturbation increment. With the adjoint state method, however, the uncertainty associated with perturbation increments is avoided.

6.5.3 SENSITIVITY ANALYSIS

To describe the sensitivity distribution of performance function $E(p)$ with respect to a distributed parameter p, we can calculate $\partial E/\partial p_i$ for each node i, where p_i is the node value of p and then draw a contour map. Using the adjoint state method to obtain all $\partial E/\partial p_i$, the computation effort is equivalent to running the simulation model twice and independent of the total number of nodes.

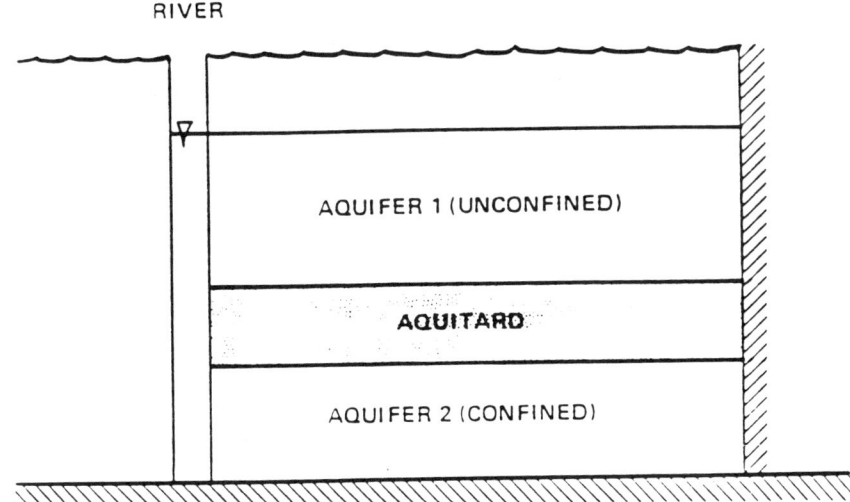

Fig. 6.5.1.   Cross-sectional View of the Leaky Aquifer System.

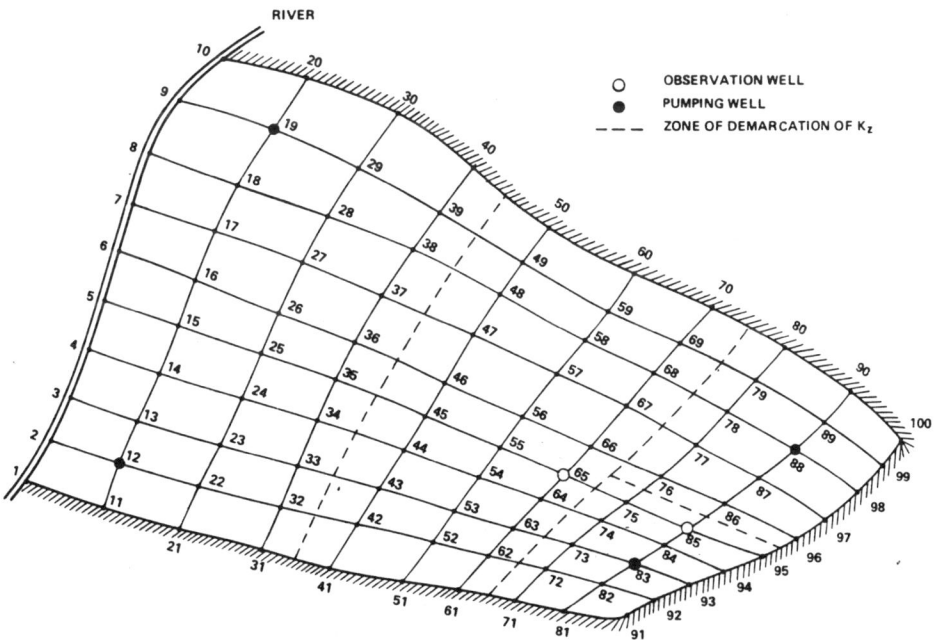

Fig. 6.5.2.   Plane View of the Leaky Aquifer System.

TABLE 6.5.1
Leakage Paramete K_z in Each Zone

	Zone			
	1	2	3	4
Value of K_z	5×10^{-4}	1×10^{-3}	1.5×10^{-3}	2×10^{-3}
Node numbers	1–40	41–70	71–75	76–80
in the zone			81–85	86–90
			91–95	96–100

TABLE 6.5.2
Parameters Used in the Leaky System

	Storage Coefficient	Hydraulic Conductivity, m/day	Thickness m
Unconfined aquifer	$S_1 = 1 \times 10^{-2}$	$K_1 = 2$	$TK_1 = 40$
Confined aquifer	$S_2 = 1 \times 10^{-4}$	$K_2 = 10$	$TK_2 = 50$

To compare the sensitivities of different parameters, we often use normalized sensitivity coefficients. For a performance function E and a parameter p, the normalized sensitivity of E with respect to p is defined as

$$s = \frac{\partial E}{\partial p} \cdot \frac{p}{E},$$ (6.5.13)

which describe the percentage change of E to a 1% change of p (Cacuci *et al.*, 1980; Sykes *et al.*, 1985).

To evaluate the efficiency of observation data in connection with parameter identification, Yeh and Sun (1990) and Sun and Yeh (1990a) introduced a new concept known as "contribution of observation R in the identification of parameter p". It is indicated by $CTB(R, p)$ as follows:

$$CTB(R, p) = \frac{\varepsilon_p}{\eta_R} \left| \int_{(D)} \frac{\partial R}{\partial p} \, dD \right|,$$ (6.5.14)

where ε_p is the maximal admissible error associated with the identified parameter p and η_R is the upper bound of the observation error associated with the observation R, and (D) is the subrigion defining parameter p.

This concept is different from the normalized sensitivity, because it is related not only to the sensitivity $\partial R/\partial p$, but also to the reliability requirement of parameter p, depicted by ε_p, and the quality of observation R, depicted by η_R.

TABLE 6.5.3
Computational Results of $(\partial h/\partial K_z)|_{(l,n,i)}$

Node Number	Influence Coefficient Method				Variational Method
	$\Delta K_z = 10^{-3}$	$\Delta K_z = 5 \times 10^{-4}$	$\Delta K_z = 10^{-4}$	$\Delta K_z = 5 \times 10^{-5}$	
54	2.61	3.48*	3.35	0.31	3.47
67	7.43	8.91*	9.77	15.26	9.25
72	7.06*	8.91	9.46	10.98	7.16
86	−38.87	−42.88*	−46.39	−47.30	−41.60

TABLE 6.5.4
Computational Results of $(\partial \phi/\partial K_z)|_{(l,n,i)}$

Node Number	Influence Coefficient Method				Variational Method
	$\Delta K_z = 10^{-3}$	$\Delta K_z = 5 \times 10^{-4}$	$\Delta K_z = 10^{-4}$	$\Delta K_z = 5 \times 10^{-5}$	
54	4.65*	9.55	7.78	0.92	4.42
67	10.68	13.67	11.44*	45.17	12.24
72	5.26	11.66	6.71*	32.35	8.34
86	14.07	17.49*	30.52	24.11	18.19

TABLE 6.5.5
Computational Results of $(\partial h/\partial K_1)|_{(\iota,n,i)}$

Node Number	Influence Coefficient Method					Variational Method
	$\Delta K_1 = 0.4$	$\Delta K_1 = 0.2$	$\Delta K_1 = 0.15$	$\Delta K_1 = 0.12$	$\Delta K_1 = 0.1$	
54	−0.0004	0.0006	0.0008	0*	0.0006	0
67	0.0007	0.0005	0.0006	0.0011*	0.0018	0.0010
72	0.0027	0.0013	0.0023	0.0031*	0.0051	0.0029
86	0.0102	0.0101	0.0112	0.0103*	0.0029	0.0104

TABLE 6.5.6
Computational Results of $(\partial \phi/\partial K_1)|_{(\iota,n,i)}$

Node Number	Influence Coefficient Method					Variational Method
	$\Delta K_1 = 0.4$	$\Delta K_1 = 0.2$	$\Delta K_1 = 0.15$	$\Delta K_1 = 0.12$	$\Delta K_1 = 0.1$	
54	0.0006*	0.0102	−0.0025	−0.0060	0.0008	0.0002
67	−0.0003	0.0002	0.0045	−0.0132	0.0022*	0.0015
72	−0.0039	0.0031*	−0.0014	−0.0093	0.0072	0.0022
86	0.0057	0.0002*	−0.0061	0.0059	0.0074	0.0011

TABLE 6.5.7
Computational Results of $(\partial h/\partial K_2)|_{(l,n,i)}$

Node Number	Influence Coefficient Method				Variational Method
	$\Delta K_2 = 4$	$\Delta K_2 = 2$	$\Delta K_2 = 0.5$	$\Delta K_2 = 0.1$	
54	0.0005	0.0006*	0.0002	0.0003	0.0006
67	0.0016*	0.0018	0.0011	0.0035	0.0017
72	0.0051	0.0051*	0.0050	0.0072	0.0052
86	0.0029*	0.0028	0.0023	0.0023	0.0030

TABLE 6.5.8
Computational Results of $(\partial \phi/\partial K_2)|_{(l,n,i)}$

Node Number	Influence Coefficient Method				Variational Method
	$\Delta K_2 = 4$	$\Delta K_2 = 2$	$\Delta K_2 = 0.5$	$\Delta K_2 = 0.1$	
54	0.0008	0.0008*	0.0007	−0.0101	0.0009
67	0.0023	0.0028*	0.0021	0.0113	0.0026
72	0.0068	0.0072*	0.0081	0.0328	0.0071
86	0.0070	0.0073*	0.0084	0.0015	0.0075

Calculating the "contribution" of each observation to each unknown parameter is very useful in designing an optimum experiment for parameter identification. In Chapter 8, we will show that when observation R is used for identifying parameter P, $CTB(R,P) > 1$ is a necessary condition.

To calculate $CTB(R,p)$ defined in (6.5.14), we can take the advantage of using the adjoint state method that in which distribution of $\partial R/\partial p$ in the flow region (Ω) can be obtained with only two simulation runs. The value of $CTB(R,p)$ can then be calculated by numerical integration.

Example 6.5.2. Let us return to Example 6.5.1 and let parameter p in (6.5.14) be the value of K_z associated with a node and R be the value of h observed at node $l = \#65$ and time t_n. Further, assume that $\varepsilon_{K_z} = 0.0001$ (1/day), $\eta_h = 0.01$ m. For each node i, contribution $CTB(h, K_z)$ can be calculated from (6.5.14), in which, the exclusive subdomain of the node is taken as (D). Using the adjoint method to complete the computation for each node, distributed contribution $CTB(h, K_z)$ can be obtained as shown in Figure 6.5.3. Parallel results for $CTB(\phi, K_z)$ is shown in Figure 6.5.4. Contributions $CTB(h, K_1)$, $CTB(h, K_2)$, $CTB(\phi, K_1)$ and $CTB(\phi, K_2)$ are shown in Figures 6.5.5 to 6.5.8. From these figures, we can

TABLE 6.5.9
Integrated contributions $CTB(h, K_z)$ and
$CTB(\phi, K_z)$ associated with observation
at node $l = 65$ and $t_n = 30$ days

	$CTB(h, K_z)$	$CTB(\phi, K_z)$
Zone 1	1.68	1.75
Zone 2	2.29	2.28
Zone 3	0.41	0.48
Zone 4	0.22	0.49

observe that the values of these contribution are less then one for all nodes. In other words, the observation R is insufficient for identifying any parameter value associated only with one node.

According to definition (6.5.14), there are several approaches to increase the value of a contribution.

(1) Increase ε_p, i.e., relax the accuracy requirement of the identified parameter p.

(2) Decrease η_R, i.e., raise the accuracy of the observation R.

(3) Enlarge region (D), i.e., reduce the dimension of parameterization such that parameter p can be associated with a larger zone.

(4) Increase the absolute value of sensitivity $\partial R / \partial p$, i.e., increase the information of observation by changing the pumping test design, such as increasing pumping rate and/or prolonging observation time.

Table 6.5.9 shows the integrated contributions $CTB(h, K_z)$ and $CTB(\phi, K_z)$, where K_z is divided into four zones as defined in Table 6.5.1.

In Sun and Yeh (1990a), readers can find numerical examples of calculating contributions for coupled groundwater flow and mass transport problems.

Exercise 6.5.1. When the regularized *OLS* criterion (4.4.4) is used for parameter identification, calculate the gradient of $E_\beta(\mathbf{k})$ by the adjoint state method.

Exercise 6.5.2. Give the adjoint equations and expressions of $\partial h_l / \partial K_z$, $\partial \phi_l / \partial K_z$, $\partial h_l / \partial K_1$, $\partial \phi_l / \partial K_1$, $\partial h_l / \partial K_2$, and $\partial \phi_l / \partial K_2$ used in Example 6.5.1.

Exercise 6.5.3. Use the concept of "contribution" defined in (6.5.14) to explain overparameterization problem.

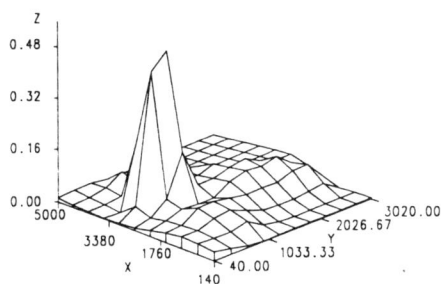

Fig. 6.5.3.   The Distribution of $CTB(h, K_z)$.

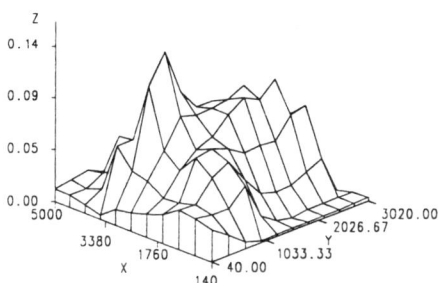

Fig. 6.5.4.   The Distribution of $CTB(\phi, K_z)$.

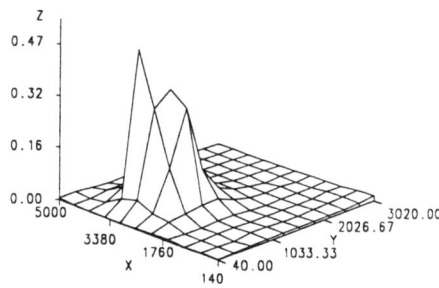

Fig. 6.5.5.   The Distribution of $CTB(h, K_1)$.

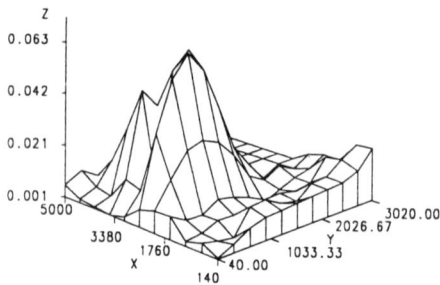

Fig. 6.5.6.   The Distribution of $CTB(h, K_2)$.

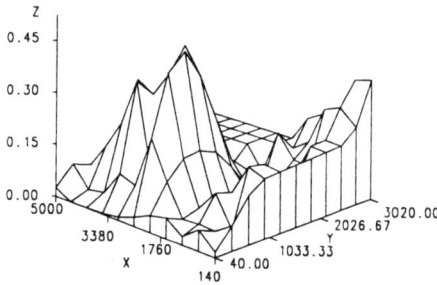

Fig. 6.5.7.   The Distribution of $CTB(\phi, K_1)$.

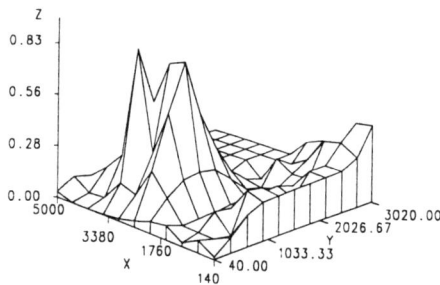

Fig. 6.5.8.   The Distribution of $CTB(\phi, K_2)$.

The Stochastic Method for Solving Inverse Problems

7.1 Defining Inverse Problems in Statistical Framework

To solve inverse problems, we must have some prior information, field observations, and a model relating the observations to unknown parameters. Since observation error and model structure error always exist and depend on some uncontrollable factors, it is necessary to consider the natural of inverse problems under the framework of stochastics.

There are three main advantages in the use of stochastic methods. First, prior information is easy to be incorporated into the statement of defining inverse problems. Second, unknown parameters and their uncertainty can be estimated at the same time. Third, the over-parameterization problem may be avoided, because unknown parameters are regarded as stochastic fields and described by only a few statistical parameters.

In this section, the inverse solution is simply defined as an information transference from measurements (or observations) to the unknown parameters. This concept has been well established in the theory of "parameter estimation" (Bard, 1974; Beck and Arnold, 1977).

7.1.1 A STATISTIC STATEMENT OF INVERSE PROBLEMS

To build a mathematical model for a real groundwater system, generally, we have three categories of information:

· prior information,

· measurement information, and

· model structure information.

In order to estimate the unknown parameters of the system as well as possible, we should use all of these categories of information and consider various errors associated with them. The problem of parameter estimation may be described as follows: by the aid of a model (with model structure errors) to transfer the measurement information (with measurement errors) from the measurement space to the parameter space to decrease the uncertainty of the estimated parameters. To make this statement to be clear, we should explain:

- · how to measure information and uncertainty,

- · how to transfer information from the measurement space to the parameter space through a model, and

- · how to estimate the uncertainty of the estimated parameters.

In the statistic framework, estimated parameters generally consist of three categories:

- · parameters that describe the statistical properties of unknown physical parameters,

- · parameters that describe the statistical properties of measurement errors, and

- · parameters that describe the statistical properties of model structure errors.

We will use β, ψ, χ to denote the three categories of parameters. In the general case, these parameters need to be estimated simultaneously. For example, it is possible to estimate the variance of head measurements simultaneously with the estimation of hydraulic conductivity. We will use a vector θ to denote the estimated parameters which may consist of all or a part of the three categories of parameters.

7.1.2 THE MEASURE OF INFORMATION AND UNCERTAINTY

Since the available information for parameter identification always includes errors which may depend on some uncontrollable factors, the true values of the identified parameters can never be known. Thus, it is reasonable to regard the unknown parameters as *random variables* or *stochastic fields*.

As well known, a continuous random vector θ is completely characterized by its *probability distribution function (pdf)*, $p(\theta)$. When $p(\theta)$ is given, we can calculate the mean value (or *mathematical expectation*) of θ by

$$\mu_\theta = E(\theta) = \int_{(\Omega)} \theta p(\theta) \, d\theta, \tag{7.1.1}$$

where (Ω) is the whole distribution space. We can also calculate the auto covariance matrix of θ by

$$\mathbf{V}_\theta = E\left[(\theta - \mu_\theta)(\theta - \mu_\theta)^T\right] = \int_{(\Omega)} (\theta - \mu_\theta)(\theta - \mu_\theta)^T p(\theta)\, d\theta, \qquad (7.1.2)$$

The *uncertainty* associated with $p(\theta)$ is measured by

$$H\left[p(\theta)\right] = -E\left[\log p(\theta)\right] = -\int_{(\Omega)} p(\theta) \log p(\theta)\, d\theta, \qquad (7.1.3)$$

while $-H\left[p(\theta)\right] = E\left[\log p(\theta)\right]$ is known as the *information content* of the distribution $p(\theta)$ (Bard, 1974).

Example 7.1.1. The *pdf* of an one-dimensional homogeneous distribution in an interval I is given by

$$p(\theta) = \begin{cases} 1/d, & \text{when } \theta \in I \\ 0, & \text{otherwise}, \end{cases} \qquad (7.1.4)$$

where d is the length of interval I. According to (7.1.3), we have

$$H(p) = -\int_I (1/d) \log (1/d)\, d\theta = \log d. \qquad (7.1.5)$$

Thus, the uncertainty of *pdf* (7.1.4) increases along with d (Figure 7.1.1).

Example 7.1.2. The *pdf* of an one-dimensional normal distribution with mean value μ and variance σ^2 is given by

$$p(\theta) = \frac{1}{\sqrt{2\pi}\,\sigma} \exp\left[-\frac{1}{2\sigma^2}(\theta - \mu)^2\right]. \qquad (7.1.6)$$

According to (7.1.3), we have

$$H(p) = -E\left[\log p\right] = \log \sigma + \frac{1}{2}(1 + \log 2\pi). \qquad (7.1.7)$$

Hence, the uncertainty associated with a normal distribution increases when its variance increases (see Figure 7.1.2).

Example 7.1.3. The result of Example 7.1.2 can be extended to the case of n-dimensional normal distribution:

$$p(\theta) = (2\pi)^{-n/2} (\det \mathbf{V}_\theta)^{-1/2} \exp\left[-\frac{1}{2}(\theta - \mu_\theta)^T \mathbf{V}_\theta^{-1}(\theta - \mu_\theta)\right]. \qquad (7.1.8)$$

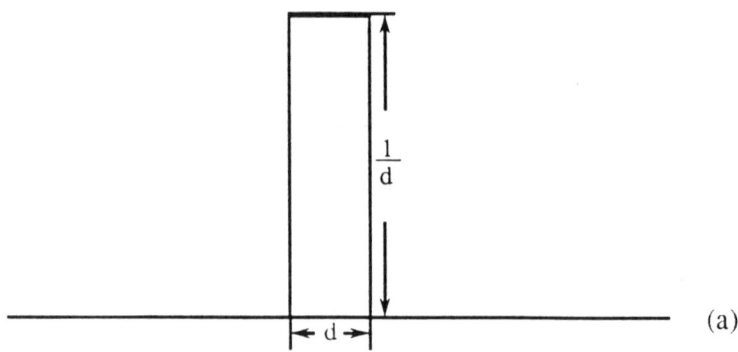

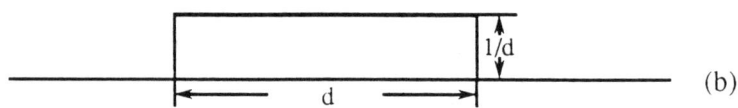

Fig. 7.1.1.   Homogeneous distribution: the distribution in (a) contains more information than that of the distribution in (b).

According to (7.1.3), its uncertainty

$$H(p) = -E\left[-\frac{n}{2}\log 2\pi - \frac{1}{2}\log\det \mathbf{V}_\theta - \frac{1}{2}(\theta - \mu_\theta)\mathbf{V}_\theta^{-1}(\theta - \mu_\theta)\right]$$

$$= \frac{n}{2}\log 2\pi + \frac{1}{2}\log\det \mathbf{V}_\theta + \frac{1}{2}T_r\mathbf{V}_\theta^{-1}\mathbf{V}_\theta$$

$$= \frac{n}{2}(1 + \log 2\pi) + \frac{1}{2}\log\det \mathbf{V}_\theta, \qquad\qquad (7.1.9)$$

where $\det \mathbf{V}_\theta$ is often called the *generalized variance* of an n-dimensional random vector θ. Equation (7.1.9) shows that the information content increases when the generalized variance decreases.

From the above examples, we see that the uncertainty defined by (7.1.3) is consistent with its intuitive explanation: a more dispersive distribution is always associated with more uncertainty. In fact, except the difference of a positive factor, (7.1.3) is the uniquely appropriate measure of the uncertainty (Bard, 1974; Tarantola, 1987).

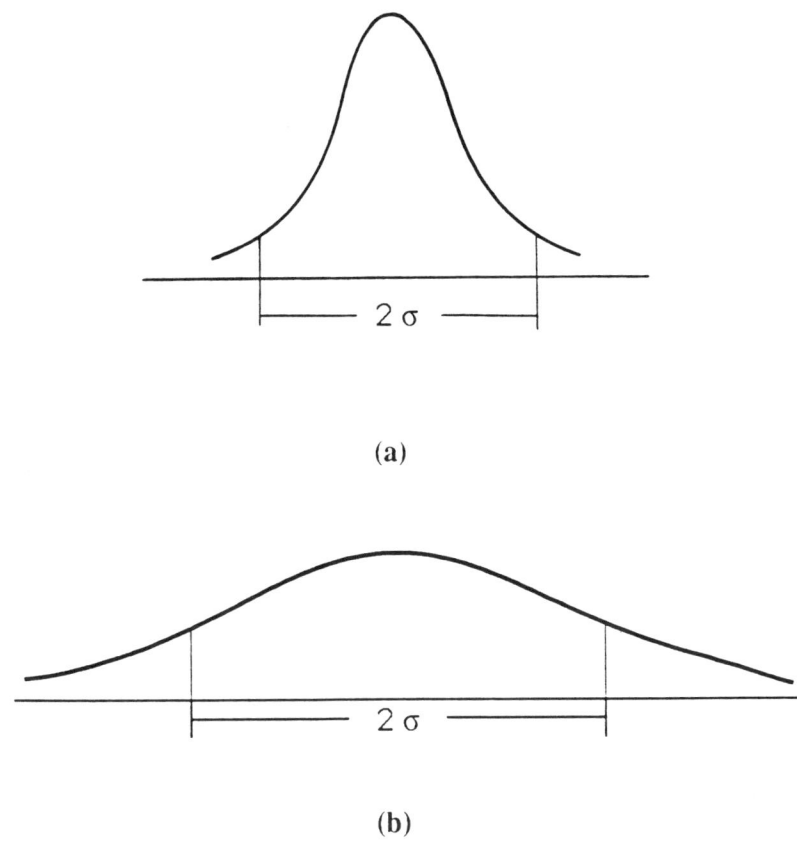

(a)

(b)

Fig. 7.1.2. Normal distribution: The distribution in (a) contains more information or less uncertainty than that of the distribution in (b).

7.1.3 PRIOR INFORMATION AND POSTERIOR DISTRIBUTION

The prior information of parameters θ can be described by a *pdf*, $p_0(\theta)$, which is called the *prior distribution* of θ. The following are some examples of prior distributions.

Example 7.1.4. Parameter θ is known exactly as θ_0. In this case, we have prior distribution

$$p_0(\theta) = \delta(\theta - \theta_0). \qquad (7.1.10)$$

where $\delta(\cdot)$ is the Dirac δ-function.

Example 7.1.5. Parameter θ is limited to the region $(V) = [\theta_{min}, \theta_{max}]$, where θ_{min} and θ_{max} are the lower bound and upper bound of θ, respectively. If there is no other information, we can assume that θ is homogeneous distribution in (V). The prior distribution is thus given by

$$
p_0(\theta) = \begin{cases} \dfrac{1}{V}, & \text{when } \theta \in (V), \\ 0, & \text{otherwise.} \end{cases} \tag{7.1.11}
$$

Example 7.1.6. If the mean value μ_θ and the covariance matrix V_θ of θ can be estimated, and there is no other information, we can assume that the prior distribution of θ is a normal distribution and given by (7.1.8).

For convenient, the normalization condition of a *pdf* is often released for the prior distribution, in other words, we do not require $\int_{(\Omega)} p_0(\theta) \, d\theta = 1$. Thus, it is permitted to assign $p_0(\theta)$ to all values of θ which are equally plausible, and assign $p_0(\theta) = 0$ to all values of θ which are entirely excluded.

After a parameter estimation procedure, i.e., to complete a transfer of information from the measurement data to parameters, we will obtain a new *pdf*, $p_*(\theta)$, which is called the *posterior distribution* of θ.

Figure 7.1.3 are two examples of comparing prior and posterior distributions. In (a), a smaller admissible interval, and in (b), a smaller variance, associated with the posterior distributions are observed.

Since some information is transferred from the measurement data to parameters, the posterior distribution has more information content or less uncertainty than the prior distribution . Generally speaking, we can use the difference

$$
I = H(p_0) - H(p_*) \tag{7.1.12}
$$

to measure the information content transferred by the parameter estimation procedure. Therefore, in the statistical framework, the parameter estimation problem becomes to find a posterior distribution.

7.1.4   THE TRANSFER OF INFORMATION

Let vector Z_D be the measured values of parameters and/or state variables in an experiment D. Their true values Z_D^* can never be known because of measurement errors. Suppose that we want to build a model $Z = M(x, \theta)$ to simulate the physical phenomenon, where vector x represents independent variables (space and/or time), θ represents model parameters. Let

$$
Z_D - M(x_D, \theta) = e_D(\theta), \tag{7.1.13}
$$

where

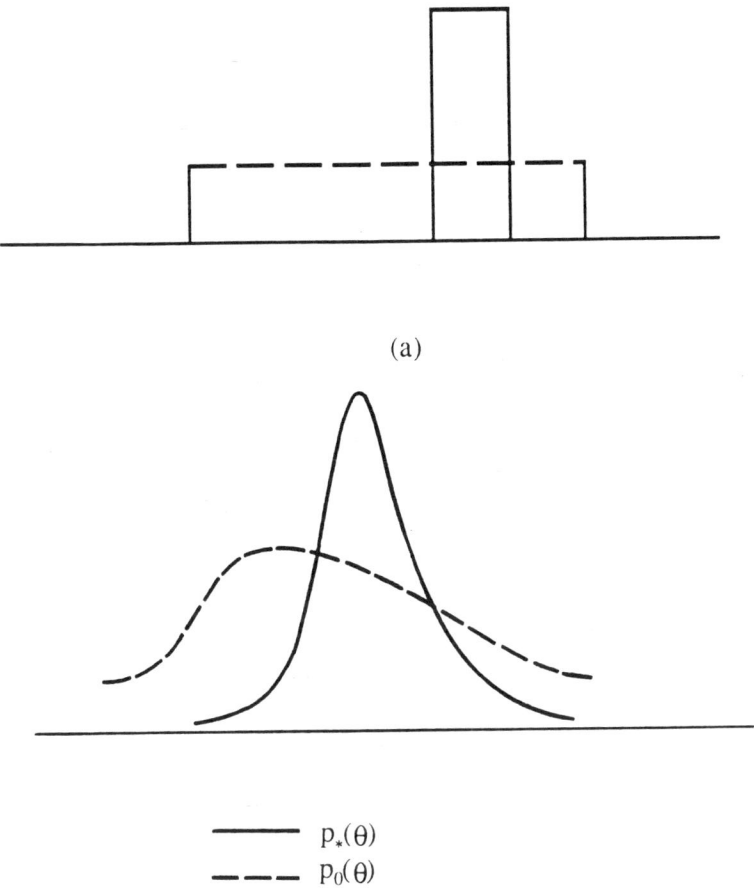

(a)

$$\overline{\hspace{2cm}} \quad p_*(\theta)$$
$$- - - - \quad p_0(\theta)$$

(b)

Fig. 7.1.3. The comparison of prior distribution and posterior distribution (a) the admissible interval is decreased; (b) the variance is decreased.

$$e_D(\theta) = (\mathbf{Z}_D - \mathbf{Z}_D^*) + (\mathbf{Z}_D^* - \mathbf{M}(\mathbf{x}_D, \theta)) \qquad (7.1.14)$$

summarizes both measurement and model structure errors. In (7.1.13) and (7.1.14), $\mathbf{x}_D$ are locations (and/or times) of measuring $\mathbf{Z}_D$.

Conditional *pdf*, $p(e_D|\theta)$, describes the distribution of errors e_D, when θ is given. It is usually named the *likelihood function* of measurements and denoted by $L(\theta)$. If the structure of model $\mathbf{M}(\mathbf{x}, \theta)$ is known, $L(\theta)$ actually represents the *pdf* of measurements , i.e., the conditional *pdf* $p(\mathbf{Z}_D|\theta)$, where $\mathbf{Z}_D$ is considered as a random vector. On the other hand, when θ is estimated from $\mathbf{Z}_D$, the posterior distribution $p_*(\theta)$ is merely the conditional *pdf* $p(\theta|\mathbf{Z}_D)$.

According to the Bayes's theorem (see Apendix B), we have

$$p(\theta|\mathbf{Z}_D) = \frac{p(\mathbf{Z}_D|\theta)p_0(\theta)}{\int_{(\Omega)} p(\mathbf{Z}_D|\theta)p_0(\theta)\, d\theta} \qquad (7.1.15)$$

or

$$p_*(\theta) = cL(\theta)p_0(\theta), \qquad (7.1.16)$$

where constant $c = (\int_{(\Omega)} p(\mathbf{Z}_D|\theta)p_0(\theta)\, d\theta)^{-1}$. Equation (7.1.16) accomplishes the transfer of information from measurements to parameters. Since prior distribution $p_0(\theta)$ is given, the problem of determining the posterior distribution now is transformed into the problem of how to define the likelihood function $L(\theta)$.

7.1.5 PROBABILITY DISTRIBUTION OF ERRORS

Let $\tilde{e}_D$ be the vector of measurement errors. In order to describe it, we have to make some assumptions on its statistical properties. Beck and Arnold (1977) presented eight standard assumptions on measurement errors. Usually, if there is no other information, we can assume that $\tilde{e}_D$ has a normal (or Gaussian) distribution with zero mean and constant covariance matrix $\tilde{\mathbf{V}}_D$, i.e.,

$$p(\tilde{e}_D) = (2\pi)^{-m/2}(\det \tilde{\mathbf{V}}_D)^{-1/2} \exp\left(-\frac{1}{2}\tilde{e}_D^T \tilde{\mathbf{V}}_D^{-1}\tilde{e}_D\right), \qquad (7.1.17)$$

where m is the dimension of $\tilde{e}_D$.

Model structure errors in groundwater modeling are often caused by

- The physical rules used to derive governing equations for a real system, such as Darcy's Law and Fick's Law, are all approximate rules.

- The conceptual model used to describe a real system may include errors. For example, the flow in the vertical direction is ignored, a fractured aquifer is regarded as a porous medium, and so forth.

- The initial and boundary conditions used for modeling a real system may include errors. For example, a boundary section is incorrectly located, a semi-permeable boundary is regarded as a impermeable boundary, and so forth.

- The parameterization used to represent distributed parameters of a real system may include errors. For example, the zonation pattern is too rough and/or incorrect.

Error vector e_D in (7.1.14) is the summation of measurement errors and model structure errors (including computational errors). Let us consider three different cases.

Case 1. The model structure errors can be ignored in comparison with the measurement errors. In these cases, e_D can be replaced by $\tilde{e}_D$, and the likelihood function may be specified as

$$L(\theta) = (2\pi)^{-m/2}(\det \tilde{\mathbf{V}}_D)^{-1/2} \times$$
$$\times \exp\left\{ -\frac{1}{2} [\mathbf{Z}_D - \mathbf{M}(\mathbf{x}_D, \theta)]^T \tilde{\mathbf{V}}_D^{-1} [\mathbf{Z}_D - \mathbf{M}(\mathbf{x}_D, \theta)] \right\}.$$

$$(7.1.18)$$

Case 2. The order of magnitude of model structure errors is the same as that of the measurement errors. If the probability distribution of model structure errors is also normal with zero mean and covariance matrix $\hat{\mathbf{V}}_D$, then the sum of these errors must be normal with zero mean and covariance matrix $\mathbf{V}_D = \tilde{\mathbf{V}}_D + \hat{\mathbf{V}}_D$. Thus, we have

$$L(\theta) = (2\pi)^{-m/2}(\det \mathbf{V}_D)^{-1/2} \times$$
$$\times \exp\left\{ -\frac{1}{2} [\mathbf{Z}_D - \mathbf{M}(\mathbf{x}_D, \theta)]^T \mathbf{V}_D^{-1} [\mathbf{Z}_D - \mathbf{M}(\mathbf{x}_D, \theta)] \right\}.$$

$$(7.1.19)$$

Case 3. The model structure errors dominate the measurement errors. In groundwater modeling, the model structure is always much simpler than that of real systems. Generally, we have to tolerate relatively large model structure errors. If the assumption of normal distribution is still reasonable for model structure errors, we can keep to use (7.1.19) as the likelihood function, else, if we can estimate the error bar η_i $(i = 1, 2, \ldots, m)$ for each component of e_D, and assume that they are independent of each other, then the distribution $p(e_D|\theta)$, or the likelihood function, is often specified as

$$L(\theta) = \left[\prod_{i=1}^{m} \frac{1}{2\eta_i} \right] \exp\left\{ -\sum_{i=1}^{m} \frac{|Z_i - M(x_i, \theta)|}{\eta_i} \right\}. \qquad (7.1.20)$$

When outliers exist in e_D, using (7.1.20) to describe the distribution of errors is better than using (7.1.19).

7.1.6 THE INVERSE SOLUTION

Substituting the likelihood function $L(\theta)$ defined in (7.1.19) or (7.1.20) into (7.1.16), we obtain the posterior distribution $p_*(\theta)$. It includes the prior information and the information transferred from measurements. In the latter, model structure information has been used. Therefore, $p_*(\theta)$ holds all available information. Various statistical properties of unknown parameters can be derived from $p_*(\theta)$. For example, the probability that the true values of the estimated parameters lie in a given range A is

$$p(\theta \in A) = \frac{1}{N} \int_A p_*(\theta)\, d\theta, \tag{7.1.21}$$

where $N = \int_{(\Omega)} p_*(\theta)\, d\theta$ is used to make $p_*(\theta)$ to be normalized.

A procedure that can generate an estimation $\hat{\theta}$ for unknown parameters θ is called an *estimator*, and $\hat{\theta}$ a *point estimation of θ*.

If we prefer to use a point estimation to represent the unknown parameters other than a range, we can select the answer from various *central estimators*, such as the mathematical expectation

$$\theta_{ME} = \mu_\theta = \frac{1}{N} \int_{(\Omega)} \theta p_*(\theta)\, d\theta, \tag{7.1.22}$$

or the *mode* of the distribution, θ_{MP}, which maximizes $p_*(\theta)$, that is

$$p_*(\theta_{MP}) = \max_\theta p_*(\theta). \tag{7.1.23}$$

The dispersion of the true parameters around a central estimation can be obtained by a dispersion estimator. For example, when the mean value μ_θ is used as the point estimation, we have the following posterior covariance matrix:

$$\mathbf{V}_* = \frac{1}{N} \int_{(\Omega)} [\theta - \mu_\theta][\theta - \mu_\theta]^T p_*(\theta)\, d\theta. \tag{7.1.24}$$

Note that an inverse problem is said to be well solved, the posterior distribution must be significantly different from its prior distribution. In other words, $p_*(\theta)$ is more concentrated to a central estimation than that of $p_0(\theta)$.

The existence of the inverse solution is usually doubtless, unless $p_*(\theta) = 0$ everywhere. This case may occur only when the prior information is contradictory with the measure information (Figure 7.1.4). If so, we have to revise the prior distribution or check whether there is something wrong in the model structure and/or measure system.

The uniqueness of $p_*(\theta)$ is evident, because it is uniquely determined by (7.1.16). However, the uniqueness of $p_*(\theta)$ does not imply the uniqueness of a central estimation. The mode of $p_*(\theta)$, θ_{MP}, may be nonunique and very sensitive with respect to the measurement errors and model structure errors (see Figure 7.1.5). The nonuniqueness problem, however, may be removed, when the prior information is increased (see Figure 7.1.6).

Exercise 7.1.1. Assume that the prior *pdf* of parameter θ is

$$p_0(\theta) = \begin{cases} 0, & -\infty < \theta < -1, \\ \theta + 1, & -1 \le \theta \le 0, \\ 1 - \theta, & 0 \le \theta \le 1, \\ 0, & 1 \le \theta < +\infty. \end{cases}$$

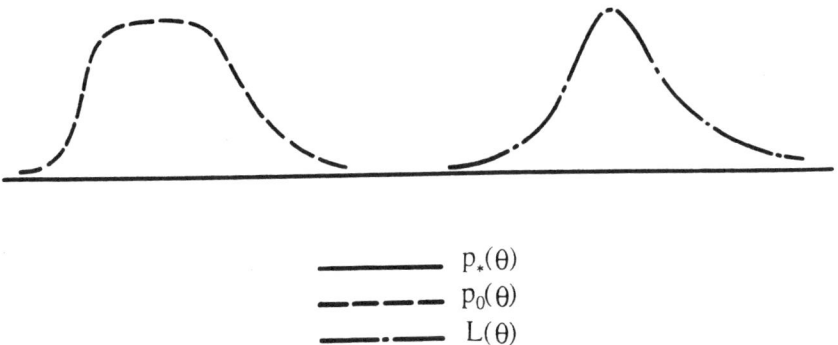

$$
\begin{array}{c}
\rule{2cm}{0.4pt} \quad p_*(\theta) \\
\text{-----} \quad p_0(\theta) \\
\text{---·---} \quad L(\theta)
\end{array}
$$

Fig. 7.1.4. $p_0(\theta)$ is not consistent with $L(\theta)$ and $p_*(\theta) \equiv 0$ is obtained.

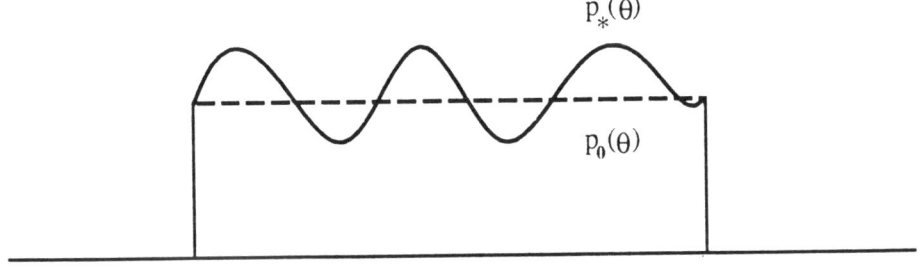

$$
\begin{array}{c}
\rule{2cm}{0.4pt} \quad p_*(\theta) \\
\text{- - -} \quad p_0(\theta)
\end{array}
$$

Fig. 7.1.5. The maximum likelihood estimation is nonunique.

After transferring information from observations, the posterior *pdf* is

$$
p_*(\theta) = \begin{cases} 0, & -\infty < \theta < -\tfrac{1}{2}, \\ 4\theta + 2, & -\tfrac{1}{2} \le \theta \le 0, \\ 2 - 4\theta, & 0 \le \theta \le \tfrac{1}{2}, \\ 0, & \tfrac{1}{2} \le \theta < +\infty. \end{cases}
$$

Calculate the uncertainty decreased by this inverse procedure.

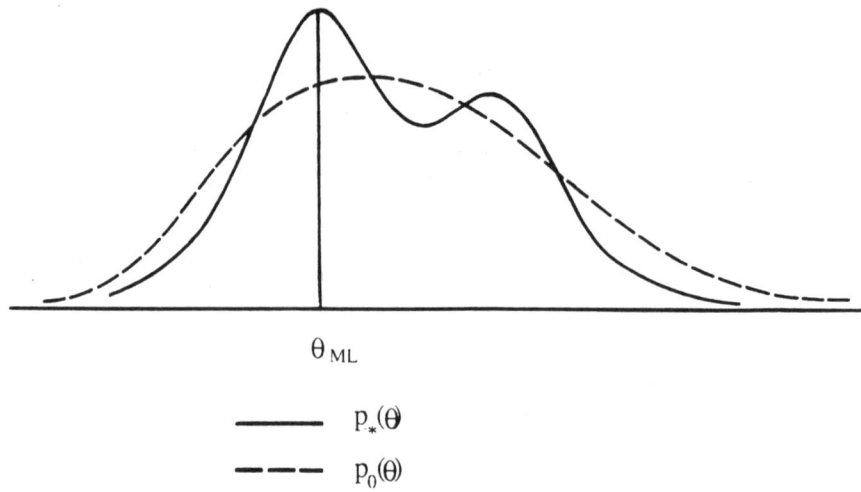

$$\theta_{ML}$$

———— $p_*(\theta)$

– – – $p_0(\theta)$

Fig. 7.1.6.   A unique θ_{MP} is obtained by the aid of the prior information.

Exercise 7.1.2.   Describe the ill-posed problem associated with inverse solutions in the statistical framework.

7.2   Statistical Properties of Estimators

All direct and indirect methods presented in Chapter 4 and Chapter 5 for parameter identification can be considered as estimators. In the last section, other estimators are introduced based on the posterior distribution. Generally, using the same $\mathbf{Z}_D$, different estimators may produce different values of the estimated parameters. In order to compare different estimators, the statistical properties of some frequently used estimators will be given in this section.

7.2.1   POINT ESTIMATION

We will use $\hat{\theta} = \mathrm{PE}(\mathbf{Z}_D)$ to denote a *point estimator*. Due to the existence of random errors, $\mathbf{Z}_D$ should be seen as a random vector. Hence, the point estimation $\hat{\theta}$ is also a random vector. Its *pdf* is referred to as the *sampling distribution* and denoted by $p_{PE}(\hat{\theta})$. If $p_{PE}(\hat{\theta})$ can be found, the mathematical expectation and covariance matrix of the estimation are determined by

$$\mu_{\hat{\theta}} = E(\hat{\theta}) = \int_{(\Omega)} \hat{\theta} p_{PE}(\hat{\theta}) \, d\hat{\theta} \tag{7.2.1}$$

and

$$\mathbf{V}_{\hat{\theta}} = E\left[(\hat{\theta} - \mu_{\hat{\theta}})(\hat{\theta} - \mu_{\hat{\theta}})^T\right]. \tag{7.2.2}$$

The following are some statistical properties defined for an estimator:

· The difference between the mean value $\mu_{\hat{\theta}}$ and the true value θ^*, i.e., $b = \mu_{\hat{\theta}} - \theta^*$, is called *the bias of the estimator*. It is a measure of "systematic" error. An estimator is said to be *unbiased*, if its bias vanishes.

· An estimator is said to be *efficient*, if its variance attains the theoretical lower bound defined by the Rao–Cramer theorem (Bard, 1974).

· An estimator is said to be *sufficient*, if it contains all information on θ, which is available in the measurements $\mathbf{Z}_D$.

· An estimator is said to be *consistent*, if the estimated parameters converge to their true values with probability 1, when the number of experiments infinitely increases.

· An estimator is said to be *robust*, if its sampling distribution is insensitive to small changes in the form of the assumed distribution.

A detailed description of these properties and relative examples can be found in Beck and Arnold (1977).

Note that the sampling distribution of an estimator not only depends upon the estimator itself, but also upon the model structure and the *pdf* of measurement errors. It can be found only under some special assumptions. For instance, model $M(\theta)$ is linear and the *pdf* of errors e_D is normal. In groundwater modeling, however, model $M(\theta)$ is generally nonlinear and the observations are very limited, the sampling distribution of an estimator can never be known, and consequently, we can not evaluate an estimator through a theoretical analysis alone.

7.2.2 MAXIMUM LIKELIHOOD ESTIMATION

We have defined an estimator in (7.1.23), which is the maximum of posterior distribution (*MPD*). In order to simplify the objective function of the optimization problem, posterior distribution $p_*(\theta)$ is often replaced by its logarithm. Thus, problem (7.1.23) is replaced by

$$\max \Phi_{\text{MPD}}(\theta), \quad \theta \in (\Theta_{\text{ad}}),$$
$$\text{where} - \Phi_{\text{MPD}}(\theta) = \log L(\theta) + \log p_0(\theta). \tag{7.2.3}$$

Due to the use of prior distribution $p_0(\theta)$, the solution of *MPD* is always meaningful in physics. In groundwater modeling, $p_0(\theta)$ is often given in the following form:

$$p_0(\theta) = \begin{cases} \dfrac{1}{\Theta_{\text{ad}}}, & \text{when } \theta \in (\Theta_{\text{ad}}), \\ 0, & \text{otherwise}, \end{cases} \tag{7.2.4}$$

where (Θ_{ad}) is the admissible domain of θ, Θ_{ad} is its measure. Equation (7.2.4) means that θ must be in (Θ_{ad}) but there is no information for selecting θ from (Θ_{ad}).

Substituting (7.2.4) into (7.2.3) yields

$$\max \Phi_{MLE}(\theta), \quad \theta \in (\Theta_{ad}),$$
$$\text{where } \Phi_{MLE}(\theta) = \log L(\theta). \tag{7.2.5}$$

Since $L(\theta) = p(e_D|\theta)$ is the likelihood function, the solution of problem (7.2.5) is called the *maximum likelihood estimation (MLE)* of parameters θ and denoted by θ_{ML}.

In order to obtain θ_{ML} by solving the nonlinear programming problem (7.2.5), the form of $L(\theta)$ must be given. As explained in Section 7.1.5, if the *pdf* of e_D is assumed to be Gaussian with zero mean and covariance matrix V_D, $L(\theta)$ will have the following form

$$L(\theta) = (2\pi)^{-m/2}(\det V_D)^{-1/2} \times$$
$$\times \exp\left\{ -\frac{1}{2} [Z_D - M(x_D,\theta)]^T V_D^{-1} [Z_D - M(x_D,\theta)] \right\}. \tag{7.2.6}$$

Remember that $\theta = (\beta, \psi)$, where β represents the unknown parameters in model M, ψ the unknown parameters in covariance matrix V_D. Using the $L(\theta)$ defined in (7.2.6), problem (7.2.5) is transferred into

$$\min \Phi_{MLN}(\theta), \quad \theta \in (\Theta_{ad}),$$
$$\text{where } \Phi_{MLN}(\theta) = \log \det V_D +$$
$$+ [Z_D - M(x_D,\theta)]^T V_D^{-1} [Z_D - M(x_D,\theta)]. \tag{7.2.7}$$

7.2.3 GENERALIZED LEAST SQUARES ESTIMATION

When the covariance matrix V_D of error e_D is known, problem (7.2.7) reduces to

$$\min \Phi_{GLS}(\beta), \quad \beta \in (\Theta_{ad}),$$
$$\text{where } \Phi_{GLS}(\beta) = [Z_D - M(x_D,\beta)]^T V_D^{-1} [Z_D - M(x_D,\beta)]. \tag{7.2.8}$$

The solution of optimization problem (7.2.8) is called the *generalized least squares (GLS) estimator*. If we assume that all components of e_D are independent of each other, then V_D must be a diagonal matrix with variances σ_i^2 ($i = 1, 2, \ldots, m$) as its diagonal elements, and problem (7.2.8) may further reduce to the *weighted least squares* (WLS) problem

$$\min \Phi_{\text{WLS}}(\beta), \quad \beta \in (\Theta_{\text{ad}}),$$
$$\text{where } \Phi_{\text{WLS}}(\beta) = [\mathbf{Z}_D - \mathbf{M}(\mathbf{x}_D, \beta)]^T \mathbf{W} [\mathbf{Z}_D - \mathbf{M}(\mathbf{x}_D, \beta)]. \tag{7.2.9}$$

The form of objective function $\Phi_{\text{WLS}}(\beta)$ in (7.2.9) coincides with that given in Chapter 4, but the weighting coefficients now are specified as $(\sigma_1^{-2}, \sigma_2^{-2}, \ldots, \sigma_m^{-2})$.

When all σ_i are equal to a constant σ, Equation (7.2.9) reduces to the *ordinary least squares* (OLS) estimation:

$$\min \Phi_{\text{OLS}}(\beta), \quad \beta \in (\Theta_{\text{ad}}),$$
$$\text{where } \Phi_{\text{OLS}}(\beta) = [\mathbf{Z}_D - \mathbf{M}(\mathbf{x}_D, \beta)]^T [\mathbf{Z}_D - \mathbf{M}(\mathbf{x}_D, \beta)]. \tag{7.2.10}$$

Therefore, the conditions of using the ordinary least squares estimation are:

(1) Errors $\mathbf{e}_D$ subject to the normal distribution with zero mean and known covariance matrix;

(2) All components of $\mathbf{e}_D$ are independent of each other;

(3) All components of $\mathbf{e}_D$ have the same variance.

If only the first two conditions are satisfied, we have to use the weighted least squares estimation (7.2.9); if only the first condition is satisfied, the generalized least squares estimation (7.2.8) should be used.

If model $\mathbf{M}(\mathbf{x}_D, \beta)$ is a linear function with respect to the unknown parameter β, then

$$\mathbf{Z}_D = \mathbf{A}\beta + \mathbf{e}_D, \tag{7.2.11}$$

where $\mathbf{A}$ is a known $m \times n$ matrix. The necessary condition for the minimization of (7.2.8) gives

$$\frac{\partial \Phi_{\text{GLS}}}{\partial \beta} = -\mathbf{A}^T \mathbf{V}_D^{-1}(\mathbf{Z}_D - \mathbf{A}\beta) = 0, \tag{7.2.12}$$

Thus, the estimated parameters can be solved as

$$\hat{\beta} = \mathbf{B}\mathbf{Z}_D, \text{ where } \mathbf{B} = (\mathbf{A}^T \mathbf{V}_D^{-1} \mathbf{A})^{-1} \mathbf{A}^T \mathbf{V}_D^{-1}. \tag{7.2.13}$$

In this situation, *GLSE* (7.2.8) is often called *multiple linear regression* (MLR). It is easy to see that *MLR* has ideally statistical properties. First, using (7.2.13) and (7.2.11), we have

$$E(\hat{\beta}) = (\mathbf{A}^T \mathbf{V}_D^{-1} \mathbf{A})^{-1} \mathbf{A}^T \mathbf{V}_D^{-1} E(\mathbf{Z}_D) = \beta, \tag{7.2.14}$$

i.e., *MLR* estimator (7.2.13) is unbiased. Second, the variance matrix of $\hat{\beta}$ is

$$
\begin{aligned}
\text{Var}(\hat{\beta}) &= E\{\left[\hat{\beta} - E(\hat{\beta})\right]\left[\hat{\beta} - E(\hat{\beta})\right]^T\} \\
&= \mathbf{B}E\{(\mathbf{Z}_D - \mathbf{A}\beta)(\mathbf{Z}_D - \mathbf{A}\beta)^T\}\mathbf{B}^T \\
&= \mathbf{B}\mathbf{V}_D\mathbf{B}^T = (\mathbf{A}^T\mathbf{V}_D^{-1}\mathbf{A})^{-1}.
\end{aligned}
\tag{7.2.15}
$$

The Gauss–Markov theorem (Beck and Arnold, 1977) shows that among all linear unbiased estimations, (7.2.13) gives the one whose variance is smallest. Thus, *GLS* is an unbiased and effective estimator, or the *best unbiased estimator*.

Note that the above conclusion is correct only when the model is linear and the *pdf* of errors is Gaussian. When the model is nonlinear, the *GLSE* can still be used as an estimator and the estimated parameters can be obtained by a nonlinear optimization method as we did in Chapter 4. The result, however, is neither unbiased nor effective.

7.2.4   GAUSSIAN PRIOR DISTRIBUTION

Now assume that the prior distribution $p_0(\beta)$ is normal, that is,

$$
p_0 = (2\pi)^{-n/2}(\det \mathbf{V}_\beta)^{-1/2} \exp\left\{ -\frac{1}{2}(\beta - \beta_0)^T\mathbf{V}_\beta^{-1}(\beta - \beta_0)\right\}, \tag{7.2.16}
$$

where β_0 is the prior estimation of β, $\mathbf{V}_\beta$ the covariance matrix. Substituting (7.2.16) and (7.2.6) into the *MPD* estimator (7.2.3), we obtain estimator

$$
\min S(\beta), \quad \beta \in \Theta_{\text{ad}},
$$
$$
\text{where } S(\beta) = [\mathbf{Z}_D - \mathbf{M}(\mathbf{x}_D, \beta)]^T \mathbf{V}_D^{-1} [\mathbf{Z}_D - \mathbf{M}(\mathbf{x}_D, \beta)] + \tag{7.2.17}
$$
$$
+ (\beta - \beta_0)\mathbf{V}_\beta^{-1}(\beta - \beta_0).
$$

If we define the generalized L_2-norm in both observation and parameter spaces, i.e., let

$$
\|\mathbf{a}\|_{\mathbf{V}_D} = \mathbf{a}^T\mathbf{V}_D^{-1}\mathbf{a}, \qquad \|\mathbf{b}\|_{\mathbf{V}_\beta} = \mathbf{b}^T\mathbf{V}_\beta^{-1}\mathbf{b}, \tag{7.2.18}
$$

where $\mathbf{a}$ and $\mathbf{b}$ are any elements in observation and parameter spaces, respectively. Then $S(\beta)$ in (7.2.17) can be rewritten as

$$
S(\beta) = \|\mathbf{Z}_D - \mathbf{M}(\mathbf{x}_D, \beta)\|_{\mathbf{V}_D}^2 + \|\beta - \beta_0\|_{\mathbf{V}_\beta}^2. \tag{7.2.19}
$$

When $\mathbf{V}_D = \sigma^2\mathbf{I}$, $\mathbf{V}_\beta = \tau^2\mathbf{I}$, it further reduces to

$$
S(\beta) = \|\mathbf{Z}_D - \mathbf{M}(\mathbf{x}_D, \beta)\|^2 + \lambda\|\beta - \beta_0\|^2, \tag{7.2.20}
$$

where $\|\cdot\|$ is the ordinary L_2-norm and $\lambda = \tau^2/\sigma^2$. Equation (7.2.20) is the objective function of regularized least squares method (RLS) that we have discussed in Chapter 4. The basic assumptions of the RLS estimator and the meaning of the regularization coefficient λ now are clearly given in the statistical framework.

When model $M(x_D, \beta)$ is a linear function with respect to β, the estimated parameters $\hat{\beta}$ can be found directly

$$\hat{\beta} = (\mathbf{A}^T \mathbf{V}_D^{-1} \mathbf{A} + \mathbf{V}_\beta^{-1})(\mathbf{A}^T \mathbf{V}_D^{-1} \mathbf{Z}_D + \mathbf{V}_\beta^{-1}\beta_0), \tag{7.2.21}$$

and the variance matrix of $\hat{\beta}$ is given by

$$\text{Var}(\hat{\beta}) = (\mathbf{A}^T \mathbf{V}_D^{-1} \mathbf{A} + \mathbf{V}_\beta^{-1})^{-1}. \tag{7.2.22}$$

7.2.5 Unknown Covariance Parameters

When the statistical parameters ψ in covariance matrix $\mathbf{V}_D$ are unknown, they may be estimated together with the unknown model parameters β by the following two-stage iterative procedure (Neuman *et al.*, 1980; Cooley, 1982; Carrera and Neuman, 1986b).

In the first stage, the model parameters β are estimated by the GLS method, where the statistical parameters ψ are given by their initial guess. In the second stage, the MLE method is used to estimate ψ, where β is replaced by the values just obtained in the first stage. The two stages are then iterated until a convergent criterion is satisfied.

Let us consider a special case that $\mathbf{V}_D$ is a diagonal matrix with unknown variances $\psi = (\sigma_1, \sigma_2, \ldots, \sigma_m)$. The problem of estimating ψ is equivalent to the determination of optimal weights in the WLS estimator. Using the two-stage iteration procedure, we have to specify a set of initial weights and solve the WLS problem to obtain the model parameter estimators $\hat{\beta}$. Let

$$\varepsilon_D = \mathbf{Z}_D - M(x_D, \hat{\beta}) \tag{7.2.23}$$

be the residuals of the estimation. The objective function of the MLE (7.2.7) can be represented as

$$\Phi(\sigma_1, \sigma_2, \ldots, \sigma_m) = \sum_{i=1}^{m} \left[2\log \sigma_i^2 + \frac{\varepsilon_{D,i}^2}{\sigma_i^2} \right], \tag{7.2.24}$$

The necessary condition of minimization gives

$$\frac{1}{\sigma_i} - \frac{\varepsilon_{D,i}^2}{\sigma_i^3} = 0, \quad (i = 1, 2, \ldots, m). \tag{7.2.25}$$

Thus, a set of estimated variances is obtained

$$\hat{\sigma}_i^2 = \varepsilon_{D,i}^2, \quad (i = 1, 2, \dots, m). \tag{7.2.26}$$

Using the updated weights to resolve the *WLS* problem, model parameters will be updated. The final solution is obtained by iteration. If $\sigma_1, \sigma_2, \dots, \sigma_m$ are the same, that is, $\mathbf{V}_D = \sigma^2 \mathbf{I}$, then we have the following estimation of σ^2:

$$\hat{\sigma}^2 = \frac{1}{m} \sum_{i=1}^{m} \varepsilon_{D,i}^2. \tag{7.2.27}$$

This estimation, however, is biased. When model $\mathbf{M}$ is linear, $E(\mathbf{e}_D) = 0$ and $\mathbf{V}_D = \sigma^2 \mathbf{I}$, an unbiased estimation of σ^2,

$$\hat{\sigma}^2 = \frac{1}{m-p} \sum_{i=1}^{m} \varepsilon_{D,i}^2 \tag{7.2.28}$$

can be derived (Beck and Arnold, 1977), where p is the dimension of β.

Substituting (7.3.28) into (7.2.15) yields

$$\mathrm{Cov}(\hat{\beta}) = \sigma^2 (\mathbf{A}^T \mathbf{A})^{-1} = \frac{R_{\mathrm{LS}}}{m-p} (\mathbf{A}^T \mathbf{A})^{-1}, \tag{7.2.29}$$

where $R_{\mathrm{LS}} = \varepsilon_D^T \varepsilon_D$ is the sum of squares of residuals. Equation (7.2.29) tells us that although R_{LS} may be decreased when the number of estimated parameters increases, the covariance $\mathrm{Cov}(\hat{\beta})$, which measures the uncertainty of the estimated parameters, may grow rapidly due to the decrease of $m - p$. Therefore, (7.2.29) gives an explanation of the overparameterization problem.

Example 7.2.1. Let us consider the identification of transmissivity T and storage coefficient S for a confined aquifer, where head observations and prior information of these parameters are available.

Suppose that there are m head observations

$$\mathbf{h}^{\mathrm{obs}} = (h_1^{\mathrm{obs}}, h_2^{\mathrm{obs}}, \dots, h_m^{\mathrm{obs}}). \tag{7.2.30}$$

The error vector between observation and model output

$$\mathbf{e}_D = \mathbf{h}^{\mathrm{obs}} - \mathbf{h}^{\mathrm{cal}}(T, S) \tag{7.2.31}$$

is assumed to be normally distributed with zero mean and covariance matrix

$$\mathbf{V}_h = \sigma_h^2 \mathbf{C}_h, \tag{7.2.32}$$

where σ_h^2 is the unknown variance, $\mathbf{C}_h$ a known matrix.

Parameters T and S are parameterized by the n_T-dimensional vector $\mathbf{T}$ and the n_S-dimensional vector $\mathbf{S}$, respectively. The prior distributions of them are assumed to be Gaussian with mean values $\mathbf{T}^\circ$, $\mathbf{S}^\circ$ and covariance matrices

$$\mathbf{V}_T = \sigma_T^2 \mathbf{C}_T, \qquad \mathbf{V}_S = \sigma_S^2 \mathbf{C}_S, \tag{7.2.33}$$

respectively, where σ_T^2 and σ_S^2 are unknown variances, $\mathbf{C}_T$ and $\mathbf{C}_S$ are known matrices.

The objective function of *MPD* estimator (7.2.7) is specified by

$$\begin{aligned}
\Phi(\boldsymbol{\theta}) = \ & m \log \sigma_h^2 + n_T \log \sigma_T^2 + n_S \log \sigma_S^2 \\
& + \sigma_h^{-2} (\mathbf{h}^{\mathrm{obs}} - \mathbf{h}^{\mathrm{cal}})^T \mathbf{C}_h^{-1} (\mathbf{h}^{\mathrm{obs}} - \mathbf{h}^{\mathrm{cal}}) \\
& + \sigma_T^{-2} (\mathbf{T} - \mathbf{T}^\circ)^T \mathbf{C}_T^{-1} (\mathbf{T} - \mathbf{T}^\circ) \\
& + \sigma_S^{-2} (\mathbf{S} - \mathbf{S}^\circ)^T \mathbf{C}_S^{-1} (\mathbf{S} - \mathbf{S}^\circ),
\end{aligned} \tag{7.2.34}$$

where

$$\boldsymbol{\theta} = (\boldsymbol{\beta}, \boldsymbol{\psi}), \qquad \boldsymbol{\beta} = (\mathbf{T}, \mathbf{S}), \qquad \boldsymbol{\psi} = (\sigma_h, \sigma_T, \sigma_S).$$

Parameters $\boldsymbol{\theta}$ can be estimated by the two-stage procedure described above. First, assume that $\boldsymbol{\psi}$ is known, and let

$$\lambda_T = \frac{\sigma_h^2}{\sigma_T^2}, \qquad \lambda_S = \frac{\sigma_h^2}{\sigma_S^2}. \tag{7.2.35}$$

Objective function $\Phi(\boldsymbol{\theta})$ then reduces to

$$\begin{aligned}
\Phi_1(\boldsymbol{\beta}) = \ & (\mathbf{h}^{\mathrm{obs}} - \mathbf{h}^{\mathrm{cal}})^T \mathbf{C}_h^{-1} (\mathbf{h}^{\mathrm{obs}} - \mathbf{h}^{\mathrm{cal}}) \\
& + \lambda_T (\mathbf{T} - \mathbf{T}^\circ)^T \mathbf{C}_T^{-1} (\mathbf{T} - \mathbf{T}^\circ) + \lambda_S (\mathbf{S} - \mathbf{S}^\circ)^T \mathbf{C}_S^{-1} (\mathbf{S} - \mathbf{S}^\circ),
\end{aligned} \tag{7.2.36}$$

which is just the objective function of regularized *GLS* method. Let us indicate the minimizer of (7.2.36) by $(\hat{\mathbf{T}}, \hat{\mathbf{S}})$.

In the second stage, we turn to determine the unknown parameters $\boldsymbol{\psi}$. Substituting $(\hat{\mathbf{T}}, \hat{\mathbf{S}})$ into (7.2.34) and using the necessary condition of minimization with respect to σ_h, we have

$$\frac{\partial \Phi}{\partial \sigma_h} = \frac{m}{\sigma_h} - \frac{R_{\mathrm{GLS}}}{\sigma_h^3} = 0, \tag{7.2.37}$$

where

$$R_{\mathrm{GLS}} = (\mathbf{h}^{\mathrm{obs}} - \hat{\mathbf{h}}^{\mathrm{cal}})^T \mathbf{C}_h^{-1} (\mathbf{h}^{\mathrm{obs}} - \hat{\mathbf{h}}^{\mathrm{cal}}),$$

$$\hat{\mathbf{h}}^{\mathrm{cal}} = \mathbf{h}^{\mathrm{cal}}(\hat{\mathbf{T}}, \hat{\mathbf{S}}).$$

The solution of (7.2.37) gives the estimate of σ_h^2:

$$\hat{\sigma}_h^2 = \frac{1}{m} R_{GLS}. \tag{7.2.38}$$

Similarly, we have

$$\hat{\sigma}_T^2 = \frac{1}{n_T}(\hat{\mathbf{T}} - \mathbf{T}^\circ)\mathbf{C}_T^{-1}(\hat{\mathbf{T}} - \mathbf{T}^\circ), \tag{7.2.39a}$$

$$\hat{\sigma}_S^2 = \frac{1}{n_S}(\hat{\mathbf{S}} - \mathbf{S}^\circ)\mathbf{C}_S^{-1}(\hat{\mathbf{S}} - \mathbf{S}^\circ). \tag{7.2.39b}$$

Note that $\hat{\sigma}_h^2$ in (7.2.38) is often replaced by its unbiased estimate

$$\hat{\sigma}_h^2 = \frac{1}{m - n_T - n_S} R_{GLS}. \tag{7.2.40}$$

To form an iterative procedure, we can replace λ_T and λ_S in (7.2.36) by $\hat{\lambda}_T = \hat{\sigma}_h^2/\hat{\sigma}_T^2$ and $\hat{\lambda}_S = \hat{\sigma}_h^2/\hat{\sigma}_S^2$, and then return to the first stage.

Let the final estimate of θ be $\hat{\theta} = (\hat{\beta}, \hat{\psi})$. In the neighboring of $\hat{\beta}$, model output $h^{cal}(\beta)$ can be linearized by taking the first order approximation of its Taylor's expression

$$\mathbf{h}^{cal}(\beta) = \mathbf{J}\beta + \mathbf{b}, \tag{7.2.41}$$

where $\mathbf{J} = \left[\partial h^{cal}/\partial \beta\right]$ is the Jacobian evaluated at $\hat{\beta}$, $\mathbf{b}$ is a constant vector depending on $\hat{\beta}$. According to (7.2.22), the covariance of $\hat{\beta}$ can be estimated approximately by

$$Cov(\hat{\beta}) \approx \left(\mathbf{J}^T \mathbf{V}_h^{-1} \mathbf{J} + \mathbf{V}_\beta^{-1}\right)^{-1}, \tag{7.2.42}$$

where

$$\mathbf{V}_h = \hat{\sigma}_h^2 \mathbf{C}_h, \qquad \mathbf{V}_\beta = \begin{bmatrix} \hat{\sigma}_T^2 \mathbf{C}_T & 0 \\ 0 & \hat{\sigma}_S^2 \mathbf{C}_S \end{bmatrix}. \tag{7.2.43}$$

Exercise 7.2.1. Describe the relationships between *MPD*, *MLE* and *GLS*.

Exercise 7.2.2. Give the conditions of using *GLS*, *WLS* and *OLS*, respectively.

Exercise 7.2.3. Give the proof of (7.2.27).

Exercise 7.2.4. Design a numerical example of using the method given in Example 7.2.1 to identify transmissivity T and storage coefficient S of a homogeneous, confined aquifer.

7.3 Parameter Estimation Based only on Parameter Observations

Up to now, all unknown physical parameters, such as hydraulic conductivities and storativities, are considered to be deterministic functions of space variables. The uncertainty is only associated with the estimated parameters and caused by insufficient and inaccurate observation data. In practice, when an aquifer is highly heterogeneous in the local scale, we can not use deterministic parameters to characterize its spatial variability, because such deterministic parameters may include too many unknown components and that may cause the inverse problem to be ill-posed. From the view of point of geostatistics, a distributed physical parameter with the variability in the small scale should be seen as a *stochastic function* or a *random field*. In this section, we will discuss how to estimate a random field by its point measurements. In other words, we will discuss how to interpolate and expolate a stochastic function.

7.3.1 SPATIAL VARIABILITY OF DISTRIBUTED PARAMETERS

When a distributed parameter, $Y(\mathbf{x})$, is considered as a stochastic function, it can be decomposed into two parts: a *trend* (or *drift*) $m(\mathbf{x})$, which describes the large-scale variability of the parameter, and a *fluctuation* $\varepsilon(\mathbf{x})$ around the drift, which describes the small-scale variability of the parameter, i.e.,

$$Y(\mathbf{x}) = m(\mathbf{x}) + \varepsilon(\mathbf{x}). \tag{7.3.1}$$

Drift $m(\mathbf{x})$ is usually assumed to be a slowly varing deterministic function depending on several undetermined coefficients β, such as a constant, a linear function, or a quadratic function. Stochastic fluctuation $\varepsilon(\mathbf{x})$ may be characterized by few statistic parameters ψ. Thus, (7.3.1) can be rewritten as

$$Y(\mathbf{x}) = m(\mathbf{x}, \beta) + \varepsilon(\mathbf{x}, \psi). \tag{7.3.2}$$

Equation (7.3.2) can be considered as a special parameterization method. The parameter estimation problem now becomes to identify $\theta = (\beta, \psi)$ and then to find such a realization $Y(\mathbf{x})$ that can approximately represent the true physical parameter.

Example 7.3.1. According to the field data reviewed by Freeze (1975), the hydraulic conductivity follows a log-normal probability distribution. This conclusion is supported by the work of Hoeksema and Kitanidis (1985a), which analyzed the data from about twenty aquifers in the United States. Thus, we have the following expression

$$\log K(\mathbf{x}) = m(\mathbf{x}) + \varepsilon(\mathbf{x}), \tag{7.3.3}$$

where the probability distribution of $\varepsilon(\mathbf{x})$ is normal with mean value $E(\varepsilon) = 0$ for any $\mathbf{x}$. The random field $\log K(\mathbf{x})$ is completely determined by its drift $m(\mathbf{x})$ and the covariance function $\mathrm{Cov}(\mathbf{x}_1, \mathbf{x}_2)$ of $\varepsilon(\mathbf{x})$. For two-dimensional problems, Hoeksema and Kitanidis (1985) used the following assumptions

$$m(\mathbf{x}) = \beta_1 + \beta_2 x + \beta_3 y, \tag{7.3.4}$$

$$\mathrm{Cov}_\varepsilon(\mathbf{x}_i, \mathbf{x}_j) = \psi_1 \delta_{ij} + \psi_2 \exp\left(-\frac{d_{ij}}{\psi_3}\right), \tag{7.3.5}$$

where (x, y) are the coordinates of point $\mathbf{x}$, δ_{ij} is the Kronecker's delta ($\delta_{ij} = 1$, if $i = j$; $\delta_{ij} = 0$ if $i \neq j$), d_{ij} is the distance between points $\mathbf{x}_i$ and $\mathbf{x}_j$, ψ_3 is named the correlation length over which the $\log K$ values become uncorrelated.

The identification of random field $Y(x, y) = \log K(x, y)$ is then reduced to the identification of $\boldsymbol{\theta} = (\boldsymbol{\beta}, \boldsymbol{\psi})$, where $\boldsymbol{\beta} = (\beta_1, \beta_2, \beta_3)$ and $\boldsymbol{\psi} = (\psi_1, \psi_2, \psi_3)$ are called *drift parameters* and *covariance parameters*, respectively.

7.3.2 Maximum Likelihood Estimation and Gaussian Conditional Mean

Suppose that there is a set of point measurements of random field $Y(\mathbf{x})$, $\mathbf{Y}_D$ are measured at a set of points $\mathbf{x}_D$. From (7.3.2), we have

$$\mathbf{Y}_D = m(\mathbf{x}_D, \boldsymbol{\beta}) + \varepsilon(\mathbf{x}_D, \boldsymbol{\psi}). \tag{7.3.6}$$

The form of (7.3.6) is exactly the same as (7.1.13), if the measurement vector $\mathbf{Z}_D$ is replaced by $\mathbf{Y}_D$, model output $\mathbf{M}(\mathbf{x}_D, \boldsymbol{\beta})$ is replaced by the drift $m(\mathbf{x}_D, \boldsymbol{\beta})$, and error vector $\mathbf{e}_D$ is replaced by the random fluctuation $\varepsilon_D = \varepsilon(\mathbf{x}_D)$. Therefore, the estimators presented in last section can also be used to estimate parameter vector $\boldsymbol{\theta} = (\boldsymbol{\beta}, \boldsymbol{\psi})$.

When the *pdf* of $\varepsilon(\mathbf{x}_D, \boldsymbol{\psi})$ is assumed to be Gaussian, the *MLE* of $\boldsymbol{\theta}$ reduces to the following optimization problem

$$\min \Phi_{MLN}(\boldsymbol{\theta}), \quad \boldsymbol{\theta} \in (\Theta_{\mathrm{ad}}),$$

$$\text{where } \Phi_{MLN}(\boldsymbol{\theta}) = \log \det \mathbf{C}_{\mathbf{Y}_D}(\boldsymbol{\psi}) + \tag{7.3.7}$$

$$+ [\mathbf{Y}_D - m(\mathbf{x}_D, \boldsymbol{\beta})]^T \mathbf{C}_{\mathbf{Y}_D}^{-1}(\boldsymbol{\psi}) [\mathbf{Y}_D - m(\mathbf{x}_D, \boldsymbol{\beta})].$$

In (7.3.7), $\mathbf{C}_{\mathbf{Y}_D}(\psi)$ is the covariance matrix of $\mathbf{Y}_D$ or $\varepsilon(\mathbf{x}_D, \psi)$. Since the form of (7.3.7) is the same as that of (7.2.7), we can use the two-stage method introduced in Section 7.2.5 to find the estimates $\hat{\beta}$ and $\hat{\psi}$ of β and ψ, respectively. The Gaussian random field $Y(x)$ is then completely determined by its mean and variance functions

$$E[Y] = m\left(x, \hat{\beta}\right), \tag{7.3.8a}$$

$$\text{Var}[Y] = C_{Y_D}\left(x, x, \hat{\psi}\right). \tag{7.3.8b}$$

The remaining problem is how to find a realization of random field $Y(x)$ to represent the unknown physical parameter. The *pdf* of the estimated parameter should be conditionalized by known measurements $\mathbf{Y}_D$. With the assumption of Gaussian distribution, the estimated value $\hat{Y}_0$, which is a realization of $Y(x)$ at x_0 and conditionalized by $\mathbf{Y}_D$, may be given by *Gaussian conditional mean*

$$\hat{Y}_0 = E[Y_0] + \mathbf{C}_{Y_0 \mathbf{Y}_D} \mathbf{C}_{\mathbf{Y}_D}^{-1} \left[\mathbf{Y}_D - m\left(\mathbf{x}_D, \hat{\beta}\right)\right]. \tag{7.3.9}$$

The variance of the estimate is given by

$$\text{Var}\left[\hat{Y}_0\right] = \text{Var}[Y_0] - \mathbf{C}_{Y_0 \mathbf{Y}_D} \mathbf{C}_{\mathbf{Y}_D}^{-1} \mathbf{C}_{Y_0 \mathbf{Y}_D}^T. \tag{7.3.10}$$

In (7.3.9) and (7.3.10), $E(Y_0)$ and $\text{Var}[Y_0]$ are unconditional mean and variance determined by (7.3.8); $\mathbf{C}_{Y_0 \mathbf{Y}_D}$ represents the cross covariance (row) matrix between Y_0 and each component of $\mathbf{Y}_D$. The proof of (7.3.9) and (7.3.10) can be found, for example, in Schweppe (1978).

Unfortunately, (7.3.10) tends to underestimate the true variance of the estimated parameter. In fact, if the random term $\varepsilon(\mathbf{x})$ in (7.3.1) had "large-scale" components, and β and ψ are estimated simultaneously, some of variability of $\varepsilon(\mathbf{x})$ may be fitted by the drift. Thus, it is impossible to distinguish between the drift and the large-scale components of the random term on the basis of measurements.

There are techniques that allow us to estimate covariance parameters without using a drift. For example, we can use a transformation to filter out the unknown drift from the measurements, such that covariance parameters ψ and drift parameters β are determined separately. A comprehensive review on this topic was given by Kitanidis (1987).

7.3.3 THE KRIGING ESTIMATOR OF WEAKLY STATIONARY FIELDS

To estimate a distributed parameter in the stochastic framework, the method presented above consists of two steps:

(1) Using a set of measurements $\mathbf{Y}_D$ to estimate the drift parameters β and covariance parameters ψ of a random field $Y(x)$, simultaneously or separately. This step can be completed by the *MLE*.

(2) Using the same set of measurements $\mathbf{Y}_D$ to find a realization $\hat{Y}(x)$ of the random field. This step can be completed by calculating the Gaussian conditional mean. For any point $\mathbf{x}_0$, $\hat{Y}(\mathbf{x}_0)$, is given by a linear combination of measurements $\mathbf{Y}_D$.

There is another approach to solve this problem, which is called *geostatistical method* or *Kriging* (Matheron, 1963). The basic idea of Kriging consists of

(1) The estimated realization $\hat{Y}(x)$ at any point $\mathbf{x}_0$ in the region of interest is assumed to be a linear combination of measurements $\mathbf{Y}_D$, i.e,

$$\hat{Y}(\mathbf{x}_0) = \sum_{i=1}^{m} \lambda_i(\mathbf{x}_0) Y_i, \qquad (7.3.11)$$

where $Y_1, Y_2, \ldots, Y_m$ are components of measurements $\mathbf{Y}_D$, $\lambda_i(\mathbf{x}_0)$ are undetermined weights.

(2) The optimal weights are then determined by the statistical structure of the random field, which can be estimated from the same set of measurements.

A random field $Y(x)$ is said to be *weakly stationary* or *second-order stationary*, if it satisfies

$$E\left[Y(\mathbf{x})\right] = \beta, \quad \text{(constant)}, \qquad (7.3.12a)$$

$$\text{Cov}\left[Y(\mathbf{x}_1), Y(\mathbf{x}_2)\right] = C(h), \qquad (7.3.12b)$$

where h is the distance between any two points $\mathbf{x}_1$ and $\mathbf{x}_2$ in the field. (7.3.12b) means that the covariance is only dependent on distance h and characterized by a function $C(h)$. For example,

$$C(h) = \psi_1 \exp\left\{-\frac{h}{\psi_2}\right\}, \qquad (7.3.13)$$

where ψ_1 and ψ_2 are covariance parameters.

Assume that β and ψ are already determined by the maximum likelihood estimation as we did in Section 7.3.2, and introduce a new field

$$Z(\mathbf{x}) = Y(\mathbf{x}) - \beta. \qquad (7.3.14)$$

It has zero mean and measurements $Z(\mathbf{x}_i) = Y(\mathbf{x}_i) - \beta$, $(i = 1, 2, \ldots, m)$.

Let $\mathbf{x}_0$ be an arbitrary point in the region of interest and the estimated $Z(\mathbf{x}_0)$, $\hat{Z}(\mathbf{x}_0)$, be a linear combination of all measurements, that is,

$$\hat{Z}(x_0) = \sum_{i=1}^{m} \lambda_i(x_0) Z(x_i) \tag{7.3.15a}$$

or in a compact form

$$\hat{Z}_0 = \sum_{i=1}^{m} \lambda_i^0 Z_i. \tag{7.3.15b}$$

The optimal weights $\lambda_1^0, \lambda_2^0, \ldots, \lambda_m^0$ may be determined by additional requirements that the estimator must be unbiased and the variance of the estimation must be minimal, i.e.,

$$E\left(\hat{Z}_0\right) = E(Z_0), \tag{7.3.16}$$

$$E\left[\left(\hat{Z}_0 - Z_0\right)^2\right] = \min_{Z} E\left[(Z - Z_0)^2\right], \tag{7.3.17}$$

where Z_0 is the true value of $Z(x_0)$ but unknown. Condition (7.3.16) is automatically satisfied in this case, because

$$E\left(\hat{Z}_0\right) = \sum_{i=1}^{m} \lambda_i^0 E(Z_i) = 0. \tag{7.3.18}$$

Condition (7.3.17) can be used to uniquely determine the weights in (7.3.15). In fact, the substitution of (7.3.15) into (7.3.17) yields

$$E\left[\left(\hat{Z}_0 - Z_0\right)^2\right] = \sum_{i=1}^{m}\sum_{j=1}^{m} \lambda_i^0 \lambda_j^0 E(Z_i Z_j) - 2\sum_{i=1}^{m} \lambda_i^0 E[Z_i Z_0] + E\left[Z_0^2\right]. \tag{7.3.19}$$

Using (7.3.12b), we have

$$E[Z_i Z_j] = E[(Y_i - \beta)(Y_j - \beta)] = \text{Cov}(Y_i, Y_j) = C(h_{ij}),$$
$$E[Z_i Z_0] = C(h_{i0}),$$
$$E[Z_0^2] = C(0),$$

where h_{ij} is the distance between points x_i and x_j. Let $C(h_{ij}) = C_{ij}$, $C(h_{i0}) = C_{i0}$, and $C(0) = C_{00}$, (7.3.19) can be rewritten as

$$E\left[\left(\hat{Z}_0 - Z_0\right)^2\right] = \sum_{i=1}^{m}\sum_{j=1}^{m} C_{ij}\lambda_i^0 \lambda_j^0 - 2\sum_{i=1}^{m} C_{i0}\lambda_i^0 + C_{00}. \tag{7.3.20}$$

This is a quadratic function of λ_i^0 $(i = 1, 2, \ldots, m)$. The necessary condition of minimum gives the following linear system

$$\sum_{j=1}^{m} C_{ij}\lambda_j^0 = C_{i0}, \quad (i = 1, 2, \ldots, m) \tag{7.3.21}$$

and the solution of (7.3.21) gives the optimal weights, which are called the *Kriging coefficients*. The variance of the estimate can be obtained by substituting (7.3.21) into (7.3.20):

$$\text{Var}\left[\hat{Z}_0 - Z_0\right] = E\left[\left(\hat{Z}_0 - Z_0\right)^2\right] = C_{00} - \sum_{i=1}^{m} C_{i0}\lambda_i^0. \tag{7.3.22}$$

Return to the original field $Y(\mathbf{x})$, we have a linear estimator

$$\hat{Y}_0 = \beta + \sum_{i=1}^{m} \lambda_i^0 (Y_i - \beta) \tag{7.3.23}$$

with estimation variance

$$\text{Var}\left[\hat{Y}_0 - Y_0\right] = C_{00} - \sum_{i=1}^{m} C_{i0}\lambda_i^0. \tag{7.3.24}$$

The effect of Kriging formulae (7.3.23) and (7.3.24) is the same as that of Gaussian conditional mean formulae (7.3.9) and (7.3.10).

A random field $Y(x)$ is said to be *increment stationary*, if the following *intrinsic hypothesis* is held

$$E\left[Y(\mathbf{x} + \mathbf{h}) - Y(\mathbf{x})\right] = 0, \tag{7.3.25a}$$
$$\text{Var}\left[Y(\mathbf{x} + \mathbf{h}) - Y(\mathbf{x})\right] = 2\gamma(\mathbf{h}), \tag{7.3.25b}$$

where $\gamma(\mathbf{h})$ is called the *variogram*. For an isotropic field, it only depends on the length h of increment $\mathbf{h}$. A weakly isotropic stationary field must be increment stationary, and since

$$C(h) = \text{Cov}\left[Y(\mathbf{x} + \mathbf{h}), Y(\mathbf{x})\right] - \beta^2,$$

we have

$$\begin{aligned}
\gamma(h) &= \frac{1}{2}E\left[(Y(\mathbf{x} + \mathbf{h}) - Y(\mathbf{x}))^2\right] \\
&= \frac{1}{2}E\left[Y^2(\mathbf{x} + \mathbf{h})\right] - E\left[Y(\mathbf{x} + \mathbf{h})Y(\mathbf{x})\right] + \frac{1}{2}E\left[Y^2(\mathbf{x})\right] \\
&= C(0) - C(h).
\end{aligned} \tag{7.3.26}$$

Thus, when $C(h)$ exists, $\gamma(h)$ must exist. On the other hand, even a finite variance $C(0)$ does not exist, variogram $\gamma(h)$ may still exist.

Now we want to find an optimal linear estimator (7.3.11) for an increment stationary field such that the unbiased condition (7.3.16) and minimum variance condition (7.3.17) are satisfied.

Condition (7.3.16) is satisfied, if and only if

$$\sum_{i=1}^{m} \lambda_i^0 = 1. \tag{7.3.27}$$

The estimation variance can be represented as

$$E\left[\left(\hat{Y}_0 - Y_0\right)^2\right] = \sum_{i=1}^{m} \sum_{j=1}^{m} \lambda_i^0 \lambda_j^0 E\left[(Y_i - Y_0)(Y_j - Y_0)\right]. \tag{7.3.28}$$

Since

$$E\left[(Y_i - Y_0)(Y_j - Y_0)\right] = \frac{1}{2} E\left[(Y_i - Y_0)^2 + (Y_j - Y_0)^2 - (Y_i - Y_j)^2\right]$$
$$= \gamma_{i0} + \gamma_{j0} + \gamma_{ij},$$

where $\gamma_{ij} = \gamma(\mathbf{x}_i - \mathbf{x}_j)$, $\gamma_{i0} = \gamma(\mathbf{x}_i - \mathbf{x}_0)$, $\gamma_{j0} = \gamma(\mathbf{x}_j - \mathbf{x}_0)$, (7.3.28) can be rewritten as

$$E\left[\left(\hat{Y}_0 - Y_0\right)^2\right] = -\sum_{i=1}^{m} \sum_{j=1}^{m} \gamma_{ij} \lambda_i^0 \lambda_j^0 + 2 \sum_{i=1}^{m} \gamma_{i0} \lambda_i^0. \tag{7.3.29}$$

The minimization of (7.3.29) subject to constraint (7.3.27) is equivalent to minimize the following expression

$$-\frac{1}{2} \sum_{i=1}^{m} \sum_{j=1}^{m} \gamma_{ij} \lambda_i^0 \lambda_j^0 + \sum_{i=1}^{m} \gamma_{i0} \lambda_i^0 - \mu \left[\sum_{i=1}^{m} \lambda_i^0 - 1\right], \tag{7.3.30}$$

where μ is an unknown Lagrange multiplier. Using the necessary condition of minimization, the minimizer $(\lambda_1^0, \lambda_2^0, \ldots, \lambda_m^0, \mu)$ of (7.3.30) can be solved from linear system

$$\sum_{j=1}^{m} \gamma_{ij} \lambda_j^0 + \mu = \gamma_{i0} \quad (i = 1, 2, \ldots, m),$$
$$\sum_{i=1}^{m} \lambda_i^0 = 1. \tag{7.3.31}$$

Thus, for an increment stationary field $Y(x)$, we have Kriging estimator

$$\hat{Y}(\mathbf{x}_0) = \sum_{i=1}^{m} \lambda_i^0 Y_i, \tag{7.3.32}$$

which is unbiased and has minimal estimation variance

$$\text{Var}\left[\hat{Y}(\mathbf{x}_0) - Y(\mathbf{x}_0)\right] = \sum_{i=1}^{m} \lambda_i^0 \gamma_{i0} + \mu. \qquad (7.3.33)$$

In (7.3.32) and (7.3.33), $\mathbf{x}_0$ is an arbitrary point in the region, λ_i^0 and μ are the solutions of (7.3.31). Note that the coefficient matrix of system (7.3.31) does not depend on $\mathbf{x}_0$. Therefore, to obtain the Kriging coefficients for different points in the region, it only needs to change the right-hand side of (7.3.31).

Now let us consider how to find the variogram for an increment stationary field. When a finite variance $C(0)$ exists, variogram $\gamma(h)$ can be obtained directly from (7.3.26), where $C(h)$ is determined by the *MLE* as described in Section 7.3.2. When $C(0)$ does not exist, $\gamma(h)$ may be estimated according to its definition (7.3.25b) and measurement data.

For a series of distance

$$0 = h_0 < h_1 < h_2 < \ldots < h_n, \qquad (7.3.34)$$

the point values of experimental variogram are defined by

$$\hat{\gamma}(h_i) = \frac{1}{2N_i} \sum_{k,j} (Y_k - Y_j)^2, \qquad (7.3.35)$$

where $Y_k = Y(\mathbf{x}_k)$, $Y_j = Y(\mathbf{x}_j)$, the summation is taken over all couples of points $(\mathbf{x}_k, \mathbf{x}_j)$ separated by a distance between $((h_{i-1} + h_i)/2, (h_i + h_{i+1})/2)$, N_i represents the number of such couples.

To find a continuous form of the variogram, we can select a theoretical model with few parameters to represent $\gamma(h)$ and then determine these parameters by the least squares method to fit the experimental variogram $\hat{\gamma}(h)$. The following are some frequently used variogram models:

Polynomial variogram:

$$\gamma(h) = \alpha h^\beta, \quad \alpha > 0, \ 0 < \beta < 2;$$

Exponential variogram:

$$\gamma(h) = \alpha\left[1 - \exp(-\beta h)\right], \quad \alpha > 0, \ \beta > 0;$$

Gaussian variogram:

$$\gamma(h) = \alpha\left[1 - \exp\left(-\beta h^2\right)\right], \quad \alpha > 0, \ \beta > 0;$$

Spherical variogram:

$$\gamma(h) = \begin{cases} \dfrac{1}{2}\alpha\left[3\dfrac{h}{\beta} - \left(\dfrac{h}{\beta}\right)^3\right], & \alpha > 0, \text{ when } h \leq \beta, \\ \alpha, & \text{when } h > \beta. \end{cases}$$

These models all satisfy

$$\gamma(0) = 0, \qquad \gamma(h) > 0, \qquad \lim_{h \to 0} \frac{\gamma(h)}{h^2} = 0, \qquad (7.3.36)$$

which are necessary conditions for a variogram function.

To check the validity of a Kriging estimator, we can use the *cross validation* technique. The error of cross validation at a measure point x_i is defined as

$$e_i = Y_i - \hat{Y}_i, \qquad (7.3.37)$$

where Y_i is the measured value at point x_i, $\hat{Y}_i$ is its Kriging estimate when Y_i itself is taken out of the set of measurements. If the Kriging estimator is acceptable, we should have:

$$\frac{1}{m} \sum_{i=1}^{m} e_i \approx 0 \quad \text{and} \quad \frac{1}{m} \sum_{i=1}^{m} \left(\frac{e_i}{\sigma_i} \right)^2 \approx 1, \qquad (7.3.38)$$

where σ_i^2 is the Kriging estimation variance. A general and complete discussion on the cross validation method used in groundwater modeling was given by Samper and Neuman (1989).

7.3.4 THE KRIGING ESTIMATOR FOR UNSTATIONARY FIELDS

When the mean value of a random field $Y(\mathbf{x})$ is not a constant, the intrinsic hypothesis (7.3.25) is no longer correct. Consequently, we can not use the variogram $\gamma(h)$ to obtain the Kriging estimation.

If the unknown parameter can be parameterized by a zonation method, i.e., the region can be divided into several zones and the unknown parameter has constant mean value in each zone, then we can assume that the unknown parameter is increment stationary in each individual zone, and thus, the Kriging formula (7.3.32) derived for increment stationary fields can be applied zone by zone.

If the drift $m(\mathbf{x}, \beta)$ can be identified first by the least squares estimate according to measurements $\mathbf{Y}_D$, then the residual field

$$Z(\mathbf{x}) = Y(\mathbf{x}) - m(\mathbf{x}, \beta) \qquad (7.3.39)$$

may be considered as a weakly stationary field with zero mean, and Kriging formula (7.3.23) can be used as an estimator.

The disadvantage of this approach is that using residual field $Z(x)$ may underestimate the random portion of the field.

We have seen that the unknown constant mean of a random field can be filtered out by considering its increment field. Similarly, one can construct the high-order increment field to filter out a non-constant mean.

Suppose that we still want to use Kriging formula (7.3.11) to estimate a unstationary field $Y(x)$ and require that the unbiased condition

$$E\left[\hat{Y}(x_0)\right] = E\left[Y(x_0)\right] \tag{7.3.40}$$

and the minimum estimation variance condition

$$\min \text{Var}\left[\hat{Y}(x_0) - Y(x_0)\right] \tag{7.3.41}$$

are satisfied. Let us consider how to determine the weights $\lambda_i(x_0)$ for an arbitrary
point x_0 in the region.

The linear combination

$$\sum_{i=0}^{m} \lambda_i^0 Y_i \quad \text{or} \quad \sum_{i=0}^{m} \lambda_i(x_0) Y(x_i), \tag{7.3.42}$$

is named a *zero-order generalized increment*, if $\lambda_0^0 = -1$ is defined and constraint
condition

$$\sum_{i=0}^{m} \lambda_i^0 = 0 \tag{7.3.43}$$

is satisfied. It is easy to prove that zero-order increment (7.3.42) is able to filter
out a constant mean. In fact, let $Y(x) = \beta + \varepsilon(x)$, where β is a constant mean,
then

$$Y(x_i) = \beta + \varepsilon(x_i), \quad i = 1, 2, \ldots, m. \tag{7.3.44}$$

Using condition (7.3.43), we have

$$\sum_{i=0}^{m} \lambda_i^0 Y(x_i) = \sum_{i=0}^{m} \lambda_i^0 [\beta + \varepsilon(x_i)] = \sum_{i=0}^{m} \lambda_i^0 \varepsilon(x_i). \tag{7.3.45}$$

Thus, constant mean β is filtered out.

Similarly, (7.3.42) is named a 2-dimensional generalized increment of order k,
if constraint conditions

$$\sum_{i=0}^{m} \lambda_i^0 x_i^p y_i^q = 0, \quad p \geq 0, \ q \geq 0, \ p + q \leq k \tag{7.3.46}$$

are satisfied, where p, q are integers, (x_i, y_i) the coordinates of point x_i. For
example, the *first-order generalized increment* requires 3 conditions:

$$\sum_{i=0}^{m} \lambda_i^0 = 0, \quad \sum_{i=0}^{m} \lambda_i^0 x_i = 0, \quad \sum_{i=0}^{m} \lambda_i^0 y_i = 0. \tag{7.3.47}$$

Besides (7.3.47), the *second-order generalized increment* requires another 3 conditions:

$$\sum_{i=0}^{m} \lambda_i^0 x_i^2 = 0, \qquad \sum_{i=0}^{m} \lambda_i^0 x_i y_i = 0, \qquad \sum_{i=0}^{m} \lambda_i^0 y_i^2 = 0. \tag{7.3.48}$$

It is easy to prove that a *k-order generalized increment* is able to filter out any expectation function $m(\mathbf{x})$ represented by a k-order polynomial $P_k(\mathbf{x})$, and the unbiased condition (7.3.40) is automatically satisfied by constraints (7.3.46), because we have

$$E[\hat{Y}_0] - E[Y_0] = E\left[\sum_{i=0}^{m} \lambda_i^0 Y_i\right] = \sum_{i=0}^{m} \lambda_i^0 E[Y_i] = \sum_{i=0}^{m} \lambda_i^0 P_k(\mathbf{x}_i) = 0.$$

$$\tag{7.3.49}$$

Now we assume that the generalized increment of order k is weakly stationary and its estimation variance can be expressed as

$$\mathrm{Var}[\hat{Y}_0 - Y_0] = E\left[\left(\sum_{i=1}^{m} \lambda_i^0 (Y_i - Y_0)\right)^2\right] = \sum_{i=0}^{m} \sum_{j=0}^{m} \lambda_i^0 \lambda_j^0 K_{ij}, \tag{7.3.50}$$

where $K_{ij} = K(h_{ij})$, h_{ij} is the distance between points $\mathbf{x}_i$ and $\mathbf{x}_j$, $K(h)$ is called the *generalized covariance of order* k. When $Y(\mathbf{x})$ itself is a weakly stationary field, $K(h)$ reduces to covariance function $C(h)$.

Once $K(h)$ is known, the Kriging weights can be obtained by the minimization of (7.3.50) subject to constraints (7.3.46).

For the case of $k = 1$, the solution of the minimization problem is given by the following Kriging equations:

$$\begin{cases} \displaystyle\sum_{j=1}^{m} \lambda_j^0 K_{ij} - \mu_1 - \mu_2 x_i - \mu_3 y_i = K_{i0}, \quad (i = 1, 2, \ldots, m) \\ \displaystyle\sum_{i=0}^{m} \lambda_i^0 = 0, \qquad \sum_{i=0}^{m} \lambda_i^0 x_i = 0, \qquad \sum_{i=0}^{m} \lambda_i^0 y_i = 0, \end{cases} \tag{7.3.51}$$

where μ_1, μ_2, μ_3 are Lagrange multipliers. The variance of the estimation is obtained by substituting (7.3.51) into (7.3.50), i.e.,

$$\mathrm{Var}\left[\hat{Y}_0 - Y_0\right] = K_{00} + \mu_1 + \mu_2 x_0 + \mu_3 y_0 - \sum_{i=1}^{m} \lambda_i^0 K_{i0}. \tag{7.3.52}$$

Note that the unknown expectation function or the drift of $Y(\mathbf{x})$ does not appear in the Kriging equations (7.3.51). We only need to determine the generalized covariance $K(h)$ from the measurements. The following are theoretical models of $K(h)$ presented by Matheron (1973):

For $k = 0$, $K(h) = a_0\delta(h) + a_1 h$,
where $a_0 \geq 0$, $a_1 \geq 0$.

For $k = 1$, $K(h) = a_0\delta(h) + a_1 h + a_3 h^3$,
where $a_0 \geq 0$, $a_1 \leq 0$, $a_3 \geq 0$.

For $k = 2$, $K(h) = a_0\delta(h) + a_1 h + a_3 h^3 + a_5 h^5$,
where $a_0 \geq 0$, $a_1 \leq 0$, $a_3 \geq (-10/3)\sqrt{a_1 a_3}$, $a_5 \leq 0$.

In the above equations, $a_0\delta(h)$ describes the nugget effect, where $\delta(h)$ is the Kronecker delta function.

The *cross validation method* mentioned in Section 7.3.3 can be used to identify the undetermined coefficients of $K(h)$. For example, let the initial guess be $k = 1$ and assume that $a_0 = 0$. Then square error e_i^2 of cross validation and estimation variance σ_i^2 obtained from (7.3.52) for each measure point x_i are both functions of undetermined coefficients a_1 and a_3, and consequently, we can set the following least squares problem

$$\min \sum_{i=1}^{m} \left[e_i^2(a_1, a_2) - \sigma_i^2(a_1, a_2) \right]^2 \tag{7.3.53}$$

subject to

$$a_1 \leq 0, \quad a_3 \geq 0$$

to obtain the values of a_1 and a_3. Problem (7.3.53) can be solved by an iterative method given in Chapter 4. $a_1 = -1$ and $a_3 = 0$ are often used as the initial estimates.

7.3.5 APPLICATIONS

The Kriging estimate method was used in hydrogeology since 1970s (Delhomme, 1978), and more applications of this method were found in recent years.

Delhomme (1979) used Kriging to estimate the transmissivity $T(x, y)$ for the Bathonian aquifer in lower Normandy, France. Random field $\log T(x, y)$ is assumed to be increment stationary and having a spherical variogram. Devary and Doctor (1982) used Kriging method to estimate a logarithm hydraulic conductivity field $\log K(x, y)$ with a few measurements of $K(x, y)$. A drift $m(x, y)$, which was assumed to be a cubic polynomial function, was filtered out first by a least squares estimator. The residuals of the estimate were regarded as the random portion $\varepsilon(x, y)$, which was assumed to be a weakly stationary field. Its covariance function

$$c(h) = 0.151 \exp\left[-0.565 \left(\frac{h}{2000} \right)^2 \right] \tag{7.3.54}$$

was obtained by fitting the experimental covariance data.

Gambolati and Volpi (1979) used Kriging method to draw contours of ground-water head distributions in the Venice aquifers, Italy. For the two upper aquifers the selected drift had the form

$$m(x, y) = A \ln \frac{Bx^2 + Cxy + (y - d)^2}{Bx^2 + Cxy + (y + d)^2} + z_0, \tag{7.3.55}$$

where d is the distance between the boundary and the center of the cone of depression, A, B and C are determined by the least squares method, and z_0 is the prescribed head on the upstream boundary.

De Marsily (1986) used the Kriging method to estimate an unstationary head field concerning an unconfined aquifer in chalk, at Aisne, France. The head measurements of 88 wells were used to determine the generalized covariance function

$$K(h) = Ah^3, \tag{7.3.56}$$

where coefficient A took different values in different zones.

In Clifton and Neuman (1982), the logarithm transmissivity field was estimated for the Avra Valley aquifer, Arizona, USA, where 148 transmissivity data were available. The aquifer was subdivided into two subregions with different spherical variograms represented by

$$\gamma(h) = \begin{cases} 0.04 + a \left[1.5 \left(\dfrac{h}{6} \right) - 0.5 \left(\dfrac{h}{6} \right)^3 \right], & 0 \leq h \leq 6, \\ b, & h > 6, \end{cases} \tag{7.3.57}$$

where $a = 0.03$, $b = 0.07$ for the northern subregion, and $a = 0.1$, $b = 0.14$ for the southern subregion. The unconditional $\log T(x, y)$ field was first conditioned by the transmissivity measurements. The result was then used as prior information in the regularized least squares criterion (7.2.17) to solve the inverse problem. Due to the use of head measurements, the estimation variance of $\log T(x, y)$ field was further decreased.

In Yeh *et al.* (1983), a Kriging interpolation was used to generate distributed head values from a few head observations. The direct method of inverse solution (Section 5.1) was then used to obtain the unknown transmissivities.

In de Marsily *et al.* (1984), an adjoint sensitivity technique was coupled with a Kriging algorithm to calibrate a flow model. The adjoint sensitivity was used to find some pilot points, where modification of the model's transmissivity obtained from the Kriging estimate or values of boundary head could directly decrease the residuals of least squares fitting. The methodology was modified and applied to the

Culebra dolomite at the Waste Isolation Pilot Plant in southeastern, New Mexico, recently by LaVenue and Pickens (1992).

Wagner and Gorelick (1989) presented a multiple realization model for groundwater quality management, in which the uncertainty associated with the hydraulic conductivity field was considered. The Kriging estimate and conditional simulation (Journel and Huijbregts, 1978) were used to generate "plausible" realizations of the hydraulic conductivity. The corresponding concentrations were then used in the management model.

Carrera *et al.* (1984) and Loaiciga (1989) used the Kriging method for optimal observation network design. This problem will be considered in next chapter.

Exercise 7.3.1. Assume that the covariance function of random field $Y(\mathbf{x})$ is given by

$$\text{Cov}(\mathbf{x}_1, \mathbf{x}_2) = 0.3 \exp(-h_{12}/100),$$

where h_{12} is the distance between $\mathbf{x}_1$ and $\mathbf{x}_2$, and the mean of $Y(\mathbf{x})$ is constant $\mu_Y = 2.0$. Write a program to generate Gaussian conditional mean for a two-dimensional rectangular region: $0 \le x \le 1000$, $0 \le y \le 500$, based on the following observations:

measure point (x, y)	measure value
(50, 50)	2.12
(100, 200)	1.96
(100, 400)	2.28
(200, 300)	1.67
(500, 100)	2.33
(700, 300)	1.44
(800, 400)	2.41
(900, 200)	2.03
(900, 400)	2.21
(950, 300)	1.82

The output is required to be given by a contour map of estimation $\hat{Y}(\mathbf{x})$.

Exercise 7.3.2. Solve the same problem as in Exercise 7.3.1 using Kriging estimation.

Exercise 7.3.3. Prove that Kriging is an exact interpolation method, i.e., the Kriging estimate associated with a measure point is exactly equal to the measurement of the point.

Exercise 7.3.4. Prove that the first-order generalized increment defined by conditions (7.3.47) can filter out any first-order polynomial mean, and the second-order generalized increment defined by conditions (7.3.47) and (7.3.48) can filter out any second-order polynomial mean.

7.4 Parameter Identification Based on Both Parameter and State Variable Observations

In Section 7.1, we have known that when the information obtained from both parameter and state variable measurements is used, the estimated parameters will be more reliable than that of using only one kind of measurement alone. In Section 7.2, we have known that the general estimate problem can be solved by the maximum likelihood estimator (*MLE*).

In this section we will discuss how to calculate the covariance functions of state observations and cross-covariance functions between parameters and state observations. These covariance functions are used not only in the *MLE*, but also in obtaining the estimated parameters by a co-Kriging technique. This methodology, which is called the stochastic inverse solution method, has been used in the identification of hydrogeological parameters in recent years.

7.4.1 STOCHASTIC PARTIAL DIFFERENTIAL EQUATIONS

When random variables are involved in a partial differential equation and/or its subsidiary conditions, the *PDE* is called a stochastic partial differential equation (*SPDE*).

The concept of *SPDE*s is very useful in groundwater modeling, because the uncertainty of hydrogeological parameters, the random nature of sink/source terms, and the inaccuracy of initial and boundary conditions can not be avoided for any real problem. For example, a transmissivity distribution of a confined aquifer may be determined by the Kriging interpolation based on some point measurements. The uncertainty in the transmissivity must cause the uncertainty in the head. The latter can be estimated from the former, when the governing equation is considered as a *SPDE*. Another example is the recharge of an unconfined aquifer. It directly depends upon the variability of precipitation. When the recharge is considered as a random function, the water table must be also a random function, and the governing equation be a *SPDE*. The third example is the solute transport in porous media. The random nature of the porous structure in every scale causes the variability of velocities and thus the variability of concentration distributions.

From the above examples, it is easy to understand that why the approach of *SPDE*s has been extensively applied in the field of groundwater modeling (Gelhar,

1976; Tang and Pinder, 1979; Dagan, 1982, 1985, 1986; Gelhar, 1986; de Marsily, 1986; Dagan, 1989).

The forward problem or the solution of a *SPDE* means to find the probability distribution of the state variable, when probability distributions of input parameters are given. However, it is generally impossible to find the whole probability distribution of state variables for any practical problem. Instead, we often limit ourselves to deal with only the first two moments, that is, the mean and covariance functions.

There are at least three methods to solve the forward problem of *SPDEs*: the spectral method, the perturbation method and the Monte-Carlo method. A simple introduction of these methods can be found in the text book of de Marsily (1986). The stochastic finite element method may be used to find a numerical solution of a *SPDE* (Ghanem and Spanos, 1991).

The spectral method can be used to obtain analytical expressions of covariance functions for simple problems. Therefore, it is often used in the theoretical study. The concept of the Monte-Carlo method is very straight and it can be used to solve any complicated field problem. The only disadvantage of Monte-Carlo method is its high computational expense. To obtain a significant statistics, several thousands of simulation runs may be needed. Yeh and Wang (1987) presented a simple example of using the Monte-Carlo method, in which uncertainties of the identified dispersivities were estimated.

Hereinafter, we will concentrate to the perturbation method and combine it with numerical solutions. Let us consider steady flow in a confined aquifer governed by

$$\frac{\partial}{\partial x}\left(e^Y \frac{\partial \phi}{\partial x}\right) + \frac{\partial}{\partial y}\left(e^Y \frac{\partial \phi}{\partial y}\right) = Q \tag{7.4.1}$$

and subject to deterministic boundary conditions, where Y is the logarithm transmissivity, ϕ the head, Q the sink/source term.

When Y is a random field, head ϕ is also a random field. Let

$$Y = F + f, \tag{7.4.2}$$

where F is the mean of Y, and f is a zero mean perturbation. Similarly, head ϕ can be represented by its expected value H plus a zero mean perturbation h, i.e.,

$$\phi = H + h. \tag{7.4.3}$$

The forward problem involves finding the expected value H and covariance $C_{\phi\phi}$ of head ϕ, as well as the cross-covariance $C_{\phi Y}$, when the mean value F and covariance C_{YY} of Y are known.

Substituting (7.4.2) and (7.4.3) into (7.4.1), replacing e^{-f} approximately by $(1 - f)$, and ignoring all second-order terms, we have

$$\left(\frac{\partial F}{\partial x} + \frac{\partial f}{\partial x}\right)\left(\frac{\partial H}{\partial x} + \frac{\partial h}{\partial x}\right) + \left(\frac{\partial^2 H}{\partial x^2} + \frac{\partial^2 h}{\partial x^2}\right)$$
$$+ \left(\frac{\partial F}{\partial y} + \frac{\partial f}{\partial y}\right)\left(\frac{\partial H}{\partial y} + \frac{\partial h}{\partial y}\right) + \left(\frac{\partial^2 H}{\partial y^2} + \frac{\partial^2 h}{\partial y^2}\right) = Qe^{-F}(1 - f).$$

(7.4.4)

Taking expectations of both sides of the equation, we obtain a *PDE* for the expected head H

$$\frac{\partial F}{\partial x}\frac{\partial H}{\partial x} + \frac{\partial^2 H}{\partial x^2} + \frac{\partial F}{\partial y}\frac{\partial H}{\partial y} + \frac{\partial^2 H}{\partial y^2} = Qe^{-F},$$

(7.4.5)

subject to the original boundary conditions. Thus, H can be obtained by a numerical method.

Subtracting (7.4.5) from (7.4.4) yields a *SPDE* relating the perturbations f and h:

$$\frac{\partial F}{\partial x}\frac{\partial h}{\partial x} + \frac{\partial F}{\partial y}\frac{\partial h}{\partial y} + \frac{\partial^2 h}{\partial x^2} + \frac{\partial^2 h}{\partial y^2} = -\frac{\partial f}{\partial x}\frac{\partial H}{\partial x} - \frac{\partial f}{\partial y}\frac{\partial H}{\partial y} - fQe^{-F}.$$ (7.4.6)

Using a finite difference or a finite element method and noting that $h = 0$ on the boundary, (7.4.6) can be discretized into an algebraic system

$$\mathbf{Ah} = \mathbf{Bf},$$

(7.4.7)

where $\mathbf{h}$ is the vector of nodal head perturbations, $\mathbf{f}$ the vector of nodal $\log T$ perturbations, $\mathbf{A}$ and $\mathbf{B}$ are coefficient matrices depending on the known mean value F and the expected head H, that is, the solution of (7.4.5). The solution of (7.4.7) is

$$\mathbf{h} = \mathbf{A}^{-1}\mathbf{Bf}.$$

(7.4.8)

Let $C_{YY} = E\left[\mathbf{ff}^T\right]$ be the covariance matrix of Y. From (7.4.8), we can obtain head covariance $C_{\phi\phi}$ matrix and head-log transmissivity cross covariance matrix $C_{\phi Y}$ as follows:

$$C_{\phi\phi} = E\left[\mathbf{hh}^T\right] = \left(\mathbf{A}^{-1}\mathbf{B}\right) C_{YY} \left(\mathbf{A}^{-1}\mathbf{B}\right)^T,$$

(7.4.9)

$$C_{\phi Y} = E\left[\mathbf{hf}^T\right] = \left(\mathbf{A}^{-1}\mathbf{B}\right) C_{YY}.$$

(7.4.10)

Note that to obtain $C_{\phi\phi}$ and $C_{\phi Y}$, $N + 1$ simulation runs are needed, where N is the number of nodes.

The above process was described by Hoeksema and Kitanidis (1984, 1985a). The same problem was also considered by Dagan (1985), in which analytical solutions were derived based on a first-order approximation and assuming that covariance C_{YY} is isotropic and exponential, that is, it has the form

$$C_{YY}\left(\mathbf{x}_i, \mathbf{x}_j\right) = \sigma_Y^2 \exp\left(-r_{ij}/l_Y\right), \tag{7.4.11}$$

where σ_Y^2 is the variance of Y, r_{ij} the distance between $\mathbf{x}_i$ and $\mathbf{x}_j$, l_Y is called the *correlation length*.

Based on the perturbation method, Mclaughlin and Wood (1988a) derived *moment equations* for a general form of governing equations in groundwater modeling. Mean value, covariance and cross-covariance functions can be obtained by solving these moment equations. The moment equation for covariance functions are defined in region $(\Omega) \times (\Omega)$ of the multiplication space, where (Ω) is the flow region. If N nodes are used in the descritization of (Ω), the number of nodes for $(\Omega) \times (\Omega)$ will be N^2. Therefore, to obtain numerical solutions for moment equations is quite expensive. A numerical example of solving moment equations can be found in Mclaughlin and Wood (1988b).

After having model output (head) covariances, confidence intervals of model output can be estimated. A confidence interval is a range that the true values of head lies within the range according to a specified probability, for example, 95% (Devore, 1987). Since the variance estimation (7.4.9) is based on first-order approximation, when we use it to estimate the uncertainty of model output of nonlinear models, parameter variances must be small. Otherwise, we have to use the Monte-Carlo method.

Cooley and Vecchia (1987) developed a method that confidence intervals of nonlinear models can be estimated by solving constrained nonlinear optimization problems. This approach was improved by incorporating hydrogeological information and calibration data into the statistical distribution of parameters (Cooley, 1993a and 1993b).

7.4.2 ESTIMATE OF STATISTICAL PARAMETERS

The inverse problem under stochastic framework is already discussed in Section 7.1. Let us consider a steady flow field and the identification of log transmissivity $Y = \ln T$. Suppose that there are m measurements of $Y(\mathbf{x})$:

$$\mathbf{Y}_D = \{Y\left(\mathbf{x}_1\right), Y\left(\mathbf{x}_2\right), \ldots, Y\left(\mathbf{x}_m\right)\} \tag{7.4.12}$$

and n observations of head $\phi(\mathbf{x})$:

$$\phi_D = \{\phi\left(\mathbf{x}_{m+1}\right), \phi\left(\mathbf{x}_{m+2}\right), \ldots, \phi\left(\mathbf{x}_{m+n}\right)\}. \tag{7.4.13}$$

The problem is how to estimate the mean value $E[Y]$ of $Y(\mathbf{x})$ and its covariance C_{YY}. In Equations (7.4.12) and (7.4.13), the subscript D means an observation design which defines $\mathbf{x}_1, \mathbf{x}_2, \ldots, \mathbf{x}_m$ as m measure points for $Y(\mathbf{x})$ and $\mathbf{x}_{m+1}, \mathbf{x}_{m+2}, \ldots, \mathbf{x}_{m+n}$ as n observation points for $\phi(\mathbf{x})$.

Assume that $E[Y]$ can be represented by parameter vector β and C_{YY} by ψ. The *MLE* algorithm (7.2.7), namely,

$$\min \Phi(\beta, \psi),$$

$$\text{where} \Phi(\beta, \psi) = \log \det C_D(\psi) + \qquad \qquad (7.4.14)$$
$$+ [\mathbf{Z}_D - M(\mathbf{x}_D, \beta)]^T C_D^{-1}(\psi) [\mathbf{Z}_D - M(\mathbf{X}_D, \beta)],$$

can be used to estimate the unknown parameters β and ψ. In (7.4.14), $\mathbf{Z}_D$ consists of $\mathbf{Y}_D$ and ϕ_D, $M(\mathbf{x}_D, \beta)$ consists of mean values of $\mathbf{Y}_D$ and ϕ_D, respectively, $C_D(\psi)$ is the covariance matrix

$$C_D(\psi) = \begin{bmatrix} C_{D,YY} & C_{D,Y\phi} \\ C_{D,Y\phi} & C_{D,\phi\phi} \end{bmatrix}, \qquad \qquad (7.4.15)$$

where

$$C_{D,YY} = E\left[\left(\mathbf{Y}_D - \overline{\mathbf{Y}}_D\right)\left(\mathbf{Y}_D - \overline{\mathbf{Y}}_D\right)^T\right], \qquad \qquad (7.4.16a)$$

$$C_{D,Y\phi} = E\left[\left(\mathbf{Y}_D - \overline{\mathbf{Y}}_D\right)\left(\phi_D - \overline{\phi}_D\right)^T\right], \qquad \qquad (7.4.16b)$$

$$C_{D,\phi\phi} = E\left[\left(\phi_D - \overline{\phi}_D\right)\left(\phi_D - \overline{\phi}_D\right)^T\right] \qquad \qquad (7.4.16c)$$

are $m \times m$, $m \times n$, $n \times m$ matrices, respectively. In above equations, $\overline{\mathbf{Y}}_D$ and $\overline{\phi}_D$ are mean values of $\mathbf{Y}_D$ and ϕ_D, respectively.

Using the notations in Section 7.4.1, these matrices can be rewritten as

$$C_{D,YY} = E\left[\mathbf{f}_D \mathbf{f}_D^T\right], \qquad \qquad (7.4.17a)$$

$$C_{D,Y\phi} = E\left[\mathbf{f}_D \mathbf{h}_D^T\right], \qquad \qquad (7.4.17b)$$

$$C_{D,\phi\phi} = E\left[\mathbf{h}_D \mathbf{h}_D^T\right], \qquad \qquad (7.4.17c)$$

where $\mathbf{f}_D$ is a zero mean perturbation vector of $Y(\mathbf{x})$ at measure points $(\mathbf{x}_1, \mathbf{x}_2, \ldots, \mathbf{x}_m)$, $\mathbf{h}_D$ is a zero mean perturbation vector of $\phi(\mathbf{x})$ at observation points $(\mathbf{x}_{m+1}, \mathbf{x}_{m+2}, \ldots, \mathbf{x}_{m+n})$. In a numerical method, the head perturbation at an observation point can be expressed approximately as a linear combination of head perturbations at its neighboring nodes determined by an interpolation algorithm. Thus, we have

$$\mathbf{h}_D = \mathbf{W}_D \mathbf{h}. \tag{7.4.18}$$

where $\mathbf{W}_D$ is a $n \times N$ weighting matrix, N the number of nodes, $\mathbf{h}$ the head perturbation vector associated with all nodes. We can assume that there is a linear relation between perturbation vectors $\mathbf{h}$ and $\mathbf{f}$ associated with all nodes:

$$\mathbf{h} = \mathbf{M}\mathbf{f}. \tag{7.4.19}$$

where $\mathbf{M}$ is an $N \times N$ matrix. For a steady flow field, for example, (7.4.8) gives $\mathbf{M} = \mathbf{A}^{-1}\mathbf{B}$. Substituting (7.4.19) into (7.4.18) and then into (7.4.17), we have

$$C_{D,YY} = E\left[\mathbf{f}_D \mathbf{f}_D^T\right], \tag{7.4.20a}$$
$$C_{D,Y\phi} = (\mathbf{W}_D\mathbf{M})\, E[\mathbf{f}\mathbf{f}_D^T], \tag{7.4.20b}$$
$$C_{D,\phi\phi} = (\mathbf{W}_D\mathbf{M})\, E[\mathbf{f}\mathbf{f}^T]\,(\mathbf{W}_D\mathbf{M})^T, \tag{7.4.20c}$$

Since the structure of C_{YY} is assumed to be known, $E[\mathbf{f}\mathbf{f}^T]$, $E[\mathbf{f}\mathbf{f}_D^T]$ and $E[\mathbf{f}_D\mathbf{f}_D^T]$ can all be represented by ψ, and thus the expression of $C_D(\psi)$ in (7.4.15) can be found. Once the expression of $C_D(\psi)$ is known, maximum likelihood estimator (7.4.14) is ready to be used and unknown parameters (β, ψ) can be estimated.

Example 7.4.1. Hoeksema and Kitanidis (1985) applied the stochastic inverse method to Jordan Aquifer of Iowa, USA. The log transmissivity field Y is assumed to be normally distributed. The following models are used to describe its mean and covariance

$$E\left[Y_i\right] = \beta_1 + \beta_2 x_i + \beta_3 y_i, \tag{7.4.21}$$
$$\mathrm{Cov}\left[Y_i, Y_j\right] = \psi_1 \delta_{ij} + \psi_2 \exp\left(-r_{ij}/l_Y\right), \tag{7.4.22}$$

where $\beta = (\beta_1, \beta_2, \beta_3)$, $\psi = (\psi_1, \psi_2)$ are undetermined statistic parameters, x_i, y_i the coordinates of measure point of Y_i, δ_{ij} the Kronecker delta. In (7.4.22), the correlation length l_Y is set to 45(Km). The variance of head measurement error, σ_ϕ^2, is considered as an additional covariance parameter ψ_3.

Using the estimation procedure given in this section with 29 head and 56 log-transmissivity measurements, Hoeksema and Kitanidis (1985b) found the following estimates for β and ψ:

$$\hat{\beta}_1 = 7.65, \qquad \hat{\beta}_2 = 0.00171\,(1/Km),$$
$$\hat{\beta}_3 = -0.00192\,(1/Km),$$
$$\psi_1 = 0.178, \qquad \psi_2 = 0.369, \qquad \psi_3 = 48.47\,(m^2).$$

7.4.3 CO-KRIGING ESTIMATE

After the statistical parameters (β, ψ) of random field $Y(\mathbf{x})$ are determined by the *MLE*, a distributed $Y(\mathbf{x})$ can be estimated for any $\mathbf{x}$ in the field by Gaussian conditional mean or Kriging estimate based on measurements $\mathbf{Y}_D$, as discussed in Section 7.3.

Now, besides $\mathbf{Y}_D$, there are some head observations ϕ_D. We should use them to improve the estimation of $Y(\mathbf{x})$, because $\phi(\mathbf{x})$ is closely related to $Y(\mathbf{x})$.

Co-Kriging is an estimation technique that can estimate two or more random fields together by using their measurements, when they are correlated.

Let $f(\mathbf{x})$ and $h(\mathbf{x})$ be zero mean random fields. There are m measurements $f(\mathbf{x}_1), f(\mathbf{x}_2), \ldots, f(\mathbf{x}_m)$ for $f(\mathbf{x})$ and n measurements $h(\mathbf{x}_1'), h(\mathbf{x}_2'), \ldots, h(\mathbf{x}_n')$ for $h(\mathbf{x})$, respectively. The measure points of $f(\mathbf{x})$ and $h(\mathbf{x})$ may be identical or different.

The co-Kriging estimation of $f(\mathbf{x})$ at any point x_0 in the field, $\hat{f}(\mathbf{x}_0)$, is a linear combination of all measurements of $f(\mathbf{x})$ and $h(\mathbf{x})$, that is,

$$\hat{f}(\mathbf{x}_0) = \sum_{i=1}^{m} \lambda_i(\mathbf{x}_0) f(\mathbf{x}_i) + \sum_{j=1}^{n} \mu_j(\mathbf{x}_0) h(\mathbf{x}_j') \tag{7.4.23}$$

or shortly written as

$$\hat{f}_0 = \sum_{i=1}^{m} \lambda_i^0 f_i + \sum_{j=1}^{n} \mu_j^0 h_j, \tag{7.4.24}$$

where λ_i^0 $(i = 1, 2, \ldots, n)$, μ_j^0 $(j = 1, 2, \ldots, m)$ are called *co-Kriging coefficients*. These coefficients can be determined by the requirements that $\hat{f}_0$ is an unbiased estimate of f_0 and the estimation variance is the minimum.

The unbiased requirement is satisfied automatically, because we have

$$E\left[\hat{f}_0 - f_0\right] = \sum_{i=1}^{m} \lambda_i^0 E[f_i] + \sum_{j=1}^{n} \mu_j^0 E[h_j] = 0. \tag{7.4.25}$$

The variance of the estimation error can be represented as

$$\mathrm{Var}\left[\hat{f}_0 - f_0\right] = E\left[\left(\sum_{i=1}^{m} \lambda_i^0 f_i + \sum_{j=1}^{m} \mu_j^0 h_j - f_0\right)^2\right]$$

$$= \sum_{i=1}^{m}\sum_{k=1}^{m} \lambda_i^0 \lambda_k^0 E[f_i f_k] + 2\sum_{i=1}^{m}\sum_{j=1}^{n} \lambda_i^0 \mu_j^0 E[f_i h_j]$$

$$+ \sum_{j=1}^{n}\sum_{l=1}^{n} \mu_j^0 \mu_l^0 E[h_j h_l] - 2\sum_{i=1}^{m} \lambda_i^0 E[f_i f_0]$$

$$-2 \sum_{j=1}^{n} \mu_j^0 E\left[h_j f_0\right] + \operatorname{Var}\left[f_0\right]. \tag{7.4.26}$$

The co-Kriging system is then derived from the necessary condition of minimization

$$\sum_{k=1}^{m} \lambda_k^0 E\left[f_i f_k\right] + \sum_{j=1}^{n} \mu_j^0 E\left[f_i h_j\right] = E\left[f_i f_0\right],$$

$$(i = 1, 2, \ldots, m); \tag{7.4.27a}$$

$$\sum_{i=1}^{m} \lambda_i^0 E\left[f_i h_j\right] + \sum_{l=1}^{n} \mu_l^0 E\left[h_j h_l\right] = E\left[h_j f_0\right],$$

$$(j = 1, 2, \ldots, n). \tag{7.4.27b}$$

The co-Kriging coefficients thus can be obtained through the solution of equations (7.4.27), and the minimal estimation variance is

$$\operatorname{Var}\left[\hat{f}_0 - f_0\right] = \operatorname{Var}\left[f_0\right] - \sum_{i=1}^{m} \lambda_i^0 E\left[f_i f_0\right] - \sum_{j=1}^{n} \mu_j^0 E\left[h_j f_0\right]. \tag{7.4.28}$$

In comparison with Kriging variance (7.3.24), the co-Kriging variance decreases an amount represented by the third term on the right-hand side of (7.4.28). It is the benefit of using the co-Kriging estimation. A general description of co-Kriging estimation and its applications in hydrogeology can be found in de Marsily (1986).

Now, return to the inverse problem presented in Section 7.4.2. Let $f = Y - F$ and $h = \phi - H$. Using co-Kriging estimation, at any point x_0 in the flow region, the log transmissivity is estimated as

$$\hat{Y}\left(x_0\right) = F\left(x_0\right) + \sum_{i=1}^{m} \lambda_i^0 \left[Y\left(x_i\right) - F\left(x_i\right)\right] + \sum_{j=m+1}^{n} \mu_j^0 \left[\phi\left(x_j\right) - H\left(x_j\right)\right]. \tag{7.4.29}$$

Hoeksema and Kitanidis (1984, 1985b), Ahmed and de Marsily (1987, 1993), Rubin and Dagan (1987a, b), and Dagan and Rubin (1988) presented some case studies of using co-Kriging estimate in groundwater modeling.

7.4.4   Adjoint State Method for *SPDEs*

Sun and Yeh (1992a) considered the identification of log hydraulic conductivity in transient flow fields, in which the covariance matrices of head observations and cross-covariance matrices between head observations and log hydraulic conductivity measurements are calculated by the adjoint state method. To obtain these matrices, the required number of simulation runs is only equal to the number of head observations plus one.

A two-dimensional transient groundwater flow in a confined aquifer is governed by the following equation

$$S\frac{\partial \phi}{\partial t} = \frac{\partial}{\partial x}\left(Km\frac{\partial \phi}{\partial x}\right) + \frac{\partial}{\partial y}\left(Km\frac{\partial \phi}{\partial y}\right) + Q, \quad (x,y) \in (\Omega), \ t \geq 0,$$

$$(7.4.30)$$

subject to appropriate initial and boundary conditions, where storage coefficient S and thickness m of the aquifer are assumed to be deterministic and known, log hydraulic conductivity $Y = \log K$ is assumed to be normally distributed. It is further assumed that the random field Y is characterized by a constant mean μ_Y and an isotropic, exponential covariance C_{YY} given in (7.4.11).

In the simplest case, random field Y is fully characterized by three statistical parameters, μ_Y, σ_Y and l_Y. In other cases, more parameters may be used to represent the mean and covariance function, or the flow region may be divided into several zones characterized by different sets of statistical parameters.

Substituting (7.4.2) and (7.4.3) into (7.4.30) yields

$$m\frac{\partial f}{\partial x}\frac{\partial H}{\partial x} + \frac{\partial}{\partial x}\left(m\frac{\partial H}{\partial x}\right) + \frac{\partial}{\partial x}\left(m\frac{\partial h}{\partial x}\right) + m\frac{\partial f}{\partial y}\frac{\partial H}{\partial y}$$
$$+ \frac{\partial}{\partial y}\left(m\frac{\partial H}{\partial y}\right) + \frac{\partial}{\partial y}\left(m\frac{\partial h}{\partial y}\right)$$
$$+ e^{-F}\left(Q - S\frac{\partial H}{\partial t} - S\frac{\partial h}{\partial t} - fQ + Sf\frac{\partial H}{\partial t}\right) = 0. \qquad (7.4.31)$$

In the derivation of above equation, we have assumed that perturbation f is small enough that $\exp(-f)$ can be substituted approximately by $(1 - f)$, and the terms $(\partial f/\partial x)(\partial h/\partial x)$, $(\partial f/\partial y)(\partial h/\partial y)$, $f(\partial h/\partial t)$ in the equation can be ignored. Recently, Loaiciga and Marino (1990) analyzed the error in dropping $(\partial f/\partial x)(\partial h/\partial x)$ and $(\partial f/\partial y)(\partial h/\partial y)$, in the case of steady flow.

Taking the expectation of (7.4.31), we obtain a *PDE* for the expected head H:

$$S\frac{\partial H}{\partial t} = e^F\left[\frac{\partial}{\partial x}\left(m\frac{\partial H}{\partial x}\right) + \frac{\partial}{\partial y}\left(m\frac{\partial H}{\partial y}\right)\right] + Q. \qquad (7.4.32)$$

Subtracting (7.4.32) from (7.4.31) yields a *SPDE* for perturbations f and h:

$$S\frac{\partial h}{\partial t} = e^F\left[\frac{\partial}{\partial x}\left(m\frac{\partial h}{\partial x}\right) + \frac{\partial}{\partial y}\left(m\frac{\partial h}{\partial y}\right)\right] + e^F m\left[\frac{\partial f}{\partial x}\frac{\partial H}{\partial x} + \frac{\partial f}{\partial y}\frac{\partial H}{\partial y}\right]$$
$$+ \left(S\frac{\partial H}{\partial t} - Q\right)f. \qquad (7.4.33)$$

To find the adjoint state equation for (7.4.33), we need a performance function

$$J(h, f) = \int_0^T \iint_{(\Omega)} R(h, f; \mathbf{x}, t) \, d\Omega \, dt, \qquad (7.4.34)$$

where T is a given time and R is a user-chosen function. The variation of performance J is

$$\delta J = \int_0^T \iint_{(\Omega)} \left(\frac{\partial R}{\partial h} \delta h + \frac{\partial R}{\partial f} \delta f \right) d\Omega \, dt. \tag{7.4.35}$$

Let $\psi(\mathbf{x}, t)$ be an arbitrary function defined on $(\Omega) \times (0, T)$ and having continuous second-order derivatives. Taking the variation of (7.4.35), multiplying the results by ψ, integrating it over $(\Omega) \times (0, T)$, and using Green's formula, we obtain

$$\int_0^T \iint_{(\Omega)} \left\{ S \frac{\partial \psi}{\partial t} + e^F \left[\frac{\partial}{\partial x} \left(m \frac{\partial \psi}{\partial x} \right) + \frac{\partial}{\partial y} \left(m \frac{\partial \psi}{\partial y} \right) \right] \right\} \delta h \, d\Omega \, dt$$

$$+ \int_0^T \iint_{(\Omega)} \left\{ e^F \left[\frac{\partial}{\partial x} \left(\psi m \frac{\partial H}{\partial x} \right) + \frac{\partial}{\partial y} \left(\psi m \frac{\partial H}{\partial y} \right) \right] \right.$$

$$\left. - \left(S \frac{\partial H}{\partial t} - Q \right) \psi \right\} \delta f \, d\Omega \, dt = 0. \tag{7.4.36}$$

Subtracting (7.4.36) from (7.4.35), and selecting ψ to satisfy

$$S \frac{\partial \psi}{\partial t} + e^F \left[\frac{\partial}{\partial x} \left(m \frac{\partial \psi}{\partial x} \right) + \frac{\partial}{\partial y} \left(m \frac{\partial \psi}{\partial y} \right) \right] = \frac{\partial R}{\partial h} \tag{7.4.37}$$

in $(\Omega) \times (0, T)$, the integral associated with δh vanishes and we obtain the functional derivative

$$\frac{\partial J}{\partial f} = \int_0^T \iint_{(\Omega)} \left\{ \frac{\partial R}{\partial f} + e^F \left[\frac{\partial}{\partial x} \left(\psi m \frac{\partial H}{\partial x} \right) + \frac{\partial}{\partial y} \left(\psi m \frac{\partial H}{\partial y} \right) \right] \right.$$

$$\left. - \left(S \frac{\partial H}{\partial t} - Q \right) \psi \right\} d\Omega \, dt. \tag{7.4.38}$$

Using (7.4.32) to reduce the right-hand side of (7.4.38) yields

$$\frac{\partial J}{\partial f} = \int_0^T \iint_{(\Omega)} \left\{ \frac{\partial R}{\partial f} + e^F m \left[\frac{\partial \psi}{\partial x} \frac{\partial H}{\partial x} + \frac{\partial \psi}{\partial y} \frac{\partial H}{\partial y} \right] \right\} d\Omega \, dt. \tag{7.4.39}$$

Therefore, once function R in (7.4.34) is chosen, we can solve (7.4.32) to obtain the expected head H, solve (7.4.37) to obtain the adjoint state ψ, and then calculate the functional derivative in (7.4.39) by a numerical integration.

Functional derivative (7.4.39) can be used to calculate the covariance matrix for the *MLE* to estimate the unknown statistical parameters. Let function R in (7.4.34) be

$$R(\mathbf{x}, t) = h(\mathbf{x}, t) \delta(\mathbf{x} - \mathbf{x}_l) \delta(t - t_k), \tag{7.4.40}$$

where $\mathbf{x}_l$ is an observation well, t_k is an observation time, and $\delta(\cdot)$ represents the Dirac δ function. From (7.4.39), for any node i, we have

$$\frac{\partial h(\mathbf{x}_l, t_k)}{\partial f_i} = \int_0^{t_k} \int\int_{(\Omega_i)} e^F m \left[\frac{\partial \psi}{\partial x} \frac{\partial H}{\partial x} + \frac{\partial \psi}{\partial y} \frac{\partial H}{\partial y} \right] d\Omega \, dt, \qquad (7.4.41)$$

where (Ω_i) is the exclusive subdomain of node i.

For a given observation time t_k, the Jacobian

$$\mathbf{J}_D(t_k) = \left[\frac{\partial \mathbf{h}_D(t_k)}{\partial \mathbf{f}} \right] \qquad (7.4.42)$$

is an $(L \times N)$ matrix, where L is the number of observation wells, N is the number of nodes, $\mathbf{h}_D(t_k)$ is the observation head perturbation vector at time t_k, and $\mathbf{f}$ is the log K perturbation vector for all nodes.

Equation (7.4.41) can be used to calculate all elements of Jacobian $\mathbf{J}_D(t_k)$. Note that only $(L+1)$ simulation runs are needed, of which one simulation run is used to obtain the expected head distribution $H(x,t)$, and the others are used to obtain the adjoint state ψ corresponding to each observation well.

Once Jacobian $\mathbf{J}_D(t_k)$ in (7.4.42) is obtained, the $(L \times L)$ covariance matrix of observation heads at time t_k can be calculated by a first-order approximation,

$$C_{D,\phi\phi}(t_k, t_k) = E\left[\mathbf{h}_D(t_k) \mathbf{h}_D^T(t_k) \right]$$
$$= \mathbf{J}_D(t_k) E\left[\mathbf{f} \mathbf{f}^T \right] \mathbf{J}_D^T(t_k), \qquad (7.4.43)$$

where the superscript T represents the transpose of a vector or a matrix, and the $(L \times L)$ cross-covariance matrix between head observation at observation time t_{k_1} and t_{k_2} can be calculated by

$$C_{D,\phi\phi}(t_{k_1}, t_{k_2}) = E\left[\mathbf{h}_D(t_{k_1}) \mathbf{h}_D^T(t_{k_2}) \right]$$
$$= \mathbf{J}_D(t_{k_1}) E[\mathbf{f} \mathbf{f}^T] \mathbf{J}_D^T(t_{k_2}). \qquad (7.4.44)$$

The $(L \times M)$ head-log K measurement cross-covariance matrix can be calculated by

$$C_{D,\phi Y}(t_k) = E\left[\mathbf{h}_D(t_k) f_D^T \right] = \mathbf{J}_D(t_k) E[\mathbf{f} f_D^T]. \qquad (7.4.45)$$

In (7.4.43), (7.4.44) and (7.4.45), the $(N \times N)$ matrix $E[\mathbf{f} \mathbf{f}^T]$ is the covariance matrix of $Y = \log K$ at all nodes which is determined by the given covariance structure (7.4.11); the $(N \times N)$ matrix $E[\mathbf{f} f_D^T]$ is the covariance matrix between Y measurements and Y of all nodes, which is a submatrix of $E[\mathbf{f} \mathbf{f}^T]$.

For transient flow, head observations at different observation times are available. There are two ways in which head observations can be used to identify parameter vector $\theta = (\mu_Y, \sigma_Y, l_Y)$.

(1) Apply the MLE sequentially, i.e., identify vector $\boldsymbol{\theta}_k$ $(k = 1, 2, \ldots, K)$ at each t_k with the head observations for that particular time period only. The $\boldsymbol{\theta}$ estimate is then obtained by averaging $\boldsymbol{\theta}_k$ over all time (Dagan and Rubin, 1988). In this case, covariance matrix C_D for time t_k in (7.4.15) is an $(L+M) \times (L+M)$ matrix

$$
C_D = \begin{bmatrix} C_{D,\phi\phi}(t_k, t_k) & C_{D,\phi Y}(t_k) \\ C_{D,\phi Y}(t_k) & C_{D,YY} \end{bmatrix},
$$

where $C_{D,\phi\phi}$ and $C_{D,\phi Y}$ are determined by (7.4.43) and (7.4.45), respectively, and $C_{D,YY}$ is the covariance matrix of Y measurements.

(2) Use head observations at all times simultaneously for parameter identification. In this case, covariance matrix C_D in (7.4.15) is an $(LK + M) \times (LK + M)$ matrix relating to all observation times,

$$
C_D = \begin{bmatrix} C_{D,\phi\phi}(t_1, t_1) & C_{D,\phi\phi}(t_1, t_2) & \cdots & C_{D,\phi\phi}(t_1, t_K) & C_{D,\phi Y}(t_1) \\ C_{D,\phi\phi}(t_2, t_1) & C_{D,\phi\phi}(t_2, t_2) & \cdots & C_{D,\phi\phi}(t_2, t_K) & C_{D,\phi Y}(t_2) \\ \vdots & \vdots & \ddots & \vdots & \vdots \\ C_{D,\phi\phi}(t_K, t_1) & C_{D,\phi\phi}(t_K, t_2) & \cdots & C_{D,\phi\phi}(t_K, t_K) & C_{D,\phi Y}(t_K) \\ C_{D,\phi Y}(t_1) & C_{D,\phi Y}(t_2) & \cdots & C_{D,\phi Y}(t_K) & C_{D,YY} \end{bmatrix},
$$

$$(7.4.46)$$

where $C_{D,\phi\phi}(t_k, t_k)$, $C_{D,\phi\phi}(t_{k_1}, t_{k_2})$ and $C_{D,\phi Y}(t_k)$ are given by (7.4.43), (7.4.44) and (7.4.45), respectively.

It is obvious that the second approach should produce better results, because the correlation described by cross covariances between different observation times is included in (7.4.46).

After the determination of statistical parameters, the measurements of both $\log K$ and head will be used again to estimate the $\log K$ field.

When co-Kriging is used to estimate $\hat{Y}_0$, and all head observations at different times are used simultaneously, we have

$$
\hat{Y}_0 = \sum_{k=1}^{K} \sum_{l=1}^{L} \mu_{l,k}(\phi_{l,k} - H_{l,k}) + \sum_{m=1}^{M} \lambda_m Y_m, \tag{7.4.47}
$$

where λ_m and $\mu_{l,k}$ are the solution of the following set of equations:

$$
\sum_{k=1}^{K} \sum_{l=1}^{L} \mu_{l,k} C_{D,\phi\phi}[\mathbf{x}_i, t_j; \mathbf{x}_l, t_k] + \sum_{m=1}^{M} \lambda_m C_{D,Y\phi}[\mathbf{x}_m, \mathbf{x}_i, t_j] =
$$

$$= C_{D,Y\phi}\left(\mathbf{x}_0; \mathbf{x}_i, t_j\right), \quad (i = 1, 2, \ldots, L; \ j = 1, 2, \ldots, K), \tag{7.4.48a}$$

$$\sum_{k=1}^{K}\sum_{l=1}^{L} \mu_{l,k} C_{D,Y\phi}\left[\mathbf{x}_i; \mathbf{x}_l, t_k\right] + \sum_{m=1}^{M} \lambda_m C_{D,YY}\left[\mathbf{x}_i, \mathbf{x}_m\right] + \nu$$
$$= C_{D,YY}\left(\mathbf{x}_i, \mathbf{x}_0\right), \quad (i = 1, 2, \ldots, M), \tag{7.4.48b}$$

$$\sum_{m=1}^{M} \lambda_m = 1. \tag{7.4.48c}$$

In (7.4.48), there are $(M + LK + 1)$ equations with $(M + LK)$ co-Kringing coefficients, λ_m and $\mu_{l,k}$, and an additional coefficient ν which is called the Lagrange multiplier.

The variance of co-Kriging estimate is given by

$$\text{Var}\left[\hat{Y}_0 - Y_0^*\right] = \sigma_Y^2 - \sum_{m=1}^{M} \lambda_m C_{D,YY}\left(\mathbf{x}_0, \mathbf{x}_m\right)$$
$$- \sum_{k=1}^{K}\sum_{l=1}^{L} \mu_{l,k} C_{D,Y\phi}\left(\mathbf{x}_0; \mathbf{x}_l, t_k\right) - \nu. \tag{7.4.49}$$

All results in this section can be extended to the case that groundwater quality observations are also involved in the inverse solution.

7.4.5 A NUMERICAL EXAMPLE

Sun and Yeh (1992a) presented a numerical example that explains how to solve inverse problems in groundwater modeling by the stochastic method described above.

Figure 7.4.1 shows a two-dimensional confined aquifer, in which $\overline{AB}$ and $\overline{CD}$ are given head boundary sections:

$$\phi|_{\overline{AB}} = 90 \text{ m}, \qquad \phi|_{\overline{CD}} = 100 \text{ m}.$$

$\overline{BEC}$ and $\overline{DFA}$ are no-flow boundary sections. The length of $\overline{AE}$ is 240 m, the length of $\overline{AF}$ is 560 m. The flow region is discretized into 168 triangular elements with 105 nodes.

Assume that random field $Y = \log k$ is characterized by a constant mean μ_Y and covariance function (7.4.11) with statistical parameters

$$\mu_Y = 3.0, \quad \sigma_Y^2 = 0.04, \quad l_Y = 200 \text{ m}.$$

Using the turning bands methods (Matheron, 1973; Mantoglow and Wilson, 1982), a realization Y^* of the random field can be generated, which is defined as the true $\log K$ field and shown in Figure 7.4.2. For any practical case, of course, Y^* is unknown.

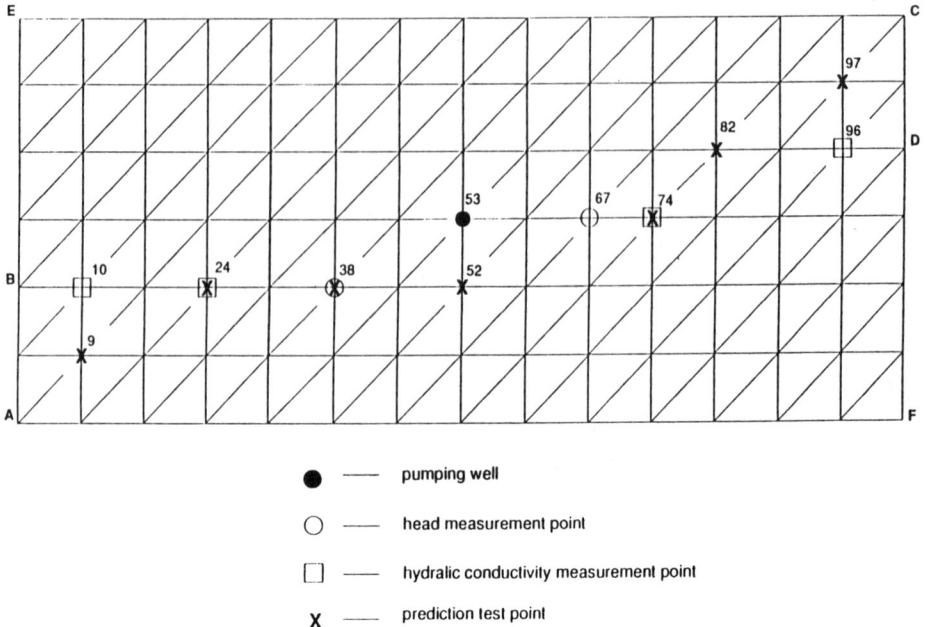

pumping well

head measurement point

hydralic conductivity measurement point

X — prediction test point

Fig. 7.4.1. Aquifer configuration.

The initial condition of groundwater flow is given in Figure 7.4.3. For simplicity, the storage coefficient and the thickness of the aquifer in Equation (7.4.30) are assumed to be constant and known as

$$S = 0.0001, \quad m = 50 \text{ m}.$$

In order to identify Y^*, an experimental design D is proposed which includes the following pumping test and data collection strategy (see Figure 7.4.1):

(1) Measure $Y = \log K$ at five nodes (10, 24, 53, 74, and 96).

(2) Pump groundwater from a well at node 53 with pumping rate $Q = 4000 \text{ m}^3/\text{day}$.

(3) Observe transient head at three observation wells located at nodes 38, 53 and 67.

(4) Collect observations at $t_1 = 0.01$ day, $t_2 = 0.1$ day and $t_3 = 4.0$ days.

We wish to predict the steady state head distribution when the pumping rate at well 53 is increased to $Q = 10,000 \text{ m}^3/\text{day}$. Let the prediction vector G be the steady state heads at nodes 9, 24, 38, 52, 53, 74, 82, and 97.

The head observations corresponding to design D can be obtained from the simulation model, in which $\exp(Y^*)$ is used as the hydraulic conductivity . The

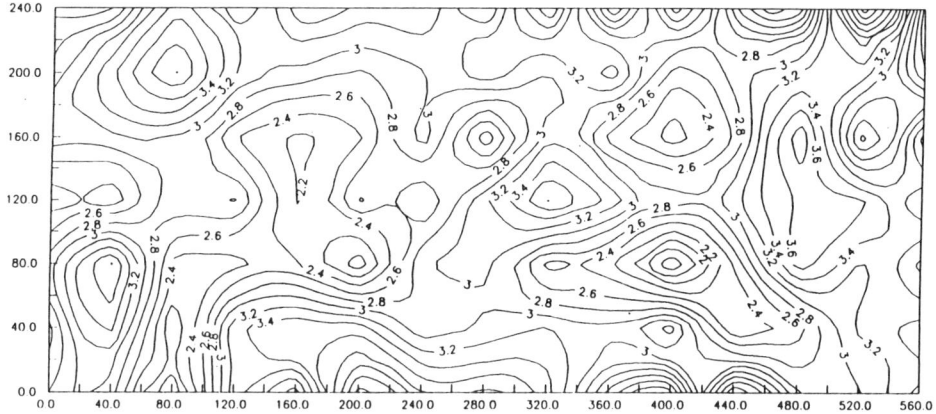

Fig. 7.4.2.   The true $\log K$ field (Y^*).

TABLE 7.4.1
$\log K$ and Head Measurements

$\log K$ measurements		Head Measurements			
Node No.	Values	Node No.	$t = 0.01$	$t = 0.1$	$t = 4.0$
10	3.807	38	98.557	92.973	91.959
24	2.330	53	97.891	92.730	91.665
53	3.165	67	99.288	94.938	93.956
74	2.916				
96	2.351				

measurements of both $\log K$ and head corresponding to design D without noise are listed in Table 7.4.1.

The procedure of the stochastic inverse solution consists of the following steps

Step 1. Use sample mean and sample variance of $\log K$ measurements as the initial estimates of statistical parameters μ_Y and σ_Y^2.

Step 2. Use the $\log K$ measurements only to estimate statistical parameters μ_Y, σ_Y^2 and l_Y through the *MLE*, and generate a $\log K$ field through Kriging;

Step 3. Use statistical parameters obtained in the last step as the initial esti-

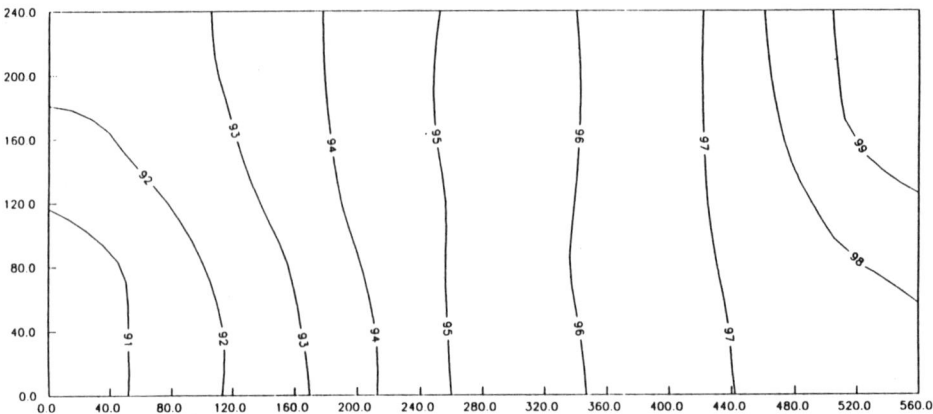

Fig. 7.4.3. The initial head distribution.

mates, and use measurements of both log K and head simultaneously to solve the stochastic inverse problem.

In this example, the multiple cell balance method (Sun and Yeh, 1983) is used to solve both the expected head equation (7.4.32) and the adjoint state equation (7.4.37). The subroutine **MCB2D** given in Appendix C can be used for this purpose with a minor modification. To obtain all elements of covariance matrix C_D in (7.4.46), only 10 simulation runs are needed.

The identified field $\hat{Y}$ is shown in Figure 7.4.4. Figure 7.4.5 show its variance distribution obtained from the co-Kriging estimation. To compare the results of inverse solution, we can use the following five criteria.

(1) The L_2-norm of the difference between Y^* and $\hat{Y}$

$$CY = \|Y^* - \hat{Y}\|_{L_2} = \left[\frac{1}{M} \sum_{m=1}^{M} \left(Y_i^* - \hat{Y}_i\right)^2\right]^{1/2}.$$

(2) The L_∞ norm of head fitting errors

$$CF_1 = \|\phi_{\text{cal}}\left(\hat{Y}\right) - \phi_{\text{obs}}\|_{L_\infty} = \max_{l,k} |\phi_{l,k}\left(\hat{Y}\right) - \phi_{l,k}|,$$

where ϕ_{obs} is the head observation vector, and $\phi_{\text{cal}}(\hat{Y})$ is the corresponding model output.

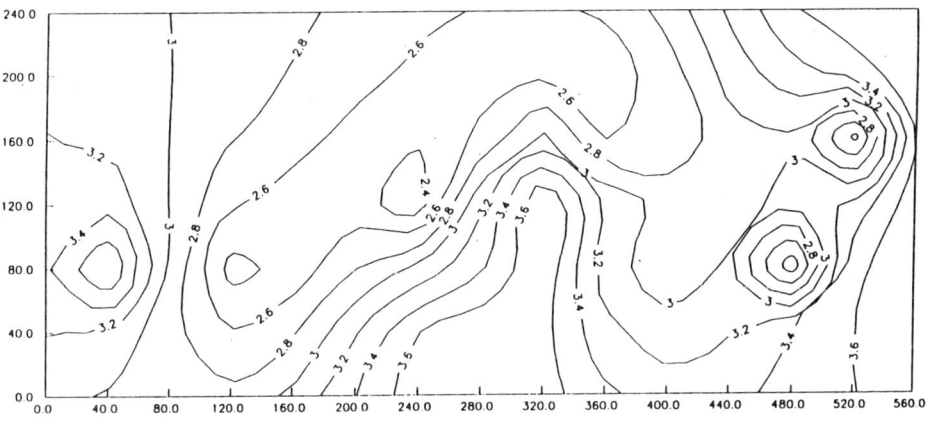

Fig. 7.4.4.   The identified log K field ($\hat{Y}$).

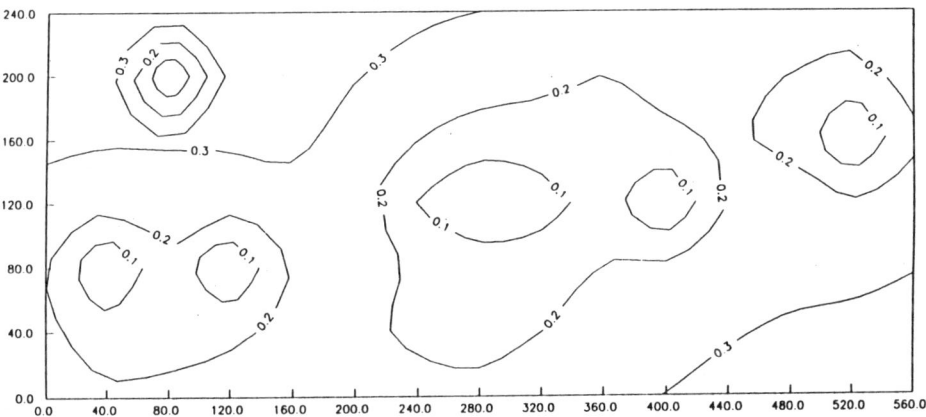

Fig. 7.4.5.   The variance distribution of identified log K field ($\hat{Y}$).

(3) The L_2 norm of the head fitting errors

$$CF_2 = \|\phi_{\text{cal}}\left(\hat{Y}\right) - \phi_{\text{obs}}\|_{L_2}.$$

(4) The L_∞ norm of the prediction errors

$$CG_1 = \|\mathbf{G}\left(Y^*\right) - \mathbf{G}\left(\hat{Y}\right)\|_{L_\infty}.$$

TABLE 7.4.2

Results of using log K measurements only and using Head measurements of one, two and three observation times simultaneously

	CY	CF_1	CF_2	CG_1	CG_2
Using log K measurements only					
	0.697	1.802	1.462	3.122	2.411
Using head measurements at $t_1 = 4.0$ days					
	0.652	1.133	0.538	1.747	0.812
Using head measurements at $t_1 = 0.1$ and at $t_2 = 4.0$ days simultaneously					
	0.613	0.556	0.306	0.927	0.547
Using head measurements at $t_1 = 0.001$, $t_2 = 0.1$ and $t_3 = 4.0$ days simultaneously					
	0.704	0.353	0.240	0.837	0.479

(5) The L_2 norm of the prediction errors

$$CG_2 = \| \mathbf{G} \left(Y^* \right) - \mathbf{G} \left(\hat{Y} \right) \|_{L_2}.$$

Table 7.4.2 is a comparison of results obtained using the log K measurements, together with the head observations of one, two, and three observation times, respectively. It shows that the utilization of head measurements of different times simultaneously can significantly decrease the head fitting errors (CF_1 and CF_2) and produce much better results for predictions (CG_1 and CG_2), although the improvement of the identified random field (CY) is not significant.

Table 7.4.3 is similar to Table 7.4.2, but using noisy data of measurements of both log K and head. We found that the stochastic inversion has good stability. This is expected, because there are only three independent statistical parameters that need to be identified.

Table 7.4.4 is a comparison of results obtained by using the head measurements of different observation times simultaneously and using them sequentially (the quasi-steady state method). The last line of this Table shows results obtained by using the head measurements simultaneously but without considering the correlation between observation times. From Table 7.4.4, we found that using the head measurements simultaneously and considering the correlation between observation times gives much better results.

Exercise 7.4.1. Give the statement of defining inverse problems when both parameter and state measurements are available.

TABLE 7.4.3

Results of using log K and Head measurements of three observation times simultaneously when measurements have noise

	CY	CF_1	CF_2	CG_1	CG_2
Retain measurements to two decimal places (noise < 0.005)					
	0.616	0.397	0.246	0.928	0.514
Retain measurements to one decimal place (noise < 0.05)					
	0.669	0.189	0.128	0.803	0.425
Add 0.1 to all measurements					
	0.626	0.352	0.217	0.605	0.328
Subtract 0.1 from all measurements					
	0.810	0.574	0.296	1.736	0.835

TABLE 7.4.4

Results of using transient Head measurements simultaneously and sequentially

	CY	CF_1	CF_2	CG_1	CG_2
Using head measurements of three observation times simultaneously					
	0.704	0.353	0.240	0.837	0.497
Using head measurements of three observation times sequentially					
	0.624	0.797	0.642	1.588	1.264
Ignoring the correlation of head measurements between observation times					
	0.801	2.239	1.464	4.132	2.856

Exercise 7.4.2. Considering a steady flow in a confined aquifer governed by

$$\frac{\partial}{\partial x}\left(e^Y \frac{\partial \phi}{\partial x}\right) + \frac{\partial}{\partial y}\left(e^Y \frac{\partial \phi}{\partial Y}\right) + R\phi - Q = 0,$$

where $R\phi$ is used to represent a leaky term and $Y = \log T$, derive the corresponding *PDE* for expected H and the *SPDE* relating $\log T$ perturbation f and head perturbation h.

Exercise 7.4.3. Use the adjoint state method given in Section 7.4.4 to calculate the covariance matrices $C_{D,\phi\phi}$ and $C_{D,\phi Y}$ for the steady flow defined in Exercise 7.4.2.

CHAPTER 8

Experimental Design, Extended Identifiabilities and Model Structure Identification

8.1 Experimental Design

The accuracy of identified model parameters and thus the reliability of model predictions depend on the quantity and quality of observation data obtained in the field according to predetermined experimental designs. A good experimental design should give enough information for model calibration while save experimental expenses. Basic ideas and methods of experimental design have been well established in statistics and extensively applied to various scientific and engineering fields (Silvey, 1980; Pázman, 1986). For groundwater modeling, however, there are some difficulties associated with the design of experiments. First, the observed state variables, such as the head and concentration, are always nonlinear functions with respect to the unknown hydrogeological parameters. Second, the model structure can never be known exactly. Third, the cost of experiments is usually expensive and the number of observation wells is very limited. Fourth, the experimental scale is generally small as compared with the aquifer scale. Therefore, in the field of groundwater modeling, special considerations should be given to the *design of experiments*.

In this section, various criteria of optimal design will be introduced based on different experimental objectives (parameter estimation, model prediction and decision making). Some new developments in this field will be also introduced.

8.1.1 EXPERIMENTAL DESIGN IN GROUNDWATER MODELING

We have given a general form of groundwater models in Equation (1.1.15), i.e.,

$$\mathbf{L}(\mathbf{u}; \mathbf{p}; \mathbf{q}) = 0$$

where **L** is a set of *PDE* operators with appropriate subsidiary conditions, **u** the state vector which can be observed at some designated points and times, **p** the parameter vector that can not be measured directly in the field, **q** the control vector whose components are natural and/or artificial excitations to the system.

An *experimental design*, *D*, generally consists of two parts.

(1) The excitation part that may include the following decisions:

  · number and locations of extraction and injection wells;

  · pumping and injection rates;

  · time periods of extraction and injection;

  · concentration of injected water;

  · artificial changes of boundary conditions.

(2) The observation part that may include the following decisions:

  · state variables to be observed;

  · number and locations of observation wells;

  · observation frequency.

When the excitation part of a design problem is predetermined and only the observation part needs to be considered, the problem is often referred to as *network design*. How to monitor the water quality around a landfill place is an example of network design.

Example 8.1.1.   Suppose that we want to design a twin-well injection-extraction tracer test for determining the dispersivity of an aquifer. In this problem, we have to make the following decisions:

  · The distance between the two wells;

  · depths of the two wells;

  · pumping and injection rates;

  · tracer concentration of injected water;

  · sampling frequency;

  · time period of the experiment.

Mercer and Faust (1980) presented a computer-aided design method for twin-well tracer test, in which appropriate ranges of design variables mentioned above were obtained by model sensitivity analysis and accessible prior information.

When we use the words "optimal design", we must have a performance E, defined by design objectives to measure all feasible designs and select the best one from them. The following design objectives are often encountered in groundwater modeling:

· Minimize the uncertainty of model structure;

· Minimize the uncertainty of model parameters;

· Maximize the reliability of model predictions;

· Minimize the risk of management decisions;

· Minimize the cost of experimental expenses;

· Minimize the time period of experiments.

Different optimal designs may be obtained when different objectives are considered. Demands of different objectives may either be complementary or conflicting. For example, increasing the sampling size can decrease the uncertainty of identified parameters and probably increase the reliability of model predictions, but cause the increase of experimental cost. Therefore, it is natural to define the optimal design problem in the framework of multiobjective decision making.

Let $\mathbf{z} = (z_1, z_2, \ldots, z_s)$ be s objectives, $\mathbf{q} = (q_1, q_2, \ldots, q_r)$ be r decision variables associated with system excitation, $\mathbf{y} = (y_1, y_2, \ldots, y_l)$ be l decision variables associated with system observation. A multiobjective experimental design problem then can be stated as follows:

$$\max \mathbf{z}(\mathbf{q}, \mathbf{y}) \qquad (8.1.1a)$$

subject to

$$\mathbf{g}(\mathbf{q}, \mathbf{y}) \leq \mathbf{b} \qquad (8.1.1b)$$

and

$$\mathbf{q} \in Q_{\mathrm{ad}}, \quad \mathbf{y} \in Y_{\mathrm{ad}}. \qquad (8.1.1c)$$

In the above equations, (8.1.1b) represents a set of constraints imposed to the decision variables. Q_{ad} and Y_{ad} are admissible regions for $\mathbf{q}$ and $\mathbf{y}$, respectively. Note that some decision variables, such as pumping rates and periods, may vary continuously. Other decision variables, however, can only take discrete values from a given set. For example, observation wells may only be selected from the existing wells.

Problem (8.1.1) is a nonlinear multiobjective programming with continuously and discontinuously varying decision variables. In practice, to solve such a general problem is too difficult. Up to date, only a few hypothetical examples and simplified cases have been reported.

8.1.2 EXPERIMENTAL DESIGN FOR PARAMETER IDENTIFICATION

Let $\mathbf{u}_D$ be the observations obtained in an experiment according to a design D. When the simulation model $u = M(\mathbf{p})$ is linear with respect to model parameters $\mathbf{p}$, observations $\mathbf{u}_D$ can be represented as

$$\mathbf{u}_D = \mathbf{A}_D \mathbf{p} + \mathbf{e}_D, \tag{8.1.2}$$

where $\mathbf{A}_D$ is a matrix. Its subscript D is used to emphasize that the matrix depends on design D. In Section 7.2.3, we have known that if the *pdf* of observation errors $\mathbf{e}_D$ is Gaussian, then the covariance matrix of estimation can be expressed as

$$\text{Cov}(\hat{\mathbf{p}}) = (\mathbf{A}_D^T \mathbf{V}_D^{-1} \mathbf{A}_D)^{-1}, \tag{8.1.3a}$$

where $\hat{\mathbf{p}}$ is the best unbiased estimate of the unknown parameters.

With further assumptions that all observation errors are independent of each other and have a constant variance σ^2, i.e., $\mathbf{V}_D = \sigma^2 \mathbf{I}$, (8.1.3a) can be reduced to

$$\text{Cov}(\hat{\mathbf{p}}) = \sigma^2 (\mathbf{A}_D^T \mathbf{A}_D)^{-1}. \tag{8.1.3b}$$

$\text{Cov}(\hat{\mathbf{p}})$ is a measure of the reliability of the estimated parameters $\hat{\mathbf{p}}$. If the objective of experimental design is to make the estimated parameters as reliable as possible, we should make $\text{Cov}(\hat{\mathbf{p}})$ as small as possible in certain sense. The "smallness" of a matrix is often described by various scalar functions of the matrix. Thus, the optimal experimental design for parameter identification, D^*, may be defined as

$$\phi\left[(\mathbf{A}_{D^*}^T \mathbf{A}_{D^*})^{-1}\right] = \min \phi\left[(\mathbf{A}_D^T \mathbf{A}_D)^{-1}\right], \quad D \in D_{\text{ad}}, \tag{8.1.4}$$

where ϕ is a scalar function with matrix as its variable, D_{ad} is a set of admissible experiments.

One of frequently used criteria is called *D-optimal design* (Silvey, 1980), where the determinant of matrices is taken as function ϕ in (8.1.4). An experimental design is D-optimal, if

$$\det(\mathbf{A}_{D^*}^T \mathbf{A}_{D^*}) = \max \det(\mathbf{A}_D^T \mathbf{A}_D), \quad D \in D_{\text{ad}}. \tag{8.1.5}$$

In Section 7.1.2, we have learned how to measure the uncertainty of the estimated parameters. According to (7.1.9), it is obvious that if the *pdf* of the estimated parameters is normal and the observations used in the estimate procedure are obtained from a D-optimal design, then the uncertainty associated with the estimated parameters will be the minimum. In other words, D-optimal design can provide the most information content to parameter estimation.

Other composite measures in addition to D-optimality are also possible. For example, a design is said to be *A-optimal* if it minimizes the trace, that is, the sum of all diagonal elements of $(A_D^T A_D)^{-1}$. A design is said to be *E-optimal*, if it minimizes the maximum eigenvalue of $(A_D^T A_D)^{-1}$. For the same problem, different optimalities may introduce slightly different designs.

The above criteria can be extended to the design of experiments for non-linear models. Let p_0 be a prior estimate of the unknown parameters. Using Taylor's expression, non-linear model $u = M(p)$ can be linearized in the neighboring of p^0. As a result, observations u_D can be represented approximately by

$$\mathbf{u}_D \approx \mathbf{J}_D \mathbf{p} + \left[\mathbf{u}_D(\mathbf{p}^0) - \mathbf{J}_D \mathbf{p}^0 \right] + \mathbf{e}_D, \tag{8.1.6}$$

where $L \times M$ matrix $\mathbf{J}_D = [\partial \mathbf{u}_D / \partial \mathbf{p}]$ is the Jacobian of observation vector $\mathbf{u}_D$ with respect to parameter vector $\mathbf{p}$, L and M are dimensions of $\mathbf{u}_D$ and $\mathbf{p}$, respectively. Using the same assumptions of deriving (8.1.3b), the estimation covariance can be approximated by

$$\text{Cov}(\hat{\mathbf{p}}) \approx \sigma^2 \left[\mathbf{J}_D^T \mathbf{J}_D \right]^{-1}. \tag{8.1.7}$$

Thus, for non-linear models, we have the following criterion of D-optimal design:

$$\max \det \left[\mathbf{J}_D^T \mathbf{J}_D \right], \quad D \in D_{\text{ad}}. \tag{8.1.8}$$

In Section 4.3, we have discussed how to calculate sensitivity matrix $\mathbf{J}_D$ with either the finite difference approximation method or the sensitivity equation method. We can also exploit the advantage that maximizing $\det \left[\mathbf{J}_D^T \mathbf{J}_D \right]$ is helpful when Gauss–Newton method is used to identify the unknown parameters.

Although the D-optimal criteria (8.1.5) for linear models and (8.1.8) for non-linear models are similar in form, there is a big difference between them: $\mathbf{A}_D$ does not depend on model parameters, but $\mathbf{J}_D$ does. When the best design depends on the model parameters to be estimated, logically, the optimal design can never be found at the design stage unless prior estimate $\mathbf{p}^0$ in (8.1.8) is very close to the true parameters, or Jacobian $\mathbf{J}_D$ is insensitive to the values of model parameters.

Generally, a *sequential design* procedure has to be used for non-linear models, in which the estimated parameters are updated after each trial and the next design is then chosen with the aid of the improved estimates. Unfortunately, the sequential design strategy is often impractical for groundwater modeling, because the required experimental time may be too long and the experimental cost may be too high.

Another alternative approach of experimental design for non-linear models is the *robust-design* strategy. It makes a design useful for all parameters in a given range. For example, design criterion

$$\max_{D \in D_{ad}} \min_{p \in P_{ad}} \det \left[\mathbf{J}_D^T(\mathbf{p}) \mathbf{J}_D(\mathbf{p}) \right] \tag{8.1.9}$$

requires to provide maximum information to a parameter vector $\mathbf{p}$ which is the most difficult one to be identified in an admissible range P_{ad}.

When there are uncertainties associated with model structure, model parameters, and observation errors, the concept of robustness is often used in the design of experiments (Steinberg and Hunter, 1984).

Using (8.1.8) as an objective, and using the cost and/or other requirements as constraints, a D-optimal design problem can be formulated. Both the objective and cost are functions of decision variables $\mathbf{q}$ and $\mathbf{y}$ defined in (8.1.1). Knopman and Voss (1988) considered the D-optimal design for a one-dimensional mass transport problem. Cleveland and Yeh (1989) used a dynamic programming algorithm to solve the D-optimal problem for designing an aquifer tracer test. In Nishikawa and Yeh (1989), the cost was taken as the objective function and the uncertainty of the identified parameters measured by $\det[(\mathbf{J}_D^T \mathbf{J}_D)^{-1}]$ was taken as a constraint, and a hypothetical example was solved by a sequential design procedure.

In practice, however, we often use the trial-and-error method to solve the D-optimal design problems. Generating numerously feasible designs and comparing them one by one with the D-optimal criterion, a best design may be found. Sometimes, we can use a heuristic technique to find a suboptimal design. For example, we can add new observation well one at a time, then check if the cost constraint is violated, if not, optimize the location and observation frequency of the new well by solving a non-linear programming problem with only a few decision variables. This procedure is repeated until enough information is obtained for parameter identification or the available budget is exceeded. When using the trial-and-error method, however, there is no guarantee that we can find the global minimum.

Hsu and Yeh (1989) formulated the experimental design problem of groundwater modeling into a non-linear mixed integer programming problem. The minimization of experimental cost was considered as the objective, while the reliability of the identified parameters, which is measured by the trace of matrix $(\mathbf{J}_D^T \mathbf{J}_D)$, was used as a constraint. In order to save computational effort, the proposed optimization problem was solved by a heuristic approach.

8.1.3 EXPERIMENTAL DESIGN FOR PREDICTION

In recent years, hydrogeologists pay more attention to the reliability of model predictions rather than the accuracy of model parameters, because the final object of groundwater modeling is the prediction of aquifer states. Since the conditions used for predictions are generally different from the conditions of experiments, the prediction equations need not be the same as the simulation equations of experiments, nor need the variables to be predicted coincide with the dependent variables of the simulation equations. For example, the flow equation may be used to simulate a pumping test for determining the hydraulic conductivity, which will be inserted into a mass transport equation to predict the arrival time of the pollutant. Let the solution of a prediction problem be $G = G(\mathbf{x}, \mathbf{p})$, where $\mathbf{x}$ represents independent space and/or time variables, $\mathbf{p}$ an M-dimensional parameter vector to be estimated. Assume that our goal is to predict a set of K values

$$\mathbf{G}_E = (G_{E,1}, G_{E,2}, \dots, G_{E,K}), \tag{8.1.10}$$

where $G_{E,k} = G(\mathbf{x}_k, \mathbf{p})$ $(k = 1, 2, \dots, K)$, $\mathbf{x}_1, \mathbf{x}_2, \dots, \mathbf{x}_K$ are given. For example, $\mathbf{G}_E$ may be the predicted values of concentration of K water supply wells at the end of a remediation process. Let the true parameters be $\mathbf{p}^*$ and $\mathbf{G}_E^* = \mathbf{G}_E(\mathbf{p}^*)$. Further, let $\hat{\mathbf{G}}_E = \mathbf{G}_E(\hat{\mathbf{p}})$, $\delta\hat{\mathbf{p}} = \hat{\mathbf{p}} - \mathbf{p}^*$, $\delta\hat{\mathbf{G}}_E = \hat{\mathbf{G}}_E - \mathbf{G}_E^*$, where $\hat{\mathbf{p}}$ is the estimated parameter vector. A Taylor series expansion up to linear terms yields

$$\delta\hat{\mathbf{G}}_E \approx \mathbf{J}_E \delta\hat{\mathbf{p}}, \tag{8.1.11}$$

where the $K \times M$ matrix $\mathbf{J}_E = [\partial\mathbf{G}_E/\partial\mathbf{p}]$ is the Jacobian of prediction vector with respect to the parameter vector. All elements of $\mathbf{J}_E$ are evaluated at $\hat{\mathbf{p}}$. The covariance matrix of prediction error is then given by

$$\begin{aligned} \text{Cov}(\hat{\mathbf{G}}_E) &= E(\delta\hat{\mathbf{G}}_E \delta\hat{\mathbf{G}}_E^T) \\ &= \mathbf{J}_E \text{Cov}(\hat{\mathbf{p}}) \mathbf{J}_E^T. \end{aligned} \tag{8.1.12a}$$

Substituting (8.1.7) into the above equation yields

$$\text{Cov}(\hat{\mathbf{G}}_E) = \sigma^2 \mathbf{J}_E (\mathbf{J}_D^T \mathbf{J}_D)^{-1} \mathbf{J}_E^T. \tag{8.1.12b}$$

When both the simulation model and prediction model are linear with respect to parameter vector $\mathbf{p}$, $\mathbf{J}_E$ only depends on prediction sampling point $\mathbf{x}_E$, and $\mathbf{J}_D$ only depends on experiment sampling points $\mathbf{x}_D$. In this case, the optimal design, which minimizes a scalar function of matrix $\mathbf{J}_E (\mathbf{J}_D^T \mathbf{J}_D)^{-1} \mathbf{J}_E^T$, may be obtained in the design stage. The diagonal elements of the matrix, $d_1, d_2, \dots, d_K$, are proportional to the variances of prediction errors. A design is said to be *G-optimal*, if it minimizes the maximal variances of prediction errors, i.e., it is the solution of the following optimization problem:

$$\min_{D} \max_{i} d_i; \quad 1 \le i \le K, \quad D \in D_{\text{ad}}. \tag{8.1.13}$$

In the theory of optimal design, it has been proved that D-optimality and G-optimality are equivalent for linear model (Kiefer and Wolfowitz, 1960). This conclusion, however, is not true for non-linear models. Strictly speaking, we could not find the optimal design for predictions in the design stage when the prediction model is non-linear.

In recent years, hydrogeologists are interested in the observation design for the case that the unknown parameters are regarded as random fields. Carrera *et al.* (1984) presented a method aimed at finding the optimal measurement points for Kriging interpolation. The optimal network was selected from a finite set of alternative networks based on the criterion of minimizing the variance of Kriging estimation.

From (7.3.33), it is easy to find out that the estimation variance does not depend on the measured values of state variables, but only on the location of measurement points. Thus, if the covariance (or variogram) is known, a optimal observation network could be determined in the design stage. On the other hand, however, the structure of covariance (or variogram) as well as its statistical parameters can only be determined by measurement information. In an example given by Carrera *et al.* (1984), the variogram used for network design was obtained by existing measurements, and a sensitivity analysis showed that the optimal network was insensitive to errors in the variogram parameters.

Loaiciga (1989) used a *mixed integer programming (MIP)* technique to solve the network design problem. Let

$$\hat{z}(\mathbf{x}_0) = \sum_{i=1}^{N} \lambda_i^0 y_i z_i \tag{8.1.14}$$

be the Kriging estimation of a weakly stationary random field, $z(\mathbf{x})$, at point $\mathbf{x}_0$. In (8.1.14), N is the total number of potential measurement points, λ_i^0 the Kriging coefficient associated with the ith observation $z_i = z(\mathbf{x}_i)$, y_i is a binary variable that takes the value 1 or 0 depending on whether or not an observation is made at the ith site $\mathbf{x}_i$. The objective of network design is to minimize the variance of the estimation error $\hat{z}_0 - z_0$. From (7.3.20), this criterion can be expressed as

$$\min_{\lambda_i^0, y_i} \sum_{i=1}^{N} \sum_{j=1}^{N} C_{ij} \lambda_i^0 \lambda_j^0 y_i y_j - 2 \sum_{i=1}^{N} C_{i0} \lambda_i^0 + C_{00} \tag{8.1.15a}$$

subject to unbiased constraint

$$\sum_{i=1}^{N} \lambda_i^0 y_i = 1 \tag{8.1.15b}$$

and cost constraint

$$\sum_{i=1}^{N} y_i c_i \leq R, \tag{8.1.15c}$$

where c_i $(i = 1, 2, \ldots, N)$ is the cost of sampling at ith site, R the available budget.

Since decision variables y_i $(i = 1, 2, \ldots, N)$ are binary, (8.1.15) is a *MIP* problem. In Loaiciga (1989), a time dependent network design problem was also considered.

In Section 7.4, we have known that the accuracy of the estimated log-hydraulic conductivity depends on the quantity and quality of both $\log K$ measurements and head observations. McLaughlin and Wood (1988a) used moment equations to quantitatively describe the connections among spatial variability of model parameters, measurement availability, and prediction accuracy. The solution of moment equation, however, requires huge computational effort.

Sun and Yeh (1992a) presented an approach for evaluating the reliability of model predictions, which is closely related to the problem of experimental design. From (8.1.12), the variance of a prediction component G_r of $\mathbf{G}_E$ can be approximately represented as

$$\text{Var}\left[G_r(\mathbf{Y}^*) - G_r(\hat{\mathbf{Y}})\right] = \sum_{i=1}^{N} \sum_{j=1}^{N} \frac{\partial G_r}{\partial Y_i} \frac{\partial G_r}{\partial Y_j} E_{ij}, \tag{8.1.16}$$

where

$$E_{ij} = E\left[(Y_i^* - \hat{Y}_i)(Y_j^* - \hat{Y}_j)\right], \tag{8.1.17}$$

$\mathbf{Y}^*$ is the true $\log K$ field, $\hat{\mathbf{Y}}$ its co-Kriging estimation, $\partial G_r/\partial \mathbf{Y}$ sensitivity coefficients evaluated at $\hat{\mathbf{Y}}$. In this situation, using the adjoint state method to calculate sensitivity coefficients is absolutely advantageous. To obtain sensitivity matrix $\partial \mathbf{G}_E/\partial \mathbf{Y}$, the total number of simulation runs is only equal to the dimension of $\mathbf{G}_E$.

Using (7.4.47), we have

$$E_{ij} = C_{D,YY}(\mathbf{x}_i, \mathbf{x}_j) - \sum_{m=1}^{M} \lambda_m^i C_{D,YY}(\mathbf{x}_i, \mathbf{x}_m)$$
$$- \sum_{k=1}^{K} \sum_{l=1}^{L} \mu_{l,k}^i C_{D,Y\phi}(\mathbf{x}_i; \mathbf{x}_l, t_k) - \sum_{m=1}^{M} \lambda_m^j C_{D,YY}(\mathbf{x}_m, \mathbf{x}_j)$$
$$- \sum_{k=1}^{K} \sum_{l=1}^{L} \mu_{l,k}^j C_{D,Y\phi}(\mathbf{x}_j; \mathbf{x}_l, t_k)$$

$$+ \sum_{m=1}^{M} \sum_{s=1}^{M} \lambda_m^i \lambda_s^j C_{D,YY}(\mathbf{x}_m, \mathbf{x}_s)$$

$$+ \sum_{m=1}^{M} \sum_{k=1}^{K} \sum_{l=1}^{L} \lambda_m^i \mu_{lk}^j C_{D,Y\phi}(\mathbf{x}_m; \mathbf{x}_l, t_k)$$

$$+ \sum_{k=1}^{K} \sum_{l=1}^{L} \sum_{m=1}^{M} \mu_{l,k}^i \lambda_m^j C_{D,Y\phi}(\mathbf{x}_m; \mathbf{x}_l, t_k)$$

$$+ \sum_{k=1}^{K} \sum_{l=1}^{L} \sum_{q=1}^{K} \sum_{p=1}^{L} \mu_{l,k}^i \mu_{p,q}^j C_{D,\phi\phi}(\mathbf{x}_l, t_k; \mathbf{x}_p, t_q), \qquad (8.1.18)$$

where λ_m^i, $\mu_{i,k}^i$, λ_m^j and $\mu_{l,k}^j$ are co-Kriging coefficients of nodes i and j, respectively. Using (7.4.48), the above expression can be reduced to

$$E_{ij} = C_{D,YY}(\mathbf{x}_i, \mathbf{x}_j) - \sum_{m=1}^{M} \lambda_m^j C_{D,YY}(\mathbf{x}_i, \mathbf{x}_m)$$

$$- \sum_{k=1}^{K} \sum_{l=1}^{L} \mu_{l,k}^j C_{D,Y\phi}(\mathbf{x}_i; \mathbf{x}_l, t_k) - \nu^j, \qquad (8.1.19)$$

where ν^j is the Lagrange multiplier of node j. Substituting (8.1.19) into (8.1.16), the variance of prediction G_r can be estimated. The reliability of prediction is explicitly connected with experimented design (D) and covariance function of $\log K$ field.

The reliability evaluation of model predictions mentioned above presents important information for the design of experiments. Assume that

(1) We have already had some $\log K$ and/or head measurements in the region of interest.

(2) These data have already been used to obtain the estimates of statistical parameters defining the covariance function.

(3) The $\log K$ field is then estimated by solving the stochastic inverse problem.

(4) The reliability of model predictions evaluated by (8.1.16) is not satisfactory.

Now, we want to add some new measurements and re-solve the inverse problem to satisfy the given accuracy requirement of model predictions. From (8.1.16), Sun and Yeh (1992a) pointed out that the variance of prediction G_r can be effectively decreased if new measurements are collected at those nodes where the absolute values of sensitivity coefficients $\partial G_r / \partial \mathbf{Y}$ are large.

Mckinney and Loucks (1992) presented a similar criterion. In their network design algorithm, new measurement locations are selected by computing the prediction variance sensitivity at a regulatory point with respect to the parameter variance at each potential measurement location. The one with the greatest sensitivity is picked as a new measurement location. It is easy to derive that

$$\frac{\partial [\text{Var} G_r]}{\partial [\text{Var} Y_j]} = \left(\frac{\partial G_r}{\partial Y_j} \right)^2 , \tag{8.1.20}$$

where $Y_j = \log K(\mathbf{x}_j)$. Mckinney and Loucks (1992) suggested to measure Y_j if $(\partial G_r / \partial Y_j)^2$ is the maximum.

Example 8.1.2. Let us return to the example given in Section 7.4.5. The sensitivity matrix of prediction vector $\mathbf{G}$ with respect to $\log K$ field can be obtained by the adjoint state method. Using these sensitivity coefficients and co-Kriging coefficients of the estimated parameter $\hat{\mathbf{Y}}$, the variance of each component of $\mathbf{G}$ can be estimated by (8.1.16). The results are listed in Table 8.1.1. It shows that the largest uncertainty of predictions is associated with the pumping well at node 53 (Figure 7.4.1), although both hydraulic conductivity and head are measured at that node.

TABLE 8.1.1
Variance Estimations of Prediction Errors

Node No.	Variance of Prediction Error
9	0.05
24	0.26
38	0.49
52	0.81
53	1.14
74	0.81
82	0.73
97	0.23

The distributed sensitivity coefficient of the predicted head at the pumping well with respect to the $\log K$ field is given in Figure 8.1.1. It clearly shows that the best sampling location is around the pumping well, but it is also important to sample at the neighboring of inflow boundary. Thus the optimal measurement network for prediction can be found prior to field experiment. However, the variances of predictions do depend on the covariance function of the unknown $\log K$ field.

Table 8.1.2 shows how the variances of predictions depend on the statistical parameters μ_Y, σ_Y^2, and l_Y. If these statistical parameters are unknown, we could not answer whether a design can provide sufficient information for required model predictions.

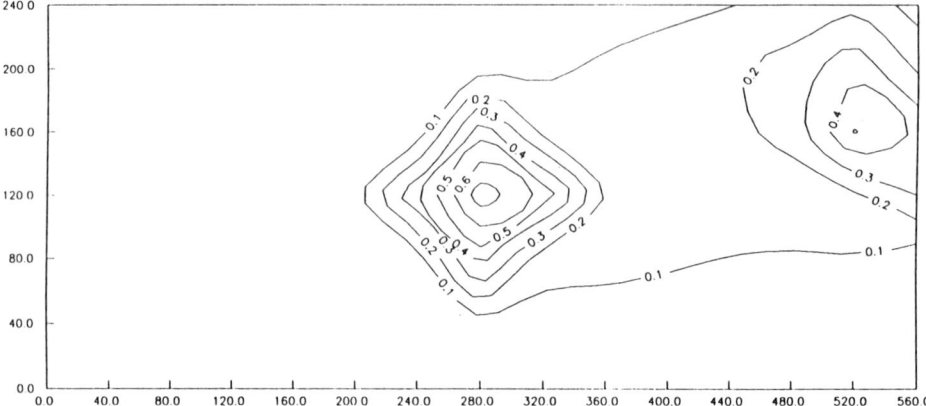

Fig. 8.1.1.   Sensitivity coefficients of predicted head at the pumping well with respect to the identified log K field.

TABLE 8.1.2

Variances Estimations of Prediction Errors With Respect to Different Statistical Parameters

Parameters			Node Numbers							
μ_Y	σ_Y^2	l_Y	9	24	38	52	53	74	82	97
3.2	0.2	200	0.04	0.21	0.41	0.67	0.93	0.62	0.54	0.17
2.8	0.4	200	0.07	0.40	0.78	1.27	1.80	1.20	1.04	0.34
3.0	0.6	200	0.10	0.62	1.21	1.97	2.76	1.84	1.58	0.51
3.0	0.4	100	0.12	0.70	1.37	2.21	3.18	2.05	1.75	0.56
3.0	0.4	400	0.04	0.24	0.48	0.79	1.11	0.75	0.65	0.20
3.2	0.6	100	0.19	1.07	2.09	3.39	4.81	3.14	2.67	0.85
3.0	0.2	400	0.02	0.11	0.22	0.35	0.49	0.33	0.29	0.09

8.1.4   EXPERIMENTAL DESIGN FOR DECISION MAKING

When a simulation model is embedded into a management problem, the decision vector **q**, i.e., the solution of the management problem, must depend on model parameters **p**. With the first order approximation, the uncertainty of management decisions and the uncertainty of model parameters are related by

$$\text{Cov}(\hat{\mathbf{q}}) = \mathbf{J}_M \text{Cov}(\hat{\mathbf{p}}) \mathbf{J}_M^T$$

$$= \sigma^2 \mathbf{J}_M (\mathbf{J}_D^T \mathbf{J}_D)^{-1} \mathbf{J}_M^T, \tag{8.1.21}$$

where matrix $\mathbf{J}_M = [\partial \mathbf{q}/\partial \mathbf{p}]$ is the Jacobian of $\mathbf{q}$ with respect to $\mathbf{p}$. All elements of $\mathbf{J}_M$ are evaluated at the estimated parameter vector $\hat{\mathbf{p}}$, and thus dependent on the observations of design D.

The optimal design problem for decision making is often formulated with two objectives:

(1) Minimize the cost of experiments.

(2) Minimize the uncertainty of management decision.

The latter may be measured by a scalar function of covariance matrix $\mathrm{Cov}(\hat{\mathbf{q}})$.

In decision theory, a cost function $c(\mathbf{p}^*, \hat{\mathbf{p}})$ may be assigned to any loss suffered because of the act of using the estimated parameter vector $\hat{\mathbf{p}}$ when the true parameter vector is $\mathbf{p}^*$. Since $\mathbf{p}^*$ is unknown, the actual cost $c(\mathbf{p}^*, \hat{\mathbf{p}})$ can not be computed. However, we can compute the *risk*, defined as the expected value of the cost of assigning $\hat{\mathbf{p}}$ to $\mathbf{p}$, i.e.,

$$R(\hat{\mathbf{p}}) \equiv E\left[c(\hat{\mathbf{p}}, \mathbf{p})\right] = \int_{(P_{\mathrm{ad}})} c(\hat{\mathbf{p}}, \mathbf{p}) p_*(\mathbf{p}) \, d\mathbf{p}, \tag{8.1.22}$$

where $p_*(\mathbf{p})$ is the posterior distribution of the unknown parameter (Section 7.1.3), which is indirectly dependent on the experimental design. Thus, we can define

$$\min(R_E + R_D) \tag{8.1.23}$$

as the objective of experimental design, where R_E is the cost of experiments, R_D the cost of risk.

The optimal aquifer remediation problem was considered by many authors in recent years. Let $\mathrm{cost}(\mathbf{q})$ be the aquifer remediation cost. It depends on decision vector $\mathbf{q}$ (locations and rates of extraction wells, for example) and some cost coefficients (extraction and treatment costs of unit volume of water, for example). The optimal remediation problem may be stated as (Ahlfeld *et al.*, 1988):

$$\min \mathrm{cost}(\mathbf{q}), \quad \mathbf{q} \in Q_{\mathrm{ad}} \tag{8.1.24a}$$

subject to

$$C_i(\mathbf{p}, \mathbf{q}) \le \overline{C}_i, \quad (i = 1, 2, \ldots, K), \tag{8.1.24b}$$

where model outputs $C_i(\mathbf{p}, \mathbf{q})$ $(i = 1, 2, \ldots, K)$ are the final concentrations of K locations of interest in the aquifer, $\overline{C}_i$ the imposed maximum admissible value of C_i.

Due to the uncertainty of the estimated model parameter **p**, we should change the deterministic constraint (8.1.24b) into the stochastic one (Wagner and Gorelick, 1987)

$$\text{Prob}\left[C_i \leq \overline{C}_i\right] \geq \pi, \tag{8.1.25}$$

where prob $\left[C_i \leq \overline{C}_i\right]$ is the probability of $C_i \leq \overline{C}_i$, π is a given reliability level.

Let us further assume that C_i is normally distributed with mean value $E[C_i]$ and standard deviation sd$[C_i]$. Under the assumption of normality, (8.1.25) becomes

$$\text{Prob}\left[\xi \leq \frac{\overline{C}_i - E\,[C_i]}{\text{sd}\,[C_i]}\right] \geq \pi, \tag{8.1.26}$$

where ξ is a standard normal random variable with zero mean and unit standard deviation. Probability inequality (8.1.26) is equivalent to

$$E\,[C_i] + F^{-1}(\pi)\text{sd}\,[C_i] \leq \overline{C}_i, \tag{8.1.27}$$

where $F^{-1}(\pi)$ is the value of the standard normal cumulative distribution corresponding to reliability level π. Equation (8.1.27) is the deterministic equivalent of chance constraint (8.1.25). The covariance matrix Cov(**C**) can be obtained from (8.1.12), namely,

$$\text{Cov}(\mathbf{C}) = \sigma^2 \mathbf{J}_E (\mathbf{J}_D^T \mathbf{J}_D)^{-1} \mathbf{J}_E^T, \tag{8.1.28}$$

where $\mathbf{J}_E = [\partial\mathbf{C}/\partial\mathbf{p}]$ is the Jacobian of final concentration vector with respect to model parameters. Since the i-th diagonal element of matrix Cov(**C**) is the variance of C_i, the optimal remediation problem with chance constraints now can be formulated as

$$\min\left[\text{cost}(\mathbf{q}) + \text{cost}(D)\right], \quad \mathbf{q} \in Q_{\text{ad}} \tag{8.1.29a}$$

subject to

$$E\,[C_i] + \sigma F^{-1}(\pi)d_i^{1/2}(D, E) \leq \overline{C}_i \quad (i = 1, 2, \ldots, K), \tag{8.1.29b}$$

where cost(D) is the cost of sampling according to design D, $d_i(D, E)$ the i-th diagonal elements of matrix $\left[\mathbf{J}_E(\mathbf{J}_D^T\mathbf{J}_D)^{-1}\mathbf{J}_E^T\right]$. Equation (8.1.29) clearly shows that the optimal remediation problem and the optimal design problem are coupled. When a few measurements are designed, the cost(D) is low but the stochastic part of constraints (8.1.29b), i.e., the second term on the left-hand side, is large. In order to satisfy these constraints, we have to decrease the deterministic part $E(C_i)$ by increasing the extraction volume. As a result, the cost(**q**) is high. On the other hand, when more measurements are designed, cost(D) is high but cost(**q**) is low, because the stochastic part of constraints (8.1.29b) is small. The solution of problem (8.1.29) gives a compromise between these two extreme cases.

In practice, problem (8.1.29) can only be solved by an iterative or sequential procedure because of its nonlinearity. Due to the complexity of the problem, only a hypothetical example was solved (Tucciarelli and Pinder, 1991).

The close relationship between management and monitoring activities has been denoted by many authors (Loaiciga and Mariño, 1987; McLaughlin and Wood, 1988a; Wagner and Gorelick, 1989; Sun and Yeh, 1990b; Andricevic and Kitanidis, 1990; Hudak and Loaiciga, 1992, 1993). Andricevic (1993) considered a coupled formulation of withdrawal and sampling design for groundwater supply models. An approach of sequential development of the sampling network coupled with the withdrawal design strategy was presented. The monitoring network guides the withdrawal decisions, and the performance of the withdrawal strategy determines how the sampling network will be expanded.

Exercise 8.1.1. Extraction-treatment-recharge is a technique of aquifer remediation. Answer the following questions:

1. What decisions should made when this technique is used to remediate an aquifer?

2. How does the hydraulic conductivity impact the remediation decisions?

3. How does an experiment for identifying the hydraulic conductivity impact the remediation decisions?

4. What are the possible objectives of experimental design for this problem?

5. What are the possible design variables for this problem?

Exercise 8.1.2. Under what conditions the A-optimality is identical to the D-optimality?

Exercise 8.1.3. Formulate a D-optimal design problem for the twin-well injection-extraction tracer test. Explain how the design variables depend on the unknown parameters.

Exercise 8.1.4. Suppose that the Dupuit's formula is applicable and our purpose is to predict the steady head in a pumping well when the pumping rate Q_E is given. A pumping test will be used to identify the hydraulic conductivity. Formulate an optimal design problem, in which the reliability of predicted head and experimental cost are considered as objectives.

8.2   Extended Identifiabilities

We have denoted that there is an inherent difficulty associated with the optimal design of experiments for nonlinear problems, i.e., the solution of optimal design depends on the values of unknown parameters. Unfortunately, the methods of sequential design and robust design may be too expensive and not feasible for groundwater modeling.

Another problem of using the D-optimal design (or the G-optimal design) is that matrix $\mathbf{J}_D^T \mathbf{J}_D$ must be non-singular, because its inverse is used in the criteria of optimal design. This requirement means that observations must be sensitive enough to each parameter component. Obviously, such an experiment may be very expensive and generally unnecessary for the purpose of building a model.

To alleviate the difficulties of D-optimal design, a concept called *sufficient design* was presented by Yeh and Sun (1984) based on the definition of an extended identifiability (δ-identifiability). This concept was further developed by Sun and Yeh (1990a), in which several kinds of extended identifiabilities relating to model predictions and decision making were defined. Experimental designs for extended identifiabilities should provide sufficient information to model calibration that a predetermined level of identifiabilities can be achieved.

In this section, we will define three kinds of extended identifiabilities: interval identifiability (*INI*), prediction equivalence identifiability (*PEI*), and management equivalence identifiability (*MEI*). The corresponding criteria of experimental design for these extended identifiabilities will be also given.

8.2.1   DEFINE PREDICTION AND DECISION SPACES

In Chapter 3, the general form of forward solutions is represented by a mapping

$$\mathbf{u} = \mathbf{M}(\mathbf{p}), \quad \mathbf{p} \in P_{\text{ad}}, \tag{8.2.1}$$

where admissible set P_{ad} is a subset of *parameter space* U_p. Corresponding to an observation design D, the model output may be represented as

$$\mathbf{u}_D = \mathbf{DM}(\mathbf{p}), \quad \mathbf{p} \in P_{\text{ad}}, \tag{8.2.2}$$

where $\mathbf{u}_D$ is a point of the *observation space* U_D.

The classical identifiability defined in Section 3.3.1 only relates to spaces U_P and U_D. It requires that mapping **DM** is an injection between the two spaces. This requirement, of course, can never be satisfied in practice because of the existence of observation error. For any practical solution of inverse problems, we have to consider a *noisy observation space* $\tilde{U}_D$ with elements

$$\tilde{\mathbf{u}}_D(\mathbf{p}) = \mathbf{u}_D(\mathbf{p}) + \eta, \tag{8.2.3}$$

where η is the observation error. The *OLS* identifiability defined in Section 4.1.2 relates spaces U_P, U_D and $\tilde{U}_D$. However, it still requires that mapping $u_D = DM(p)$ be injective and continuous. An experimental design that causes all $p \in P_{ad}$ to be *OLS* identifiable may be very expensive if it exists. To escape from the ill-posed problem of inverse solution, we should not limit ourselves in only considering parameter and observation spaces.

In practice, the ill-posed problem may not kill the result of inverse solution. When we build a model for a practical problem, what we care is the reliability of model applications, other than the accuracy of model parameters. To explain this concept quantitatively, it is convenient to introduce two other spaces: *prediction space* U_E and *decision space* U_J. The former is determined by a prediction problem E and the latter by a management problem J.

For each parameter $p \in P_{ad}$, the solution of prediction problem E generates a point $g_E(p)$ in the prediction space U_E. When we solve the inverse problem, more than one $\hat{p} \in P_{ad}$ may be obtained from $\tilde{u}_D$ (non-uniqueness), and these $\hat{p}$'s may or may not be close to the true parameter p^* (instability). However, if all $g_E(\hat{p})$ are close to $g_E(p^*)$ in a satisfactory extent, we would like to accept the result of inverse solution in spite of its non-uniqueness and instability. Obviously, less observations may be sufficient for satisfying such a requirement.

Similarly, for each parameter $p \in P_{ad}$, the solution of management problem J gives a set of decisions $q_J(p)$ which is a point in the decision space U_J. If all $q_J(\hat{p})$ are close to $q_J(p^*)$ in a satisfactory extent, we think that the result of inverse solution is acceptable. This is the basic idea of defining extended identifiabilities.

Figure 8.2.1 shows the relationships between spaces $U_P, U_D, \tilde{U}_D, U_E$ and U_J. Generally, these spaces are assumed to be Bananch spaces. However, if we only deal with discretized models, all of them are finite dimensional.

Since the components of vectors p, u_D, g_E and q_J may have different dimensions and orders of magnitudes, it is better to use the generalized least squares norm for these spaces. Let C_P, C_D, $\tilde{C}_D$, C_E and C_J be given weighting matrices, which are assumed to be symmetric and positive definite. We define norms:

$$\|\cdot\|_{C_P} = (\cdot^T C_P^{-1} \cdot)^{1/2}, \qquad \|\cdot\|_{C_D} = (\cdot^T C_D^{-1} \cdot)^{1/2},$$

$$\|\cdot\|_{\tilde{C}_D} = (\cdot^T \tilde{C}_D^{-1} \cdot)^{1/2}, \qquad \|\cdot\|_{C_E} = (\cdot^T C_E^{-1} \cdot)^{1/2}, \qquad (8.2.4)$$

$$\|\cdot\|_{C_J} = (\cdot^T C_J^{-1} \cdot)^{1/2}$$

for spaces $U_P, U_D, \tilde{U}_D, U_E$ and U_J, respectively.

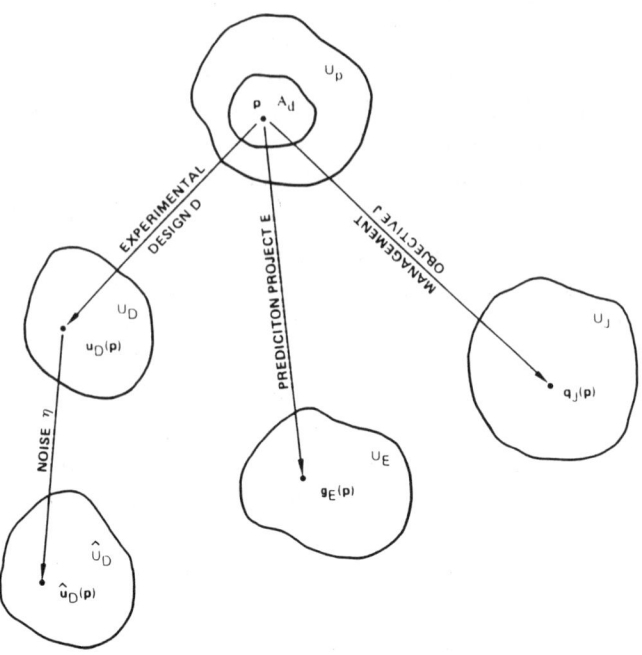

Fig. 8.2.1. Parameter space, observation space, prediction space, and decision space.

8.2.2 INTERVAL IDENTIFIABILITY AND EXPERIMENTAL DESIGN

The *output least squares identifiability (OLSI)* defined in Section 4.1.2 may be expressed as follows: for a parameter $p_0 \in P_{ad}$, if there is an experimental design D, such that the generalized least squares problem

$$\min_{\mathbf{p}} S_D(\mathbf{p}, \mathbf{p}_0), \quad \mathbf{p} \in P_{ad}$$

$$S_D(\mathbf{p}, \mathbf{p}_0) = \|\mathbf{u}_D(\mathbf{p}) - \tilde{\mathbf{u}}_D(\mathbf{p}_0)\|_{\tilde{\mathbf{C}}_D} \tag{8.2.5}$$

has a unique solution and the solution continuously depends on observations $\tilde{\mathbf{u}}_D(\mathbf{p}_0)$, then p_0 is said to be *OLS identifiable* with respect to design D and weighting matrix $\tilde{\mathbf{C}}_D$.

In practice, the uniqueness of the inverse solution may be relaxed if the identified parameters are not "too far" from the true parameter. The *interval identifiability (INI)* is defined as follows:

For a parameter $p_0 \in P_{ad}$ and given weighting matrices $\mathbf{C}_P$ and $\mathbf{C}_D$, if there is a design D and a number $\delta > 0$, such that

$$\|\mathbf{u}_D(\mathbf{p}) - \mathbf{u}_D(\mathbf{p}_0)\|_{\mathbf{C}_D} < \delta \quad \text{implies} \quad \|\mathbf{p} - \mathbf{p}_0\|_{\mathbf{C}_P} < 1 \tag{8.2.6}$$

for all $\mathbf{p} \in P_{ad}$, then p_0 is said to be δ-interval identifiable with respect to design D and weighting matrices $\mathbf{C}_P$ and $\mathbf{C}_D$.

Condition (8.2.6) is equivalent to the following statement:

$$\|u_D(p) - u_D(p_0)\|_{C_D} \geq \delta > 0 \text{ for all } p \in P_{ad} \text{ and } \|p - p_0\|_{C_p} \geq 1.$$

$$(8.2.7)$$

Thus, in order to test the *INI*, we can calculate the minimum

$$S_D(p_0) = \min_{p \in P_{ad}} \|u_D(p) - u_D(p_0)\|_{C_D} \qquad (8.2.8a)$$

subject to

$$\|p - p_0\|_{C_p} \geq 1. \qquad (8.2.8b)$$

If $S_D(p_0) \geq \delta > 0$, then p_0 is δ-interval identifiable.

The set $\{p| \|p - p_0\|_{C_p} < 1\}$ is a neighborhood of parameter p_0. In fact, it is a high-dimensional "ellipse" with its center at p_0. The shape and size of the ellipse can be controlled through the selection of weighting matrix C_P. For example, if

$$|p^{(1)} - p_0^{(1)}| < \varepsilon_1, |p^{(2)} - p_0^{(2)}| < \varepsilon_2, \ldots, |p^{(m)} - p_0^{(m)}| < \varepsilon_m$$

are required, where $p^{(1)}, p^{(2)}, \ldots, p^{(m)}$ and $p_0^{(1)}, p_0^{(2)}, \ldots, p_0^{(m)}$ are components of p and p_0, respectively, then it is sufficient to take

$$C_P = \begin{bmatrix} \varepsilon_1^2 & & & 0 \\ & \varepsilon_2^2 & & \\ & & \ddots & \\ 0 & & & \varepsilon_m^2 \end{bmatrix}. \qquad (8.2.9)$$

The first advantage of the *INI* is its ability to deal with the observation error η. If the upper bound $\bar{\eta}$ of $\|\eta\|_{\tilde{C}_D}$ and the upper bound $\bar{\delta}$ of residuals of the least squares problem (8.2.5) can be estimated, and a design D can be found such that parameter p_0 is δ-interval identifiable, where $\delta \geq \bar{\delta} + \bar{\eta}$, then for any $p \in P_{ad}$, we can conclude that

$$\|u_D(p) - \tilde{u}_D(p_0)\|_{\tilde{C}_D} < \bar{\delta} \text{ implies } \|p - p_0\|_{C_p} < 1. \qquad (8.2.10)$$

In fact, condition $\|u_D(p) - \tilde{u}_D(p_0)\|_{\tilde{C}_D} < \bar{\delta}$ implies

$$\|u_D(p) - u_D(p_0)\|_{\tilde{C}_D} \leq \|u_D(p) - \tilde{u}_D(p_0)\|_{\tilde{C}_D} + \|\tilde{u}_D(p_0) - u_D(p_0)\|_{\tilde{C}_D}$$
$$< \bar{\delta} + \bar{\eta}.$$

$$(8.2.11)$$

Since p_0 is $(\bar{\delta} + \bar{\eta})$-interval identifiable, $\|p - p_0\|_{C_p} < 1$ can be obtained directly from (8.2.11) and the definition (8.2.6) of interval identifiability.

The second advantage of the *INI* is that a design D is already included in its definition. If $S_D(\mathbf{p}_0)$ in (8.2.8) is not less than $(\bar{\delta} + \bar{\eta})$, design D must be sufficient for the interval identifiability of $\mathbf{p}_0 \in P_{ad}$ with respect to $\mathbf{C}_P$ and $\tilde{\mathbf{C}}_D$. When a design D is insufficient, we can modify it through increasing pumping rates, extending pumping period, adding new observation wells and so on, until $S_D(\mathbf{p}_0) \geq \bar{\delta} + \bar{\eta} > 0$ is achieved. Since the true parameter is unknown in the design stage, the sufficiency of a design D should be tested by condition

$$S_D = \min_{\mathbf{p}_0 \in P_{ad}} S_D(\mathbf{p}_0) \geq \bar{\delta} + \bar{\eta}, \tag{8.2.12}$$

where P_{ad} may be estimated by prior information.

An important problem is how to determine $\bar{\delta}$ and $\bar{\eta}$ in (8.2.12). According to the definition of $\bar{\delta}$, we have to estimate the residuals of the least squares problem (8.2.5). Let $\hat{\mathbf{p}}$ be a minimizer of the problem. We then have

$$S_D(\hat{\mathbf{p}}, \mathbf{p}_0) \leq S_D(\mathbf{p}_0, \mathbf{p}_0) = \|\mathbf{u}_D(\mathbf{p}) - \tilde{\mathbf{u}}(\mathbf{p}_0)\|_{\tilde{\mathbf{C}}_D} = \|\eta\|_{\tilde{\mathbf{C}}_D} \leq \bar{\eta}. \tag{8.2.13}$$

Therefore, a conservation estimate of $\bar{\delta}$ is $\bar{\delta} \equiv \bar{\eta}$. For the observation error η, we only need to know the upper bound of its norm $\|\eta\|_{\tilde{\mathbf{C}}_D}$, which is independent of its statistical distribution.

The third advantage of the *INI* is that when observations obtained from an experiment are sufficient for $2\bar{\eta}$-interval identifiability, the optimization procedure of solving the inverse problem can be simplified. Once a parameter $\tilde{\mathbf{p}}$ is found in an iteration step such that $\|\mathbf{u}_D(\tilde{\mathbf{p}}) - \mathbf{u}_D(\mathbf{p}_0)\|_{\tilde{\mathbf{C}}_D} < \bar{\eta}$, the optimization procedure can be stopped immediately because $\tilde{\mathbf{p}}$ must satisfy $\|\tilde{\mathbf{p}} - \mathbf{p}_0\|_{\mathbf{C}_P} < 1$, and thus it is a satisfactory approximation solution. The convergency of the least squares problem (8.2.5) is not necessary.

Now let us consider the algorithm of testing the sufficiency of a design D for the *INI*. As explained above, first, we have to solve non-linear programming problem (8.2.8) for each $\mathbf{p}_0 \in P_{ad}$, then solve another non-linear programming problem (8.2.12). This computation is too complicated to carry out. Fortunately, for a practical problem, we can often find such a $\mathbf{p}_0$ from physics that it is the most difficult one to be identified, and we only need to solve (8.2.8) for this $\mathbf{p}_0$. In Sun and Yeh (1990b), problem (8.2.8) is replaced by an unconstrained optimization problem

$$\min_{\mathbf{p} \in P_{ad}} F(\mathbf{p}) = \|\mathbf{u}_D(\mathbf{p}) - \mathbf{u}(\mathbf{p}_0)\|_{\tilde{\mathbf{C}}_D}^2 + \gamma g(\mathbf{p}), \tag{8.2.14}$$

where

$$g(\mathbf{p}) = \max\{1 - \|\mathbf{p} - \mathbf{p}_0\|_{\mathbf{C}_P}^2, 0\} \tag{8.2.15}$$

is the penalty function and γ the penalty factor. Since function $F(\mathbf{p})$ has the form of sum of function squares, the Gauss–Newton method can be used to solve this problem. If the parameter space is discretized and its dimension is low, however, we can use the trial-and-error method to find the minimum of problem (8.2.8).

Sun and Yeh (1990b) presented a necessary condition and a sufficient condition for testing the *INI* without solving any optimization problem. In Section 6.5.3, we have introduced a concept called "the contribution of observation R to the identification of parameter p". According to (6.5.14) and (6.5.16), the contribution of observation u_D^i to the identification of parameter p^j can be represented as

$$\mathrm{CTB}(u_D^i, p^j) = \frac{\varepsilon_j}{\eta_i} \left| \int_{(\Omega_j)} \frac{\partial u_D^i}{\partial p^j} \, d\Omega \right|, \tag{8.2.16}$$

where u_D^i is a component of $\mathbf{u}_D(\mathbf{p})$; p^j, a component of $\mathbf{p}$, is associated with subregion (Ω_j); ε_j is the accuracy requirement of *INI* for p^j, which can be seen as the element ε_j of matrix $\mathbf{C}_P$ in (8.2.9); η_i is the upper bound of observation error associated with u_D^i.

A necessary condition for the *INI* is that there exists at least one observation u_D^i for each component p^j of the unknown parameter whose contribution is larger than 1. In fact, let us assume that $\mathrm{CTB}(u_D^i, p^j) < 1$ for all u_D^i. For each u_D^i we can find two parameters, $\mathbf{p}_1$ and $\mathbf{p}_2$, with only the jth component being different, and satisfying

$$\left| u_D^i(\mathbf{p}_1) - \tilde{u}_D^i \right| \le \eta_i \text{ and } \left| u_D^i(\mathbf{p}_2) - \tilde{u}_D^i \right| \le \eta_i,$$

$$\text{but } \left| u_D^i(\mathbf{p}_1) - u_D^i(\mathbf{p}_2) \right| > \eta_i. \tag{8.2.17}$$

Equation (8.2.17) means that the observation u_D^i is unable to differentiate $\mathbf{p}_1$ and $\mathbf{p}_2$. On the other hand, however, we have

$$\eta_i < \left| u_D^i(\mathbf{p}_1) - u_D^i(\mathbf{p}_2) \right| = \mathrm{CTB}(u_D^i, p^j) \frac{\eta_i}{\varepsilon_j} \left| p_1^j - p_2^j \right| < \frac{\eta_i}{\varepsilon_j} \left| p_1^j - p_2^j \right|. \tag{8.2.18}$$

It gives $\left| p_1^j - p_2^j \right| > \varepsilon_j$, i.e., p^j is not interval identifiable. Thus, the proposition is proved. For a given design D, we can calculate contributions of all observations to all components of the unknown parameter and test whether the necessary condition is satisfied. If it is not, design D is insufficient for the *INI*. For non-linear models, since contributions depend on unknown parameters, we have to find a $\mathbf{p}_0 \in P_{\mathrm{ad}}$ such that the observations are the most insensitive, and then evaluate the contributions to $\mathbf{p}_0$.

Now, let us give a sufficient condition for the *INI*. Suppose that the upper bounds $|p^j - p_0^j|$ can be estimated and denoted by b_j $(j = 1, 2, \ldots, m)$, if there is at least one observation u_D^i for each p^j such that the following inequality is satisfied:

$$\mathrm{CTB}(u_D^i, p^j) > 2 + \sum_{k \neq j}^{m} \mathrm{CTB}(u_D^i, p^k) \frac{2b_k}{\varepsilon_k} \quad (j = 1, 2, \ldots, m), \tag{8.2.19}$$

then for any $\mathbf{p}_1, \mathbf{p}_2$ that satisfy (8.2.17), we have

$$\left| p_1^j - p_2^j \right| < \varepsilon_j \quad (j = 1, 2, \ldots, m), \tag{8.2.20}$$

that is, the unknown parameter $(\mathbf{p}_0)$ is interval identifiable.

In fact, from (8.2.17), we have

$$2\eta_i \geq |u_D^i(\mathbf{p}_1) - u_D^i(\mathbf{p}_2)| = \eta_i \left| \sum_{k=1}^{m} \mathrm{CTB}(u_D^i, p^k) \frac{p_1^k - p_2^k}{\varepsilon_k} \right|$$

$$\geq \eta_i \mathrm{CTB}(u_D^i, p^j) \frac{|p_1^j - p_2^j|}{\varepsilon_j} - \eta_i \sum_{k \neq j}^{m} \mathrm{CTB}(u_D^i, p^k) \frac{2b_k}{\varepsilon_k}. \tag{8.2.21}$$

Using condition (8.2.19), the *INI* condition (8.2.20) can be obtained directly from (8.2.21). In practice, the sufficient condition (8.2.19) may be satisfied in the following situations:

· The observation u_D^i is very sensitive to p^j in comparison with other components ($\mathrm{CTB}(u_D^i, p^j)$ is large);

· All p^k ($k \neq j$) have been estimated with high accuracy (b_k is small);

· u_D^i is insensitive to p^k ($k \neq j$) ($\mathrm{CTB}(u_D^i, p^k)$ is small).

An example of using the sufficient condition to test the sufficiency of a design D for *INI* will be given in Section 8.2.4.

8.2.3   PREDICTION EQUIVALENCE IDENTIFIABILITY AND EXPERIMENTAL DESIGN

The *INI* defined in last section allows us to design an experiment to guarantee that the identified parameters must fall into a predetermined range described by the weighting matrix $\mathbf{C}_P$. Obviously, if we limit ourselves only in parameter estimation, then we do not know how to assign the range. As explained in Section 8.2.1, the determination of $\mathbf{C}_P$ must depend on the accuracy requirements of model applications.

Suppose that there is a prediction problem E. The prediction space U_E is formed by prediction vectors $\mathbf{q}_E(\mathbf{p})$, $\mathbf{p}_0 \in P_{\mathrm{ad}}$; and the accuracy requirements of the prediction problem are characterized by a given weighting matrix $\mathbf{C}_E$. We define the prediction equivalence identifiability (*PEI*) as follows:

For a parameter $p_0 \in P_{ad}$, if there exists a design D and a number $\delta > 0$, such that

$$\|u_D(p) - u_D(p_0)\|_{C_D} < \delta \text{ implies } \|g_E(p) - g_E(p_0)\|_{C_E} < 1$$

(8.2.22)

for all $p_0 \in P_{ad}$, then the parameter p_0 is said to be *δ-prediction equivalence identifiable* with respect to design D and weighting matrices C_E and C_D.

Equation (8.2.22) is equivalent to the following statement:

$$\|u_D(p) - u_D(p_0)\|_{C_D} \geq \delta > 0$$
$$\text{for all } \{p | p \in P_{ad} \text{ and } \|g_E(p) - g_E(p_0)\|_{C_E} \geq 1\}.$$

(8.2.23)

The following set

$$\Omega_E(p_0, C_E) = \{p | p \in P_{ad} \text{ and } \|g_E(p) - g_E(p_0)\|_{C_E} < 1\}$$

(8.2.24)

is called the *prediction equivalence set* of parameter p_0 with respect to weighting matrix C_E. Condition (8.2.23) means that when parameter $p \notin \Omega_E(p_0, C_E)$, the model output $u_D(p)$ must be significantly different from the model output $u_D(p_0)$. If $p \in \Omega_E(p_0, C_E)$, however, the differentiation between $u_D(p)$ and $u_D(p_0)$ may be obscured. A design D satisfying this requirement may be inexpensive.

Similar to the *INI*, when observation error exists, condition (8.2.23) should be replaced by

$$\|u_D(p) - u_D(p_0)\|_{\hat{C}_D} \geq 2\bar{\eta} \text{ for all } \|g_E(p) - g_E(p_0)\|_{C_E} \geq 1.$$

(8.2.25)

Using the first-order Taylor series expansion of $g_E(p)$ around p_0, we have

$$g_E(p) = g_E(p_0) + J_E(p - p_0),$$

(8.2.26)

where $K \times M$ matrix $J_E = [\partial g_E / \partial p]$ is the Jacobian of K-dimensional prediction vector g. All partial derivatives in J_E are evaluated at p_0. Using (8.2.23), we can obtain

$$\begin{aligned}
\|g_E(p) - g_E(p_0)\|_{C_E}^2 &= [g_E(p) - g_E(p_0)]^T C_E^{-1} [g_E(p) - g_E(p_0)] \\
&\approx [J_E(p - p_0)]^T C_E^{-1} [J_E(p - p_0)] \\
&= (p - p_0)^T J_E C_E^{-1} J_E (p - p_0) \\
&= \|p - p_0\|_{\hat{C}_P}^2,
\end{aligned}$$

(8.2.27)

where

$$\hat{C}_P = J_E^T C_E^{-1} J_E.$$

(8.2.28)

Therefore, the prediction equivalence set $\Omega_E(\mathbf{p}_0, \mathbf{C}_E)$ can be approximately replaced by an interval equivalence set $\{\mathbf{p} | \mathbf{p} \in P_{\text{ad}}, \|\mathbf{p} - \mathbf{p}_0\|_{\hat{\mathbf{C}}_P} < 1\}$. In other words, the problem of *PEI* is approximately reduced to a problem of *INI*.

The sufficiency of a design D for *PEI* can be tested by solving the following non-linear programming problem:

$$S_D = \min_{\mathbf{p}, \mathbf{p}_0} \|\mathbf{u}_D(\mathbf{p}) - \mathbf{u}_D(\mathbf{p}_0)\|_{\mathbf{C}_D}; \quad \mathbf{p} \in P_{\text{ad}}, \ \mathbf{p}_0 \in P_{\text{ad}} \qquad (8.2.29a)$$

subject to

$$\|\mathbf{g}_E(\mathbf{p}) - \mathbf{g}_E(\mathbf{p}_0)\|_{\mathbf{C}_E} \geq 1. \qquad (8.2.29b)$$

If $S_D \geq 2\bar{\eta}$, design D must be sufficient for *PEI*. In fact, when the identified parameter $\hat{\mathbf{p}}$ satisfies $\|\mathbf{u}_D(\hat{\mathbf{p}}) - \tilde{\mathbf{u}}_D\|_{\tilde{\mathbf{C}}_D} < \bar{\eta}$, where $\tilde{\mathbf{u}}_D$ is the observations obtained by experiment D, it must satisfy the prediction accuracy requirement, i.e., $\|\mathbf{g}_E(\hat{\mathbf{p}}) - \mathbf{g}_E(\mathbf{p}^*)\|_{\mathbf{C}_E} < 1$, where $\mathbf{p}^* \in P_{\text{ad}}$ is the unknown true parameter.

It is important to denote that if both $\mathbf{p}$ and $\mathbf{p}_0$ are considered as unknown variables, the optimation problem (8.2.29) can be solved in the design stage. Unfortunately, this problem is too difficult to solve. To test the sufficiency of a design D for *PEI*, we often transform it into an *INI* problem and then use the necessary condition and sufficient condition of *INI* to find the answer. The optimal design of experiments can be found from all sufficient designs for the *PEI* based on an additional criterion J, for example, the cost. Thus, the optimal design problem for prediction now is formulated as follows

$$J(D^*) = \min_D J(D), \quad D \in D_{AS}, \qquad (8.2.30)$$

where D_{AS} is the set of all admissible and sufficient designs for the *PEI*.

Example 8.2.1. A numerical example given by Yeh and Sun (1984) clearly explains the concept of *PEI*.

Figure 8.2.2 shows a two-dimensional confined aquifer. Its length and width are about 8000 and 5000 m, respectively. Its boundary section ABC is a river with a constant head $H_R = 100$ m. The rest of the boundary is impermeable. The flow region is divided into two zones. Their transmissivities are noted by $T^{(1)}$ and $T^{(2)}$, respectively. The true values of $T^{(1)}$ and $T^{(2)}$ are unknown, but their bounds are given as

$$500 \text{ m}^2/\text{d} \ \leq T^{(1)} \leq 1000 \text{ m}^2/\text{d},$$
$$1000 \text{ m}^2/\text{d} \leq T^{(2)} \leq 2000 \text{ m}^2/\text{d}. \qquad (8.2.31)$$

In addition, let the initial head be 100 m in the entire region and the storage coefficient S be 0.0002.

The parameter space $\mathbf{U}_P$ in this example is a two-dimensional space with points $\mathbf{T} = (T^{(1)}, T^{(2)})$. The admissible set P_{ad} is a rectangle, R, defined by (8.2.31).

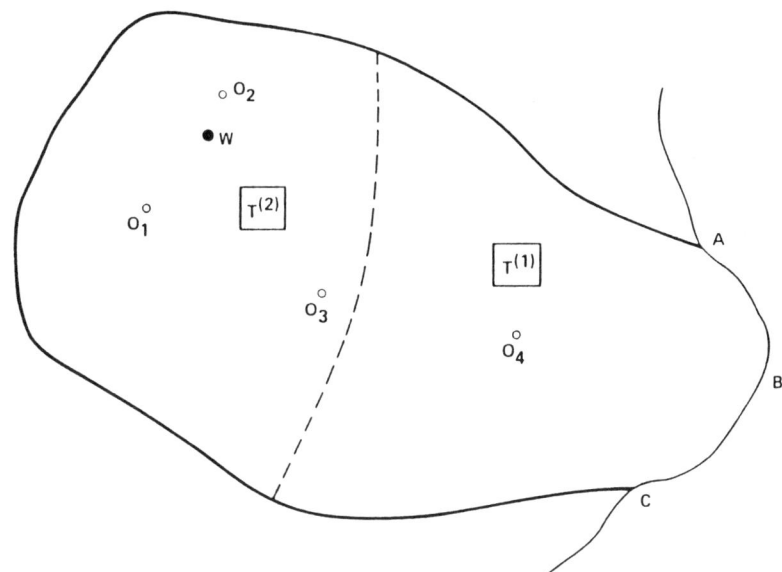

Fig. 8.2.2. Aquifer configuration.

The *PEI* Problem. A predetermined pumping well is located at point W in Figure 8.2.2. The prediction problem is to find the maximum drawdown, or the steady head h_E in the well when the pumping rate is 10,000 m³/d. It is required that the error of prediction does not exceed 3 m. Thus, the prediction vector $\mathbf{g}_E$ in this problem has only one component, h_E, the weighting matrix $\mathbf{C}_E$ only has one element $\varepsilon = 3$ m. The general requirement of *PEI* now reduces to $|h_E(\hat{\mathbf{T}}) - h_E(\mathbf{T}_0)| < 3$ m, where $\mathbf{T}_0$ and $\hat{\mathbf{T}}$ are the true and estimated transmissivities, respectively.

For any transmissivity $\mathbf{T}$, the prediction problem can be solved by a numerical method. Consequently, the prediction equivalence set

$$\Omega_E(\mathbf{T}_0, \varepsilon) = \{\mathbf{T}|\, \mathbf{T} \in R \text{ and } |h_E(\mathbf{T}) - h_E(\mathbf{T}_0)| < 3\}$$

can be found for any $\mathbf{T}_0 \in R$.

The shaded areas in Figure 8.2.3 (a)–(d) show $\Omega_E(\mathbf{T}_0, \varepsilon)$ for $\mathbf{T}_0 = (500, 1000)$, $\mathbf{T}_0 = (500, 2000)$, $\mathbf{T}_0 = (700, 1500)$, and $\mathbf{T}_0 = (1000, 2000)$, respectively. The boundary of $\Omega_E(\mathbf{T}_0, \varepsilon)$ consists of curves $B_1(\mathbf{T}) = h_E(\mathbf{T}) - h_E(\mathbf{T}_0) - \varepsilon = 0$, $B_2(\mathbf{T}) = h_E(\mathbf{T}) - h_E(\mathbf{T}_0) + \varepsilon = 0$, and a part of boundary of R.

Our purpose is to find sufficient designs for the *PEI* such that the estimated $\hat{\mathbf{T}}$ must be in $\Omega_E(\mathbf{T}_0, \varepsilon)$ for any $\mathbf{T}_0 \in R$.

Experimental Design for the *PEI*. Let us consider a design D_0 of pumping test defined by (See Figure 8.2.2):

(1) One pumping well located at point W;

(2) Pumping rate $Q_D = 3000$ (m³/d);

(3) Pumping period $t_D = 10$ days;

(4) Four observation wells located at O_1, O_2, O_3 and O_4;

(5) Six observation times: $t_{obs} = 0.5, 1.0, 2.0, 4.0, 7.0, 10.0$ (days).

Suppose that the model structure error can be ignored and the observation error is always less than 0.1 (m). Due to the simplification of this example, optimization problem (8.2.29) can be solved directly. Its minimizer is $T_0 = (500, 1000)$, and $T = (597.5, 1000)$. Its minimum is $S_{D_0} = 0.142$. These results tell us: (1) $T_0 = (500, 1000)$ is the most difficult parameter to be identified in R; (2) design D_0 is insufficient for the *PEI*, because $S_{D_0} < 2\bar{\eta} = 0.2$.

Thus, we have to modify the design to obtain more information for parameter identification. We found that increasing the number of observation wells and observation frequencies does not significantly increase the value of S_D. To increase the value of S_D we can either increase the pumping period t_D or increase the pumping rate Q_D. For example, S_D will be larger than 0.2, when $Q_D = 3000$ (m³/d) and $t_D = 12$ days, or, $Q_D = 3800$ (m³/d) and $t_D = 10$ days.

The Optimal Pumping Test Design. The optimal location of pumping well is point W, that is, the same location of pumping well defined in the prediction problem. Only one observation well located adjacent to the pumping well is enough. For an optimal design, we only need to determine the optimal pumping period and pumping rate.

Figure 8.2.4 shows the set D_{AS} of all admissible and sufficient designs. Its boundary is composed of three parts: curve PQ satisfying $S_D = 0.2$, line segment PR expressing the maximum admissible pumping period, and line segment RQ expressing the maximum admissible pumping rate. The optimal design, D^*, which is determined by a cost criterion $J(D)$, should be on the curve PQ.

Parameter Identification and Verification. Assume that the optimal pumping test design D^* has been determined as $Q_D = 3800$ (m³/d) and $t_D = 10$ days. The head in the pumping well will be observed eight times:

$t_{obs} = 0.5, 1.0, 2.0, 3.0, 4.5, 6.0, 8.0, 10.0$ (days).

We can verify that this design is sufficient for any $T_0 \in R$. For example, let the true transmissivity be $T_0 \in R = (750, 1500)$. Using the simulation model to simulate the design, we can obtain head values of the pumping well at the designed

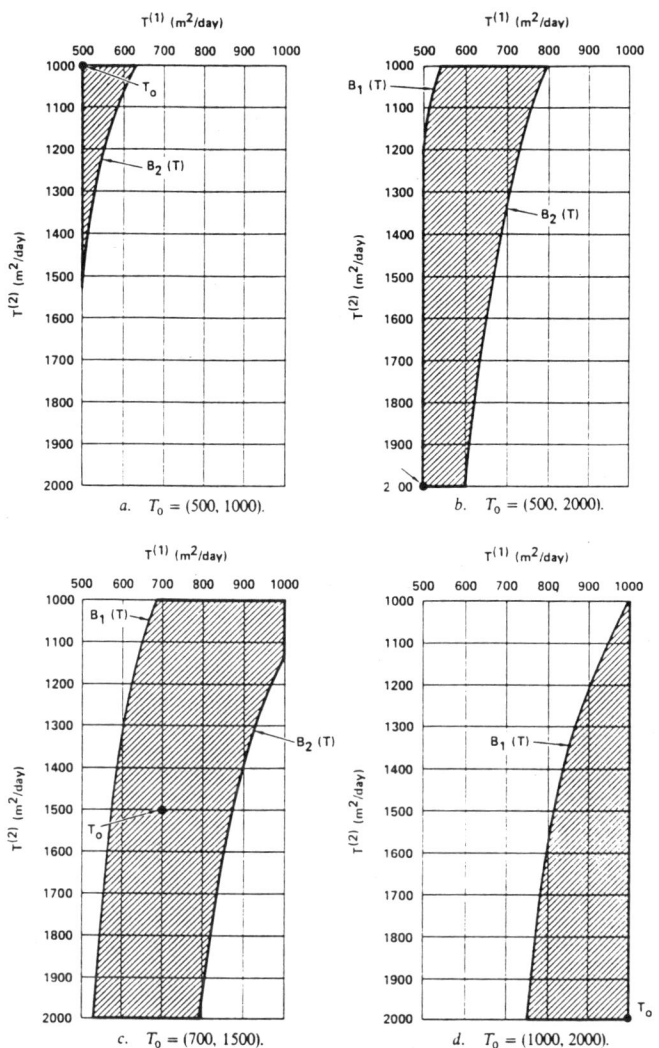

Fig. 8.2.3. The ε equivalent set of parameter T_0 for design X_0.

observation times. After adding observation error ($\overline{\eta} = 0.1$ m) to these values, a set of "observations" is obtained (Table 8.2.1). With these observations, the inverse solution is $\hat{\mathbf{T}} = (722, 1561)$. It is obviously different from the true parameter $\mathbf{T}_0$. However, using the prediction model we can find $h_E(\hat{\mathbf{T}}) = 78.68$ (m) and $h_E(\mathbf{T}_0) = 79.03$ (m). Thus

$$|h_E(\hat{\mathbf{T}}) - h_E(\mathbf{T}_0)| = 0.35 \text{ (m)} < 3 \text{(m)},$$

i.e., $\hat{\mathbf{T}}$ is indeed prediction equivalent with $\mathbf{T}_0$.

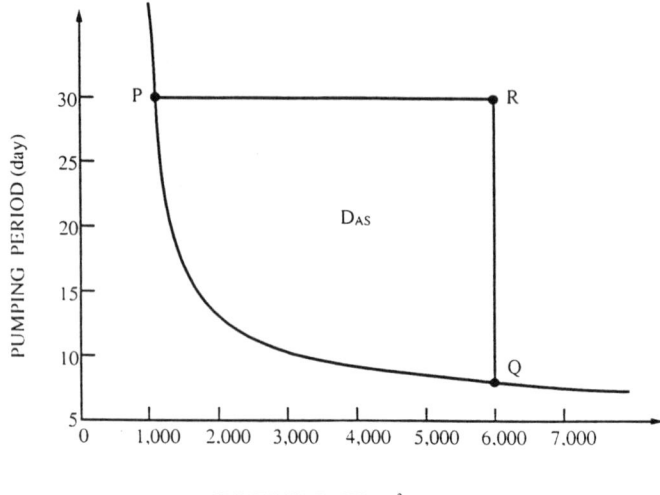

PUMPING RATE (m³/day)

Fig. 8.2.4.   Admissible set of pumping test design.

TABLE 8.2.1
Generated Observations

	Observed Heads at Well W		
t_{obs}	Noise $\bar{\eta} = 0$	Noise $\bar{\eta} = 10$ cm	Noise $\bar{\eta} = 30$ cm
0.5	98.46	98.56	98.66
1.0	97.90	97.80	97.60
2.0	97.07	97.17	97.37
3.0	96.39	96.49	96.69
4.5	95.56	95.46	95.26
6.0	94.88	94.98	94.58
8.0	94.19	94.29	94.49
10.0	93.66	93.56	93.36

Note that this design is quite conservative because it is required to be sufficient for all $T_0 \in R$. Therefore, for a particular parameter T_0, a prediction equivalent parameter can still be found for even a higher noise level. When $T_0 = (750, 1500)$, $\bar{\eta} = 0.3$ (m), the inverse solution is $\hat{T} = (636, 1676)$, and the corresponding prediction head is $h_E(\hat{T} = 77.26$ (m). Since $|h_E(\hat{T}) - h_E(T_0)| = 1.77$ (m) < 3 m, $\hat{T}$ is still a prediction equivalent parameter with T_0.

8.2.4 MANAGEMENT EQUIVALENCE IDENTIFIABILITY AND EXPERIMENTAL DESIGN

If the purpose of building a model is to find an optimal decision for a given management problem J rather than a prediction problem E, then it is only necessary to consider the error of the optimal decision variables $q_J(p)$ caused by the uncertainty of the identified parameter p. Analogous to the *PEI*, we define management equivalence identifiability (*MEI*) as follows:

For a parameter $p_0 \in P_{ad}$, if there exists a design D and a number $\delta > 0$, such that

$$\|u_D(p) - u_D(p_0)\|_{C_D} < \delta \text{ implies } \|q_J(p) - q_J(p_0)\|_{C_J} < 1 \qquad (8.2.32)$$

for all $p_0 \in P_{ad}$, then the parameter p_0 is said to be *δ-management equivalence identifiable* with respect to design D and weighting matrices C_J and C_D. Similarly, (8.2.32) is equivalent to condition

$$\|u_D(p) - u_D(p_0)\|_{C_D} \geq \delta > 0$$
$$\text{for all } \{p|p \in P_{ad} \text{ and } \|q_J(p) - q_J(p_0)\|_{C_J} \geq 1\}. \qquad (8.2.33)$$

If there exists observation error, $\delta \geq 2\overline{\eta}$ is required. The following set

$$\Omega_J(p_0, C_J) = \{p| p \in P_{ad} \text{ and } \|q_J(p) - q_J(p_0)\|_{C_J} < 1\} \qquad (8.2.34)$$

is named a *management equivalence set* of parameter p_0 with respect to weighting matrix C_J. To find an accurate expression for $\Omega_J(p_0, C_J)$ is very difficult. An approximated estimation of $\Omega_J(p_0, C_J)$, however, can be obtained by the Taylor series expansion. Replacing g_E by q_J in (8.2.27), we have

$$\|q_J(p) - q_J(p_0)\|_{C_J}^2 \approx \|p - p_0\|_{\overline{C}_P}^2, \qquad (8.2.35)$$

where

$$\overline{C}_P = J_J^T C_J^{-1} J_J \qquad (8.2.36)$$

and $J_J = [\partial q_J / \partial p]$ is the Jacobian of decision vector q_J with respect to parameter vector p. All elements of J_J are evaluated at p_0. Thus, the set $\Omega(p_0, C_J)$ is approximately replaced by the set $\{p| p \in P_{ad} \text{ and } \|p - p_0\|_{\overline{C}_p} < 1\}$, and the problem of *MEI* is approximately reduced to the problem of *INI*.

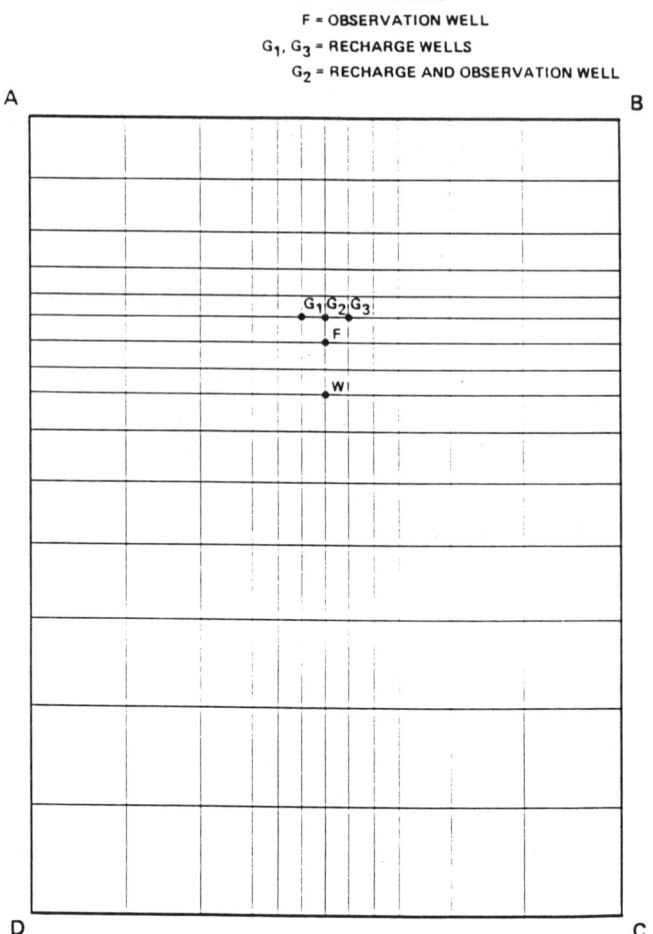

Fig. 8.2.5. The flow region and its discretization.

Example 8.2.2. (Sun and Yeh, 1990b). *A management problem of contaminated sources.*

A hypothetical aquifer is bounded between two rivers (see Figure 8.2.5). The head and concentration at boundary sections $\overline{AB}$ and $\overline{CD}$ (rivers) are given. $\overline{AC}$ and $\overline{BD}$ are no-flow boundary sections. Suppose that we want to use this aquifer as a natural dilution pool to dispose the wastewater. Three wastewater recharge wells are located at points G_1, G_2, and G_3 (nodes 71, 72, and 73). The recharge rate is 1200 m³/d (400 m³/d for each recharge well). The concentration of the recharged water is $C' = 10^3$ g/m³. A pumping well is located at point W (node 111). Because of pumping, the clean water flows from the rivers into the aquifer and dilutes the recharged wastewater.

A management problem is to determine the minimum pumping rate required so that the concentration of the mixed water does not exceed a given value $\overline{C} = 250$ g/m^3 anywhere in the aquifer. This problem can be formulated as follows:

$$q = \min W \tag{8.2.37}$$

subject to

$$C(x, y, t) \leq \overline{C},$$

$$(x, y) \in (\Omega), \quad t > 0,$$

where W is the pumping rate. The concentration distribution $C(x, y, t)$ is determined by the following coupled two-dimensional flow-mass transport equations:

$$S \frac{\partial h}{\partial t} = \nabla \cdot (Km\nabla h) - W + Q, \tag{8.2.38a}$$

$$\frac{\partial(\theta C)}{\partial t} = \nabla \cdot (D\theta \nabla C) - \nabla \cdot (V\theta C) - \frac{W}{m}C + \frac{Q}{m}C' \tag{8.2.38b}$$

with the following initial and boundary conditions:

$$h|_{t=0} = h_0, \quad C|_{t=0} = C_0, \quad h|_{\overline{AB}} = h_{b_1}, \quad h|_{\overline{CD}} = h_{b_2},$$

$$\frac{\partial h}{\partial n}\bigg|_{\overline{AC}} = 0, \quad \frac{\partial h}{\partial n}\bigg|_{\overline{BD}} = 0, \quad C|_{\overline{AB}} = C_{b_1}, \quad C|_{\overline{CD}} = C_{b_2},$$

$$\frac{\partial C}{\partial n}\bigg|_{\overline{AC}} = 0, \quad \frac{\partial C}{\partial n}\bigg|_{\overline{BD}} = 0,$$

where $h_0 = h_{b_1} = h_{b_2} = 100$ m, and $C_0 = C_{b_1} = C_{b_2} = 0$. Assume that the aquifer is homogeneous and isotropic and S, θ, and m are known: $S = 10^{-4}$, $\theta = 0.05$, and $m = 50$ m, but K, α_L and α_T are unknown. For simplicity, we further assume that $\alpha_T = \alpha_L/3$. Therefore, only two constants K and α_L need to be identified, that is, $\mathbf{p} = (K, \alpha_L)$. The prior information is represented by the admissible set $P_{ad} = \{\mathbf{p}|\, 10 \leq K \leq 40; 35 \leq \alpha_L \leq 110\}$.

Now, we can present a *MEI* problem. Suppose that the error of the management decision q in (8.2.37) is required to be less than 2000 m^3/d. If the true parameter is $\mathbf{p}_0$, then a parameter $\mathbf{p}$ which is management equivalent with $\mathbf{p}_0$ must satisfy

$$|q(\mathbf{p}) - q(\mathbf{p}_0)| < 2000. \tag{8.2.39}$$

TABLE 8.2.2
The distribution of function $q(K, \alpha_L)$ (dimension: m³/day)

α_L (m)	K (m/day)				
	10	20	30	40	50
30	21512	21500	21496	21495	21490
40	18115	18105	18101	18100	18098
60	13785	13776	13772	13765	13760
80	11185	11177	11172	11163	11156
100	9485	9478	9470	9464	9453
120	8317	8310	8302	8295	8283

Reduce to the Interval Identifiability. All parameters **p** satisfying (8.2.39) form the management equivalent set. According to the previous discussion, it can be approximately replaced by an interval in the parameter space. After solving a series of management problem (8.2.37) for different parameters, we obtained Table 8.2.2. From this table we found that q is sensitive to α_L but not to K.

Using the finite difference approximation to the data listed in Table 8.2.2, we can obtain the sensitivity coefficients $\partial q/\partial K$ and $\partial q/\partial \alpha_L$, as shown in Tables 8.2.3 and 8.2.4. Since we have

$$|q(\mathbf{p}) - q(\mathbf{p}_0)| \approx \left|\frac{\partial q}{\partial K}\Delta K + \frac{\partial q}{\partial \alpha_L}\Delta \alpha_L\right| \leq \left|\frac{\partial q}{\partial K}\right||\Delta K| + \left|\frac{\partial q}{\partial \alpha_L}\right||\Delta \alpha_L|,$$

(8.2.40)

and $|\partial q/\partial K| \leq 2$ for any $\mathbf{p}_0 \in P_{ad}$ (Table 8.2.3), and $|\Delta K| \leq 30$ is always less than 30, the *MEI* requirement (8.2.39) will be satisfied if

$$\left|\frac{\partial q}{\partial \alpha_L}\right||\Delta \alpha_L| < 2000 - 60 = 1940$$

or

$$|\alpha_L - \alpha_L^0| < \varepsilon(\alpha_L^0), \text{ where } \varepsilon(\alpha_L^0) = 1940\left|\frac{\partial q}{\partial \alpha_L}\right|_{\alpha_L^0}^{-1}.$$

(8.2.41)

Using Table 8.2.4, $\varepsilon(\alpha_L^0)$ can be obtained for different α_L^0, as shown in Table 8.2.5. The value of $\varepsilon(\alpha_L^0)$ increases when α_L^0 increases. Thus, the *MEI* problem is approximately reduced to an *INI* problem.

Experimental Design for the *MEI*. Let us consider an experimental design D_0 consisting of (See Figure 8.2.5):

TABLE 8.2.3
The distribution of partial derivative $\partial q/\partial K$

α_L	K			
	15	25	35	45
30	−1.2	−0.4	−0.1	−0.5
40	−1.0	−0.4	−0.1	−0.2
60	−0.9	−0.4	−0.7	−0.5
80	−0.8	−0.4	−0.9	−0.7
100	−0.7	−0.8	−0.6	−1.1
120	−0.7	−0.8	−0.7	−1.2

TABLE 8.2.4
The distribution of partial derivative $\partial q/\partial \alpha_L$

α_L	K				
	10	20	30	40	50
35	339.7	339.5	339.5	339.5	339.2
50	216.5	216.4	216.4	216.7	216.9
70	130.0	130.0	130.0	130.1	130.2
90	85.0	85.0	85.1	85.0	85.2
110	58.9	58.4	58.4	58.5	58.5

(1) A recharge well is located at point G_2 with recharge rate $Q_D = 150 \text{ m}^3/\text{day}$. The concentration of recharge water is $C_D' = 10^3 \text{ g/m}^3$;

(2) A pumping well is located at point W with pumping rate $W_D = 2000 \text{ m}^3/\text{day}$;

(3) Two observation wells located at points G_2 and F, with observation times: $t_{\text{obs}} = 1, 2, 3$ days. The total number of observation data is six.

For any fixed K, define function $\|u_D(p) - u_D(p_0)\|^2_{C_D}$ as

TABLE 8.2.5
The values of $\varepsilon(\alpha_L^0)$ and $\delta^2(\alpha_L^0)$

Function	Values of α_L^0				
	35	50	70	90	110
$\varepsilon(\alpha_L^0)$	5.7	8.9	14.8	22.8	32.9
$\delta^2(\alpha_L^0)$	288.8	220.3	241.5	269.6	269.4

$$S_D^2(\alpha_L, \alpha_L^0) = \frac{1}{6} \sum_{l=1}^{6} \left[C_l(\alpha_L) - C_l(\alpha_L^0) \right]^2. \tag{8.2.42}$$

Corresponding to different $\alpha_L^0 \in (35, 110)$, it is easy to obtain the values of

$$\delta = \min_{\alpha_L} S_D(\alpha_L, \alpha_L^0),$$

subject to

$$|\alpha_L - \alpha_L^0| \geq \varepsilon^0(\alpha_L^0).$$

The results are listed in Table 8.2.5. The smallest $\delta^2 = 220.3$ corresponds to $\alpha_L^0 = 50$. If the upper bound of observation error norm is $\bar{\eta} = 5$ g/m^3, then condition $\delta > 2\bar{\eta}$ is satisfied for all $\alpha_L^0 \in (35, 110)$. Hence, according to (8.2.33), design D_0 is a sufficient design for the *MEI*.

Note that the values of δ^2 in Table 8.2.5 are very close for different parameters. This fact shows an important advantage of the *MEI*, that the difficulty caused by nonlinearity may be avoided in the experimental design. Parameter $\alpha_L^0 = 110$ is the most difficult one to be identified, but at the same time, it is allowed to have the maximum estimate error from the point of view of *MEI*.

The sufficiency of the design D_0 can be also derived from the sufficient condition of *INI*. In this example, condition (8.2.19) reduces to

$$\frac{2 + \text{CTB}(C, K)(|\Delta K|/\varepsilon_K)}{\text{CTB}(C, \alpha_L)} < 1, \tag{8.2.43}$$

where

$$\text{CTB}(C, K)|\Delta K|/\varepsilon_K = \frac{1}{\bar{\eta}} \left| \frac{\partial C}{\partial K} \right| |\Delta K|,$$

$$\text{CTB}(C, \alpha_L) = \frac{1}{\bar{\eta}} \left| \frac{\partial C}{\partial \alpha_L} \right| \varepsilon(\alpha_L).$$

The problem of justifying the sufficiency of a design now is reduced to the calculation of $\partial C/\partial K$ and $\partial C/\partial \alpha_L$. Since the value of $\text{CTB}(C, K)$ is very small everywhere, $\text{CTB}(C, K)|\Delta K|/\varepsilon_K < 0.1$ is always correct. Thus, condition (8.2.43) can be replaced by $\text{CTB}(C, \alpha_L) > 2.1$. The value of $\text{CTB}(C, \alpha_L)$ depends on various design parameters, such as Q_D, W_D, t_D and C_D'. After completing a series of sensitivity analyses (see Sun and Yeh, 1990b), a design, D^*, may be obtained. It is the same as D^0, except decreasing W_D to 1000 (m^3/d), Q_D to 100 (m^3/d) and t_D to 1.2 days. D^* gives less information but is still sufficient for the *MEI*.

Parameter Identification and Verification. Let the true parameter values be $K^0 = 20$ (m/d), $\alpha_L^0 = 50$ (m). Using a numerical model to simulate experimental design D^*, we can obtain model output $u_D(p_0)$. "Observations" $\tilde{u}_D$ are then generated by adding observation error $\eta = (\eta_h, \eta_C)$ to $\tilde{u}_D$.

The estimated parameters $\hat{K}$ and $\hat{\alpha}_L$ are the solution of optimization problem

$$\min_{K, \alpha_L} \|u_D(K, \alpha_L) - \tilde{u}_D\|_{\tilde{C}_D}^2 = \frac{w_h}{3} \sum_{l=1}^{3} \left(\frac{h_l - \tilde{h}_l}{\sigma_h} \right)^2 + \frac{w_C}{3} \sum_{l=1}^{3} \left(\frac{C_l - \tilde{C}_l}{\sigma_C} \right)^2$$

(8.2.44a)

subject to

$$10 \le K \le 40, \qquad 35 \le \alpha_L \le 110. \tag{8.2.44b}$$

The Gauss–Newton method is used to solve (8.2.44), where $w_h = 1$, $w_c = 100$, $\sigma_h = 0.01$ and $\sigma_C = 1.0$, are used. The results are listed in Tables 8.2.6 to 8.2.8.

Table 8.2.6 shows that no matter if the noise is Gaussian or not, so long as $\eta_C \le 5$, the identified parameters K and α_L always satisfy $|K - 20| < 30$, $|\alpha_L - 50| < 8.9$, that is, they are management equivalent with the true parameters.

Table 8.2.7 shows that, regardless of the values of the initial estimates of the unknown parameters in the set P_{ad}, the identified parameters always satisfy the requirements $|K - 20| < 30$ and $|\alpha_L - 50| < 8.9$.

Table 8.2.8 shows that no matter what the true parameters are, the identified parameters are always management equivalent with the true parameters.

From Tables 8.2.6 to 8.2.8, we also find that the optimization procedure generally can be terminated after only one or two iteration runs, when the residual $\delta < \bar{\eta}$ is achieved. The instability problem that is often encountered in the use of the Gauss–Newton method is naturally avoided when the observation data are obtained from a sufficient design.

For a set of identified parameters, for example, $K = 27.58$ and $\alpha_L = 43.95$ in Table 8.2.5, we can obtain $q(K, \alpha_L) = 17,125$. Since $q(20, 50) = 15,650$, the requirement $|q(p) - q(p_0)| = 1475 < 2000$ is satisfied.

Exercise 8.2.1. Under what assumptions, weighting matrix $\tilde{C}_D$ can reduce to a diagonal matrix? Calculate $\bar{\eta}$ when the observation error for ith observation is η_i and $\tilde{C}_D$ is a diagonal matrix with diagonal elements σ_i^2 $(i = 1, 2, \ldots, L)$.

Exercise 8.2.2. When the number of prediction objectives is large than one, assign the weighting matrix C_E in the definition of *PEI*.

Exercise 8.2.3. Prove that when condition (8.2.25) is satisfied, the identified parameter $\hat{p}$ must be in $\Omega_E(p_0, C_E)$, provided that $\|u_D(\hat{p}) - \tilde{u}_D(p_0)\|_{\tilde{C}_D} < \bar{\eta}$.

TABLE 8.2.6
The results of parameter identification with different noise distributions

	Noise			Iteration 1			Iteration 2		
	η_h	η_C	$\bar{\eta}^2$	K	α_L	$\bar{\delta}^2$	K	α_L	$\bar{\delta}^2$
Truncation	0.05	5.0	2525	18.55	46.10	127.5*	23.03	48.16	2.8
$\tilde{\mathbf{u}}_D = \mathbf{u}_D + \eta$	0.05	5.0	2525	27.67	38.88	6617.4	27.58	43.95	228.4*
$\tilde{\mathbf{u}}_D = \mathbf{u}_D - \eta$	0.05	5.0	2525	10.0	58.28	2237.1*	14.33	55.59	233.4

<1> For this table, the true parameter values are $K = 20$ and $\alpha_L = 50$, and all initial estimates are $K = 30$ and $\alpha_L = 70$.

<2> * means the optimization procedure can be terminated at this iteration.

Exercise 8.2.4. What are the weighting matrices $\mathbf{C}_P$, $\tilde{\mathbf{C}}_D$ and $\mathbf{C}_J$ used in Example 8.2.2?

Exercise 8.2.5. Assume that the Dupuit's formula of steady flow in a confined-aquifer is applicable for predicting the drawdown in a pumping well ($r_w = 0.2$ m), but transmissivity T and the radius of influence R are unknown. The prediction problem is to find the drawdown S_w in the pumping well when pumping rate is $Q_w = 10^4$ m^3/day. The prediction error is required to be less than 3 m. The prior information shows that $200 \leq T \leq 500$ m^2/day, $500 \leq R \leq 1000$ m. Design an experiment for the *PEI* (The observation error is assumed to be less than 0.1 m).

8.3 Model Structure Identification

We have mentioned in Chapter 2 that model calibration should include two contents: adjusting model structure and identifying model parameters. So far, we mainly limit ourselves to the latter, that is, the solution of inverse problems, and assume that the model structure error can be ignored.

In practice, however, model structure errors often dominate measurement and computation errors. In most cases, model failure is caused by using an incorrect conceptual model. In Section 7.1.5, we have mentioned various structure errors that may occur in groundwater modeling. When the quantity and quality of observed data are insufficient, an incorrect conceptual model may fit these observations in a satisfactory extent, but give unacceptable prediction results.

Since the structure of a real aquifer is usually very complicated, various simplification assumptions are employed when building an applicable model. However, we should consider the following problem: what complexity level of the conceptual model is appropriate for a given aquifer? This problem, Unfortunately, is very

TABLE 8.2.7

The results of parameter identification with different initial estimates

Initial Estimation		Iteration 1			Iteration 2		
K	α_L	K	α_L	$\bar{\delta}^2$	K	α_L	$\bar{\delta}^2$
40	110	10	32	56240.5	16.07	46.02	992.1*
40	35	14.57	49.26	962.8*	20.28	47.42	212.0
10	35	16.07	46.39	770.4*	20.87	48.26	54.29
10	110	16.15	35	38934.1	21.45	46.30	904.7*

<1> For this table, the true parameter values are $K = 20$ and $\alpha_L = 50$, and the noise level is the same as that in Table 8.2.5.
<2> * means the optimization procedure can be terminated at this iteration.

TABLE 8.2.8

The results of parameter identification with different true parameter values

True Parameters		Noise			Initial Estimation		Iteration 1			Iteration 2		
K	α_L	η_h	η_C	$\bar{\eta}^2$	K	α_L	K	α_L	$\bar{\delta}^2$	K	α_L	$\bar{\delta}^2$
15	35	0.05	5.0	2525	20	60	10.00	44.96	2990.5	12.43	40.26	112.5*
35	50	0.05	5.0	2525	20	60	25.56	55.18	788.8*	26.30	57.26	99.8
15	70	0.05	5.0	2525	20	60	10.00	82.34	2932.9	12.43	79.22	118.3*
15	90	0.05	5.0	2525	20	60	15.31	81.86	139.8*	17.29	82.17	75.9
35	110	0.05	5.0	2525	20	60	25.98	94.24	5447.6	29.95	99.34	305.5*

<1> * means the optimization procedure can be terminated at this iteration.

difficult to answer. Up to date, only a few works are related to this topic in the field of groundwater modeling.

In this section, we will consider how to identify the parameter structure and how to estimate the model structure error. Two new conceptions which called *structure equivalence models for prediction (SEMP)* and *structure equivalence models for management (SEMM)* will be introduced. We will also consider how to design experiments for discriminating different model structures when they are not equivalent.

8.3.1 IDENTIFICATION OF PARAMETER STRUCTURE

An inherent difficulty associated with the zonation method of parameterization is the determination of zonation pattern. Although we may have some geologic and hydrogeological information, it is still difficult to divide the flow region into several homogeneous zones. We can not avoid the pattern problem by increasing the number of zones, because increasing the dimension of parameterization may seriously decrease the reliability of the identified parameters. (Shah *et al.*, 1978; Yeh and Yoon, 1981). We have given an explanation of overparameterization problem in Section 7.2.5. However, equation (7.2.29) does not give us an optimal dimension of zonation. Carrera and Neuman (1986a,c) recommended the following four criteria for determining the dimension of zonation and comparing different zonation patterns

$$\text{AIC} = S + 2M, \tag{8.3.1a}$$
$$\text{BIC} = S + M \ln L, \tag{8.3.1b}$$
$$\phi = S + 2M \ln(\ln L), \tag{8.3.1c}$$
$$d_M = S + M \ln(L/2\pi) + \ln |F_M|, \tag{8.3.1d}$$

where S is the value of log-likelihood function, for example, $\Phi_{MLN}(\theta)$ in (7.2.7), or $S(\beta)$ in (7.2.17); M the dimension of parameterization; L the number of measurement data; F_M is the Fisher information matrix. A zonation pattern is the optimal one if it minimizes a selected criteria. Increasing M may cause a decrease of S, but at the same time the second term of each of these criteria increases. When L is given, these criteria tend to select the simpler model in structure. Once M is selected, the best zonation pattern is determined completely by the minimization of S.

Sun and Yeh (1985) were the first to consider the identification of *parameter structure* for groundwater modeling. It is easy to prove that when the pattern of zonation is incorrect, the values of the identified parameters are also incorrect. In Sun and Yeh (1985), the problem of identifying the parameter structure is formulated as a problem of combinatorial optimization.

For a discrete model, a distributed parameter $p(\mathbf{x})$ is replaced first by a vector $\mathbf{p}_N = (p_1, p_2, \ldots, p_N)^T$, where N is the total number of nodes. To solve the inverse problem, $\mathbf{p}_N$ is replaced further by a vector $\mathbf{p}_M$ with lower dimension ($M \ll N$) through a parameterization method. The relationship between $\mathbf{p}_N$ and $\mathbf{p}_M$ can be written as

$$\mathbf{p}_N = \mathbf{G}\mathbf{p}_M, \tag{8.3.2}$$

where $N \times M$ matrix $\mathbf{G}$ is called the *structure matrix*. It depends on what parameterization is used. For example, when the zonation method is used, $\mathbf{G}$ is a diagonal matrix with elements either 1 or 0. The location of non-zero elements describes the pattern of zonation. When the finite element interpolation method is used, $\mathbf{G}$ depends on locations of nodes and basis functions.

Now, suppose that criterion (7.2.20) is used for identifying parameter $\mathbf{p}_N$, i.e.,

$$\begin{aligned} S(\mathbf{p}_N) &= [\mathbf{Z}_D - \mathbf{M}_D(\mathbf{p}_N)]^T [\mathbf{Z}_D - \mathbf{M}_D(\mathbf{p}_N)] \\ &\quad + \lambda(\mathbf{p}_N - \mathbf{p}_N^0)^T (\mathbf{p}_N - \mathbf{p}_N^0). \end{aligned} \tag{8.3.3}$$

Replacing nonlinear model output $\mathbf{M}_D(\mathbf{p}_N)$ by its first order approximation and substituting (8.3.2) into (8.3.3), we have

$$\begin{aligned} S(\mathbf{G}\mathbf{p}_M) &\approx (\mathbf{a}_0 - \mathbf{J}_D\mathbf{G}\mathbf{p}_M)^T (\mathbf{a}_0 - \mathbf{J}_D\mathbf{G}\mathbf{p}_M) \\ &\quad + \lambda(\mathbf{p}_M - \mathbf{p}_M^0)^T \mathbf{G}^T \mathbf{G}(\mathbf{p}_M - \mathbf{p}_M^0), \end{aligned} \tag{8.3.4}$$

where

$\mathbf{a}_0 = \mathbf{Z}_D - \mathbf{M}_D(\hat{\mathbf{p}}_M^0) + \mathbf{J}_D\hat{\mathbf{p}}_M^0$;
$\mathbf{p}_M^0$ – a guess of $\mathbf{p}_M$ based on prior information;
$\hat{\mathbf{p}}_M^0$ – the current value of $\mathbf{p}_M$ in the iteration;
$\mathbf{J}_D$ – $L \times N$ Jacobian matrix of observations with respect to $\mathbf{p}_N$. It can be calculated by the adjoint state method.

The updated solution is then given by

$$\hat{\mathbf{p}}_M = \left[\mathbf{G}^T\mathbf{J}_D^T\mathbf{J}_D\mathbf{G} + \lambda\mathbf{G}^T\mathbf{G}\right]^{-1} (\mathbf{G}^T\mathbf{J}_D^T\mathbf{a}_0 + \lambda\mathbf{G}^T\mathbf{G}\mathbf{p}_M^0). \tag{8.3.5}$$

The above equation clearly shows that $\hat{\mathbf{p}}_M$ depends on structure matrix $\mathbf{G}$. In other words, different parameter structures generate different parameter estimations. Thus, we can find a $\hat{\mathbf{G}}$, such that $S(\hat{\mathbf{G}}\hat{\mathbf{p}}_M)$ in (8.3.4) is the minimum, where $\hat{\mathbf{p}}_M$ is the optimal parameter associated with the optimal parameter structure $\hat{\mathbf{G}}$.

Using the distance weighting parameterization method given in Section 3.3.1, the elements of matrix $\mathbf{G} = [g_{im}]_{N \times M}$ are determined by

$$g_{im} = \omega_{i,m}\sigma_{i,m} / \sum_{m=1}^{M} \omega_{i,m}, \quad m \neq k, \tag{8.3.6a}$$

$$g_{ik} = 1 - \sum_{m \neq k} g_{im}, \tag{8.3.6b}$$

where $\omega_{i,m}$, $\sigma_{i,m}$ are defined in (3.3.12). Thus, matrix **G** is parameterized by α, β and coordinates of all basis points. We can systematically change α, β and locations of basis points to search the minimum of $S(\mathbf{G}\mathbf{p}_M)$. In Sun and Yeh (1985), all basis points are selected from nodes. Different structures are obtained by moving basis points from nodes to nodes. Since different combinations of basis points cause different combinations of coefficients in (8.3.4), the problem of identifying parameter structure then becomes a problem of combination optimization. According to the criteria given in (8.3.1), however, M can not be large, and thus, for a practical problem, it is possible to move the basis points one at a time to find the optimal parameter structure.

Example 8.3.1. (Sun and Yeh, 1985). Let us consider a confined aquifer shown in Figure 8.3.1. Section (ABC) of its boundary is a river, where the head is given as $H_B = 100$ m. The others are no-flow boundaries. There are two pumping wells W_1 and W_2 and five observation wells (O_1 to O_5) in this region. The pumping rates are $Q_{w_1} = Q_{w_2} = 10,000$ m^3/day. The pumping period is 5 days.

A finite element approximation with 169 nodes and 288 elements is used to solve the forward problem and its adjoint problem. These "fine" nodes are further combined into 25 "coarse" nodes for the solution of inverse problems, that is, $N = 25$. The purpose of using fine nodes is to improve the accuracy of the computed sensitivity matrix $\mathbf{J}_D$.

Case 1. Zonation. Assume that the true pattern of zonation is given in Figure 8.3.2. There are two zones with true transmissivity values $T_1^* = 1000$ m^2/day and $T_2^* = 500$ m^2/day, respectively. The observations are obtained by adding noises to the solution of the simulation model (Table 8.3.1). The probability distributions of observation noises are assumed to be independent of each other and normal with mean value $E = 0$ and deviation $\sigma = 0.5$ m.

The initial estimates include $M = 2$, two basis points located at nodes #7 and #18, and $T_1^0 = T_2^0 = 700$ m^2/day. The piecewise transmissivity can be described by taking $\alpha = 2.0$ and $\beta = 4.0$. Without noise, the true pattern of zonation is obtained after two iteration runs with estimated transmissivity values $\hat{T}_1 = 994$ m^2/day and $\hat{T}_2 = 499$ m^2/day that are almost identical to their true values. With noise, the true pattern of zonation can still be obtained after only two iterations, but the estimated transmissivity values will change to $\hat{T}_1 = 929$ m^2/day and $\hat{T}_2 = 559$ m^2/day.

In this case, if an incorrect patter of zonation is used for parameter identification, true values of transmissivity can never be found, no matter if with or without observation errors.

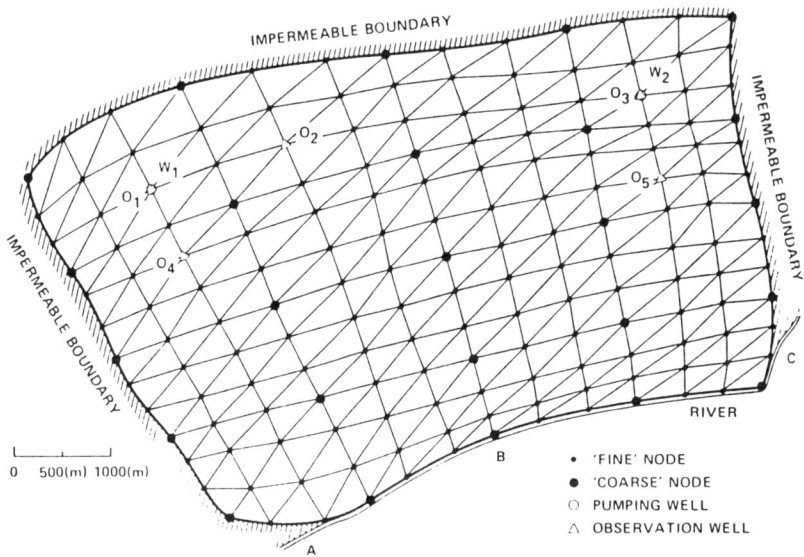

Fig. 8.3.1. Aquifer configuration and finite element discretization.

Case 2. Continuous Distribution. Suppose that the true transmissivity belongs to the continuous type. The true nodal values are shown in Figure 8.3.3. Corresponding observations and noises are given in Table 8.3.2. We tried to use different dimensions M of parameterization, from $M = 1$ to $M = 6$. For $M = 1$, assuming the aquifer is homogeneous, we obtain $\hat{T}_1 = 1035$ m^2/day. The results for $M = 2, 4, 6$ are given in Table 8.3.3. The optimal solution is associated with $M = 4$. Figure 8.3.2b shows the distribution of $\hat{T}_N$. It is much better than using any zonation method. From Table 8.3.3, we also find out that using $M = 6$ not only wastes computational efforts, but gets inferior results.

8.3.2 STRUCTURE EQUIVALENT MODELS AND MODEL STRUCTURE REDUCTION

In the last section, the optimal dimension of parameterization, or generally speaking, the complexity level of model structure, is determined completely by available observations, neither the physical structure of the real system nor model applications are considered. In fact, the physical structure and model application should be the primary factors of determining a model structure. In this section, we will consider how to find an appropriate complexity level of model structure based on model applications.

Let $\mathbf{M}_A$ and $\mathbf{M}_B$ be different model structures. Their parameters, $\mathbf{p}_A$ and $\mathbf{p}_B$, may belong to different parameter spaces. The admissible sets of $\mathbf{p}_A$ and $\mathbf{p}_B$ are

TABLE 8.3.1
True Heads, Noise, and Observations of Case 1

Observation Well	True Heads	Noise $(E = 0, \sigma = 0.5)$	Observations
O_1	90.38	−0.90	89.48
	86.51	0.01	86.52
	83.88	0.69	84.57
	81.99	−0.15	81.84
	80.64	0.09	80.73
O_2	96.00	0.87	96.87
	92.64	0.03	92.67
	90.15	0.91	91.06
	88.36	0.46	88.82
	87.06	−0.37	86.69
O_3	89.92	−0.87	89.05
	86.38	−0.45	85.93
	84.12	−0.71	83.41
	82.58	−0.26	82.32
	81.49	0.17	81.66
O_4	94.67	−0.69	93.98
	91.21	0.41	91.62
	88.75	0.19	88.94
	86.99	−0.32	86.67
	85.72	0.59	86.31
O_5	93.89	0.24	94.13
	90.64	0.19	90.83
	88.52	0.01	88.53
	87.07	−0.41	86.66
	86.05	0.26	86.31

denoted by $P_{A,\text{ad}}$ and $P_{B,\text{ad}}$, respectively. In addition, let D be an experimental design. For any $\mathbf{p}_A \in P_{A,\text{ad}}$, let $\hat{\mathbf{p}}_{BA} \in P_{B,\text{ad}}$ be the solution of the following optimization problem:

$$\min_{\mathbf{p}_B} \|\tilde{\mathbf{u}}_D(\mathbf{p}_A) - \mathbf{u}_D(\mathbf{p}_B)\|_{\tilde{\mathbf{C}}_D}, \quad \mathbf{p}_B \in P_{B,\text{ad}}, \tag{8.3.7}$$

where $\tilde{\mathbf{u}}_D(\mathbf{p}_A) = \mathbf{u}_D(\mathbf{p}_A) + \boldsymbol{\eta}$, $\mathbf{u}_D(\mathbf{p}_A)$ is the model output, $\boldsymbol{\eta}$ the observation error. Other notations are the same as that in Section 8.2. Equation (8.3.7) means that $\hat{\mathbf{p}}_{BA}$ is the best fitting parameter when M_B is used to match the "observations" of $M_A(\mathbf{p}_A)$.

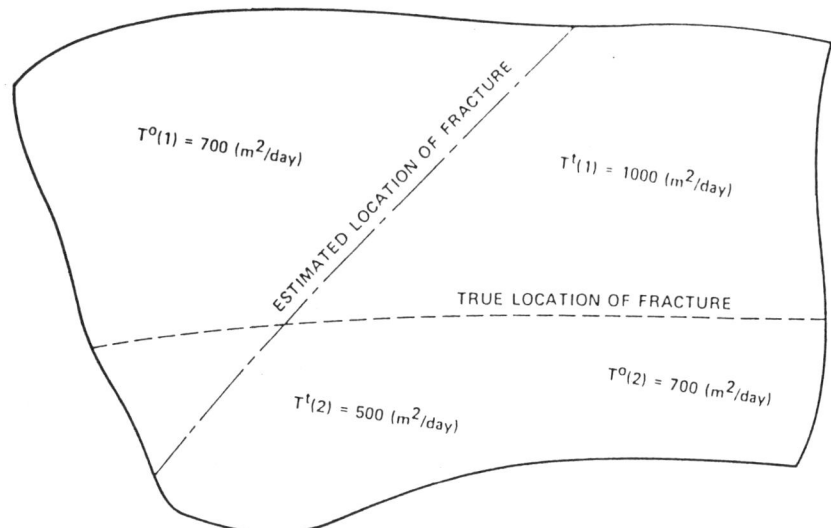

Fig. 8.3.2. The zonation pattern identification of Case 1.

Furthermore, assume that $\mathbf{g}_E$ is a prediction vector. If

$$\|\mathbf{g}_E(\mathbf{p}_A) - \mathbf{g}_E(\hat{\mathbf{p}}_{BA})\|_{\mathbf{C}_E} < 1, \tag{8.3.8}$$

then we say that model structure $\mathbf{M}_A$ can be replaced by model structure $\mathbf{M}_B$ for prediction $\mathbf{g}_E$. If $\mathbf{M}_B$ can also replaced by $\mathbf{M}_A$, then we say that structures $\mathbf{M}_A$ and $\mathbf{M}_B$ are *equivalent for prediction* $\mathbf{g}_E$.

To judge whether $\mathbf{M}_A$ can be replaced by $\mathbf{M}_B$, we introduce the following definition of *model structure error*. The minimum

$$E_{BA}(\mathbf{p}_A) = \min_{\mathbf{p}_B} \{\|\mathbf{u}_D(\mathbf{p}_A) - \mathbf{u}_D(\mathbf{p}_B)\|_{\mathbf{C}_D}^2 +$$
$$\lambda^2 \|\mathbf{g}_E(\mathbf{p}_A) - \mathbf{g}_E(\mathbf{p}_B)\|_{\mathbf{C}_E}^2\}^{1/2}, \quad \mathbf{p}_B \in P_{B,\text{ad}} \tag{8.3.9}$$

is called the structure error associated with parameter $\mathbf{p}_A$ when $\mathbf{M}_A$ is replaced by $\mathbf{M}_B$. Actually, $E_{BA}(\mathbf{p}_A)$ is the residual when $\mathbf{M}_B$ is used to fit both $\mathbf{u}_D(\mathbf{p}_A)$ and $\mathbf{g}_E(\mathbf{p}_A)$. In (8.3.9), λ is a weighting factor. The minimizer of (8.3.9) is denoted by $\mathbf{p}_{BA}$. Generally speaking, different structure errors may be obtained for different $\mathbf{p}_A \in P_{A,\text{ad}}$. The maximum structure error of using $\mathbf{M}_B$ to replace $\mathbf{M}_A$ is then defined as

$$\overline{E}_{BA} = \max_{\mathbf{p}_A} E_{BA}(\mathbf{p}_A), \quad \mathbf{p}_A \in P_{A,\text{ad}} . \tag{8.3.10}$$

Now we can propose the following sufficient condition for model sturcture replacement: if there exists a design D, such that

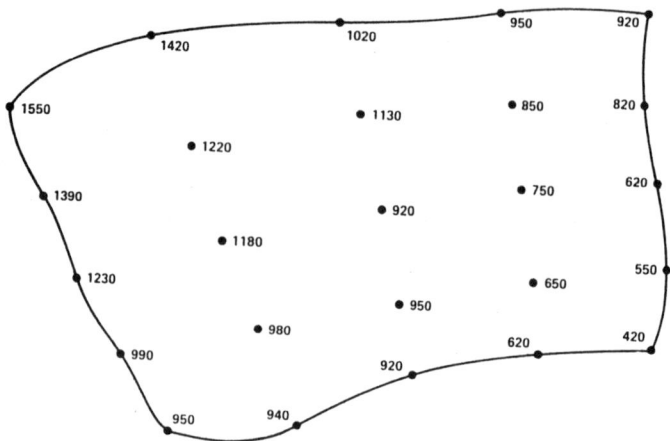

Fig. 8.3.3.   The true distribution of transmissivity for Case 2.

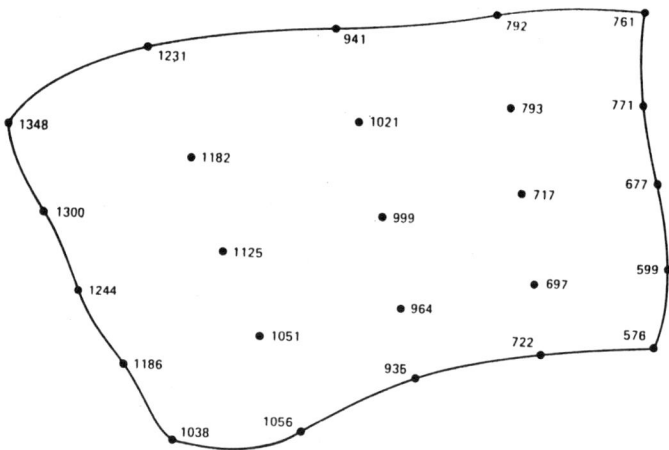

Fig. 8.3.4.   The identified transmissivity distribution for Case 2, where $M = 4$, $\alpha = 1.1$, $\beta = 1.0$.

$$\|\mathbf{u}_D(\mathbf{p}_{B,1}) - \mathbf{u}_D(\mathbf{p}_{B,2})\|_{\mathbf{C}_D} > 2(\overline{E}_{BA} + \overline{\eta}) \qquad (8.3.11)$$

stands for all $\mathbf{p}_{B,1} \in P_{B,\mathrm{ad}}$ and $\mathbf{p}_{B,2} \in P_{B,\mathrm{ad}}$ satisfying

$$\|\mathbf{g}_E(\mathbf{p}_{B,1}) - \mathbf{g}_E(\mathbf{p}_{B,2})\|_{\mathbf{C}_E} \geq 1 - \frac{1}{\lambda}\overline{E}_{BA}, \qquad (8.3.12)$$

then we conclude that $\mathbf{M}_A$ can be replaced by $\mathbf{M}_B$ for prediction $\mathbf{g}_E$.
    In fact, we have

$$\|\mathbf{g}_E(\mathbf{p}_A) - \mathbf{g}_E(\hat{\mathbf{p}}_{BA})\|_{\mathbf{C}_E} \leq \|\mathbf{g}_E(\mathbf{p}_A) - \mathbf{g}_E(\mathbf{p}_{BA})\|_{\mathbf{C}_E} +$$

TABLE 8.3.2
True Heads, Noise and Observations of Case 2

Observation Well	True Heads	Noise $(E = 0, \sigma = 0.5)$	Observations
O_1	93.13	−0.90	92.23
	90.39	0.01	90.40
	88.59	0.69	89.28
	87.36	−0.15	87.21
	86.53	0.09	86.62
O_2	96.83	0.87	97.70
	94.36	0.03	94.39
	92.64	0.91	93.55
	91.46	0.46	91.92
	90.66	−0.37	90.29
O_3	90.85	−0.87	89.98
	87.88	−0.45	87.43
	86.14	−0.71	85.43
	85.03	−0.26	84.77
	84.31	0.17	84.14
O_4	95.97	−0.69	95.28
	93.48	0.41	93.07
	91.8	0.19	91.61
	90.65	−0.32	90.23
	89.86	0.59	90.45
O_5	94.64	0.24	94.88
	91.95	0.19	92.14
	90.34	0.01	90.35
	89.31	−0.41	88.90
	88.64	0.26	89.90

$$+ \|g_E(\mathbf{p}_{BA}) - g_E(\hat{\mathbf{p}}_{BA})\|_{\mathbf{C}_E}. \qquad (8.3.13)$$

Definitions (8.3.9) and (8.3.10) yield

$$\|g_E(\mathbf{p}_A) - g_E(\mathbf{p}_{BA})\|_{\mathbf{C}_E} < \frac{1}{\lambda}\overline{E}_{BA}. \qquad (8.3.14)$$

On the other hand,

$$\|\mathbf{u}_D(\mathbf{p}_{BA}) - \mathbf{u}_D(\hat{\mathbf{p}}_{BA})\|_{\mathbf{C}_D} \leq \|\mathbf{u}_D(\mathbf{p}_{BA}) - \mathbf{u}_D(\mathbf{p}_A)\|_{\mathbf{C}_D}$$
$$+ \|\mathbf{u}_D(\mathbf{p}_A) - \tilde{\mathbf{u}}_D(\mathbf{p}_A)\|_{\mathbf{C}_D} +$$

TABLE 8.3.3

Results of Case 2

Parameter Dimension	Coefficients α and β	Initial Location of Basis Point (Node Number)	Estimated Value of Parameter (m²/day)	Optimum Location of Basis Point (Node Number)	Optimum Value of Parameter (m²/day)	$\|T_N^* - T_N^t\|$	Least Square Error E
$M = 2$	$\alpha = 1.0$	7	1200	3	1024	0.70×10^3	4.701
	$\beta = 4.0$	19	700	15	632		
$M = 4$	$\alpha = 1.1$	7	1200	7	1182		
	$\beta = 1.0$	9	850	4	792	0.53×10^3	4.604
		17	1000	23	935		
		19	650	24	722		
$M = 6$	$\alpha = 1.4$	6	1400	6	1301		
	$\beta = 1.0$	7	1200	1	592		
		9	850	10	1557	1.30×10^3	5.321
		17	1000	23	702		
		19	650	14	566		
		25	500	5	945		

$$+\|\tilde{u}_D(p_A) - u_D(\hat{p}_A)\|_{C_D}$$
$$+\|u_D(\hat{p}_A) - u_D(\hat{p}_{BA})\|_{C_D}$$
$$\leq 2(\overline{E}_{BA} + \overline{\eta}),\qquad(8.3.15)$$

where $\hat{p}_A \in P_{A,\mathrm{ad}}$ is the best parameter fitting observations $\tilde{u}_D(p_A)$. With (8.3.11) and (8.3.12), Equation (8.3.15) gives

$$\|g_E(p_{BA}) - g_E(\hat{p}_{BA})\|_{C_E} < 1 - \frac{1}{\lambda}\overline{E}_{BA}.\qquad(8.3.16)$$

Substituting (8.3.14) and (8.3.16) into (8.3.13), (8.3.8) is obtained.

Condition (8.3.11) means that if there is a design D which can provide sufficient information to overcome both model structure error and observation error, then it is possible to replace a model structure by another model structure. The sufficient condition defined by (8.3.11) and (8.3.12), actually, is a *PEI* condition for model structure M_B. When model structure error $\overline{E}_{BA} = 0$, (8.3.11) and (8.3.12) reduce to the *PEI* condition (8.2.25). Thus, the discussion given above extends the concept of *PEI* to the case that both model structure error and observation error are involved.

Generally, structure error $\overline{E}_{BA}$ of replacing M_A by M_B and structure error $\overline{E}_{AB}$ of replacing M_B by M_A are not equal. There are two cases:

· M_A and M_B are incomparable;

· M_B is a simplification of M_A.

For the former, to verify the equivalence of M_A and M_B, we have to calculate both $\overline{E}_{BA}$ and $\overline{E}_{AB}$. For the latter, however, since we always have $\overline{E}_{AB} = 0$, it is only required to calculate $\overline{E}_{BA}$.

When the dimension of p_A is large, it is difficult to obtain $\overline{E}_{BA}$ by solving the maximum–minimum problem (8.3.10). In practice, it is possible to guess one or several values of p_A from physics, with which the maximum structure error may be associated. If so, we only need to solve problem (8.3.9).

Structure equivalent models for decision making can be defined similarly.

8.3.3 EXPERIMENTAL DESIGN FOR MODEL DISCRIMINATION

When model structures M_A and M_B are not equivalent for prediction, and existing data are not enough for determining which one is better, we have to collect more information to discriminate them.

For example, there may exist two conceptual models for simulating a confined aquifer. One is based on inflow coming from a boundary section, while the other on leakage from an overlaying aquifer. We may design a pumping test to discriminate them. Another example is the mass transport in a fractured aquifer. The candidate models may include

· A double porosity model which considers the dispersion in both fractures and porous matrices.

· An equivalent porous medium model.

· A pure convection model which ignores the impact of dispersion.

A tracer test may tell us which model should be selected, if they are not equivalent for prediction.

In the field of groundwater modeling, discrimination problem was first considered by Knopman and Voss (1988). The objective of optimal design for model discrimination is to maximize the information of model structure based on certain criteria. A multiple objective design problem which involves model discrimination, parameter estimation and sampling cost was considered by Knopman and Voss (1989), and Knopman *et al.* (1991). Usunoff *et al.* (1992) presented an applicable approach that allows us to judge whether a design is sufficient for discriminating two or more given models.

Let us consider the case of discriminating two model structures $\mathbf{M}_A$ and $\mathbf{M}_B$. On the analogy of defining the model structure error, we can define the distance between $\mathbf{M}_A$ and $\mathbf{M}_B$ with respect to design D as follows:

$$d_D(\mathbf{M}_A, \mathbf{M}_B) = \min\{d_{DA}, d_{DB}\}, \tag{8.3.17a}$$

where

$$d_{DA} = \min_{\mathbf{p}_A} \min_{\mathbf{p}_B} \{\|\mathbf{u}_0 - \mathbf{u}_0(\mathbf{p}_B)\|_{\mathbf{C}_0}^2 + \lambda^2 \|\mathbf{u}_D(\mathbf{p}_A) - \mathbf{u}_D(\mathbf{p}_B)\|_{\mathbf{C}_D}^2\}^{1/2}, \tag{8.3.17b}$$

$$d_{DB} = \min_{\mathbf{p}_B} \min_{\mathbf{p}_A} \{\|\mathbf{u}_0 - \mathbf{u}_0(\mathbf{p}_A)\|_{\mathbf{C}_0}^2 + \lambda^2 \|\mathbf{u}_D(\mathbf{p}_A) - \mathbf{u}_D(\mathbf{p}_B)\|_{\mathbf{C}_D}^2\}^{1/2}. \tag{8.3.17c}$$

In (8.3.17), $\mathbf{u}_0$ represents the existing data, $\mathbf{C}_0$ is a weighting matrix defined for the observation space of existing data, λ a weighting factor. In (8.3.17), d_{DA} is the minimum residual when $\mathbf{M}_B$ is used to fit all existing data and the model output of $\mathbf{M}_A$, d_{DB} has the similar meaning. $d_D(\mathbf{M}_A, \mathbf{M}_B)$ measures the ability of discriminating $\mathbf{M}_A$ and $\mathbf{M}_B$ by existing data and the data that will be obtained after design D is realized.

The optimal design problem for model discrimination now can be stated as

$$\min_D \mathrm{Cost}(D) \tag{8.3.18a}$$

subject to

$$d_D(\mathbf{M}_A, \mathbf{M}_B) \geq \varepsilon, \tag{8.3.18b}$$

where ε is the lowest requirement of discrimination. Since $d_D(\mathbf{M}_A, \mathbf{M}_B)$ is difficult to calculate, (8.3.18) may not make much sense for practical problems.

Usunoff *et al.* (1992) concentrated on how to evaluate the discrimination capability of a given design, that is, to answer whether it will indeed discriminate among the alternative models once the experiment is performed. For simplification, Usunoff *et al.* (1992) assumed that model parameters can only take values from a discrete subset of the admissible set. Moreover, parameter sets leading to significantly different predictions of the experiment are treated as different conceptual models. With these assumptions, we only need to answer the following questions:

· Based on the experiment, can we reject $\mathbf{M}_B$ as wrong if $\mathbf{M}_A(\mathbf{p}_A)$ is the true model?

· Otherwise, can we reject $\mathbf{M}_A$ as wrong if $\mathbf{M}_B(\mathbf{p}_B)$ is the true model?

Instead of calculating the distance $d_D(\mathbf{M}_A, \mathbf{M}_B)$ in (8.3.17a), we only need calculate

$$d_D\left[\mathbf{M}_A(\mathbf{p}_A), \mathbf{M}_B(\mathbf{p}_B)\right] = \min\left[d_{DA}(\mathbf{p}_A), d_{DB}(\mathbf{p}_B)\right], \qquad (8.3.19)$$

that is, the distance between $\mathbf{M}_A(\mathbf{p}_A)$ and $\mathbf{M}_B(\mathbf{p}_B)$. In (8.3.19),

$$d_{DA}(\mathbf{p}_A) = \min_{\mathbf{p}_B}\left\{\|\mathbf{u}_0 - \mathbf{u}_0(\mathbf{p}_B)\|_{\mathbf{C}_0}^2 + \lambda\|\mathbf{u}_D(\mathbf{p}_A) - \mathbf{u}_D(\mathbf{p}_B)\|_{\mathbf{C}_D}^2\right\}^{1/2},$$
$$\qquad (8.3.20a)$$

$$d_{DB}(\mathbf{p}_B) = \min_{\mathbf{p}_A}\left\{\|\mathbf{u}_0 - \mathbf{u}_0(\mathbf{p}_A)\|_{\mathbf{C}_0}^2 + \lambda\|\mathbf{u}_D(\mathbf{p}_A) - \mathbf{u}_D(\mathbf{p}_B)\|_{\mathbf{C}_D}^2\right\}^{1/2}.$$
$$\qquad (8.3.20b)$$

Equations (8.3.20) are obtained from (8.3.17) by assuming that: in (8.3.17b), $\mathbf{p}_A$ is given, and in (8.3.17c), $\mathbf{p}_B$ is given. To calculate $d_{DA}(\mathbf{p}_A)$, we only need:

(1) Solving the forward problem with model $\mathbf{M}_A(\mathbf{p}_A)$ to obtain $\mathbf{u}_D(\mathbf{p}_A)$;

(2) Solving the inverse problem with model structure $\mathbf{M}_B$ to obtain a fitting parameter $\hat{\mathbf{p}}_B$, in which, existing data and $\mathbf{u}_A(\mathbf{p}_A)$ are used as "observations".

The value of $d_{DB}(\mathbf{p}_B)$ can be calculated simultaneously. When distance $d_D[\mathbf{M}_A(\mathbf{p}_A), \mathbf{M}_B(\mathbf{p}_B)]$ in (8.3.19) is larger than ε, the lowest limit of discrimination, then we conclude that design D is acceptable for discriminating $\mathbf{M}_A(\mathbf{p}_A)$ and $\mathbf{M}_B(\mathbf{p}_B)$. The value of ε depends on the norm of observation errors. Generally, we require that $\varepsilon \geq 2\bar{\eta}$.

The approach given above can be used to discriminate any discrepancy in model structure. An interesting application can be found in Usunoff *et al.* (1992).

8.3.4 A GEOLOGICAL PARAMETERIZATION APPROACH

Now let us consider how to make the model structure to be consistent with the real structure of an aquifer. A new kind of parameterization method, which is called the *geological parameterization approach*, is presented recently by Sun *et al.* (1994b). It allows us to incorporate all well-log data and other geological information into the solution of three-dimensional inverse problems in groundwater modeling. With this method, the real structure of an aquifer can be simulated in detail without increasing the dimension of parameterization.

For each well constructed, there may be a record of what geological materials were found in drilling the well and at what depth. This record is called a well-log. Along the vertical direction from groundsurface to bedrock, a well log can be described by the following two sets of numbers:

$$l_1, l_2, \ldots, l_K; \quad k_1, k_2, \ldots, k_K, \tag{8.3.21}$$

where $l_j(j = 1, 2, \ldots, K)$ represents the thickness of jth geological layer which is formed by geological material k_j. An example of well-logs is given in Table 8.3.4. Besides the use of well-logs, there are some other geophysical techniques, such as the gravity method, magnetic method and etc., that may help us to find more information about the geological structure. Let us consider how to use this kind of information directly in the identification of parameters.

Three-dimensional groundwater flow in a confined aquifer is governed by

$$\frac{\partial}{\partial x}\left(K_x \frac{\partial \phi}{\partial x}\right) + \frac{\partial}{\partial y}\left(K_y \frac{\partial h}{\partial y}\right) + \frac{\partial}{\partial z}\left(K_z \frac{\partial \phi}{\partial z}\right) = S_0 \frac{\partial \phi}{\partial t} + Q, \tag{8.3.22}$$

where $\phi(x, y, z, t)$ is the piezometric head (L); K_x, K_y and K_z are components of hydraulic conductivity tensor in principle directions (L/T). We will assume that $K_x = K_y = K_H$, and use K_V to represent K_z. In Equation (8.3.22) S_0 is the specific storativity ($1/L$), Q the sink/source term. To obtain a solution of (8.3.22), appropriate initial and boundary conditions must be supplemented.

Three-dimensional flow problems can be solved by various numerical methods. The mixed finite difference — multiple cell balance method (*FD-MCB*) uses two-dimensional MCB discretization in the horizontal direction and finite difference discretization in the vertical direction. The general form of (*FD-MCB*) method for solving coupled groundwater flow and contaminant transport problems can be found in Sun (1989a). In the FD-MCB method, the flow region is first divided into a number of horizontal layers. These layers are then subdivided into a number of triangular prisms so that the corresponding nodes in each layer is located along a single vertical line. The six vertices of each triangular prism are taken as nodes (Figure 8.3.5). Next, a secondary division is introduced to form exclusive subdomains for all nodes. As shown in Figure 8.3.6, the exclusive subdomain of node ($\mathbf{x}_n$) is a multiangular prism surrounding the node.

TABLE 8.3.4
An Example of Well-Logs

Depth (m)		Geological material
From	To	
0	24	top soil and sand
24	34.50	sandy clay
34.50	42	clay
42	45.50	sand
45.50	56.50	sandy clay
56.50	60	fine sand
60	78.50	sandy clay
78.50	89	fine sand
89	94	coarse sand and gravel
94	99.50	fine sand
99.50	104.50	coarse sand
104.50	110	blue clay
110	121	gravel
121	131	sand
131	143	clay streaks w/sand
143	154	sand
154	164.50	sandy clay
164.50	175.50	packed fine sand
175.50	186.50	sand
186.50	198	sand and clay
198	203.50	coarse sand
203.50	209.50	fine packed sand
209.50	218	coarse sand
218	220	hard packed sand and clay

FD-MCB discretization equations can be derived directly from the local mass balance over exclusive subdomains of all nodes and time steps.

The first step of the geological parameterization method is to estimate the local geological structure in each exclusive subdomain. There are two approaches that can serve this purpose:

· If the considered aquifer has continuous layer structure, we can use the computer contouring techniques developed in mathematical geology (Davis, 1986), including Kriging and Kalman filtering (Rajaram and Mclaughlin, 1990), to estimate the elevations of formation surfaces for each geological

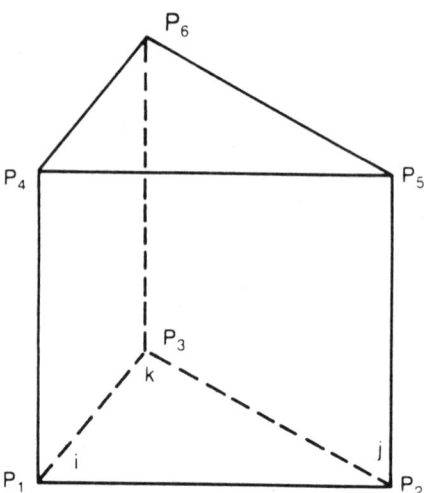

Fig. 8.3.5.   Three-dimensional triangular prism element.

layer, that is, to find the global structure of the aquifer. Then, the local
structure associated with each exclusive subdomain can be determined. If the
aquifer contains discontinuities such as faults, inner boundaries defined by
faults should be considered.

· Directly estimate the thickness of each geological material in each exclusive
  subdomain by an interpolation/extrapolation method. For example, we can
  use the two-dimensional moving averages method (Davis, 1986) or Kriging
  to complete the estimate for all exclusive subdomains in the same horizontal
  layer.

In any approach mentioned above, well logs are used as control points of estimation. Other geological information, including the judgment of an experienced
geologist, may be incorporated into the estimation process through introducing
some estimated well logs. In Sun *et al.* (1994), the Kriging method discussed in
Section 7.3 is used to obtain the local structure estimate for all exclusive subdomains, i.e.,

$$l_j(\mathbf{x}_n), \quad n = 1, 2, \ldots, N; \; j = 1, 2, \ldots, J, \tag{8.3.23}$$

where N is the total number of nodes, J the total number of materials.

The second step of the geological parameterization method is to generate a
parameterization algorithm. Let

$$K_1, K_2, \ldots, K_J; \quad S_1, S_2, \ldots, S_J \tag{8.3.24}$$

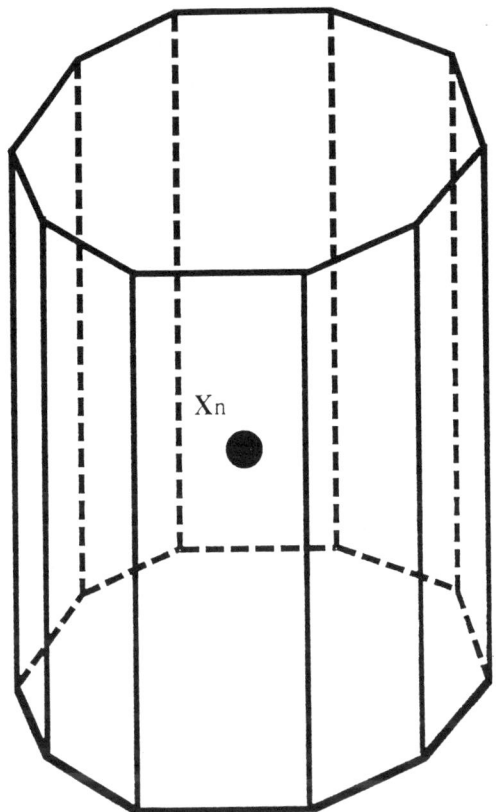

Fig. 8.3.6. Exclusive subdomain of a node $\mathbf{x}_n$.

be values of hydraulic conductivity and specific storativity associated with J kinds of geological materials, respectively. The distributed parameters K_H, K_V and S_0 then can be obtained for each node $\mathbf{x}_n$

$$K_H(\mathbf{x}_n) = \sum_{j=1}^{J} [l_j(\mathbf{x}_n)K_j] / l_n, \qquad (8.3.25)$$

$$K_V(\mathbf{x}_n) = l_n / \sum_{j=1}^{J} [l_j(\mathbf{x}_n)/K_j], \qquad (8.3.26)$$

$$S_0(\mathbf{x}_n) = \sum_{j=1}^{J} [l_j(\mathbf{x}_n)S_j] / l_n, \qquad (8.3.27)$$

where $l_n = \sum_{j=1}^{J} l_j(\mathbf{x}_n)$ is the total height of the exclusive subdomain of $\mathbf{x}_n$. These expressions can be easily derived from Darcy's Law. Thus, distributed parameters K_H, K_V and S_0 are represented by $K_1, K_2, \ldots, K_J$ and $S_1, S_2, \ldots, S_J$ in the whole aquifer. The duty of parameterization is completed.

To find $K_1, K_2, \ldots, K_J$ and $S_1, S_2, \ldots, S_J$, we can solve the inverse problem with head observations. Since these parameters have certain physical meaning, two types of constraints may be available:

· Range constraints

$$K_{j,\min} < K_j < K_{j,\max}, \quad (j = 1, 2, \ldots, J). \tag{8.3.28}$$

· Order constraints

$$K_{j_1} < K_{j_2} < \cdots < K_{j_J}; \quad S_{i_1} < S_{i_2} < \cdots < S_{i_J}, \tag{8.3.29}$$

where $(j_1, j_2, \ldots, j_J)$ and $(i_1, i_2, \ldots, i_J)$ are different arranges of $(1, 2, \ldots, J)$. These constraints may help to improve the ill-posedness of inverse solution.

Numerical examples and a case study of using the proposed geological parameterization method can be found in Sun *et al.* (1994).

Conclusion

We have introduced basic concepts, theories and methods of inverse problems in groundwater modeling. Both deterministic and stochastic approaches were considered. The basic problem that runs through the whole book is how to overcome or avoid the ill-posedness of inverse solutions. The fundamental way out is to gain sufficient observations, both in quantity and quality, based on predetermined criteria and model applications. It is also important to incorporate all existing geological and hydrogeological information into the inverse solution procedure.

Before ending this book, we would like to suggest a general procedure of constructing mathematical models for groundwater modeling. This procedure is something different from the traditional approach.

Step 1. Collect existing geological and hydrogeological information.

Step 2. Present conceptual models based on the collected data. These conceptual models may have different dimensions, structures and boundary conditions.

Step 3. Calibrate conceptual models by existing data. Some of them may be rejected during the calibration procedure. Note that any model M_B with parameters p_B should not be rejected when it satisfies

$$\|u[M_A(p_A)] - u[M_B(p_B)]\|_{U_D} \leq \varepsilon + \eta,$$

where model M_A with parameters p_A is the calibration result, u the model output, $\|\cdot\|_{U_D}$ the norm defined in the observation space. On the right-hand side of the above equation, $\varepsilon = \|u[M_A(p_A)] - u^{obs}\|_{U_D}$ is the norm of residuals, η the norm of observation errors. Thus, we may have a set of models (models with the same structure but different parameter values are considered as different models) such that any one in the set should not be rejected by existing data and prior information.

Step 4. Present a prediction (or management) model as well as its accuracy requirement based on the projected model applications.

Step 5. Let g be the solution of the prediction (or management) problem. Since g depends on model structure and model parameters, different values of g may be obtained using different simulation models. The distribution and uncertainty of g, however, can be estimated by the model set defined in Step 3. Usually, we can transfer the prediction (or management) uncertainty into a cost of prediction (or management) risk. If the risk cost is unacceptable, we have to gain more data to decrease the "size" of the model set.

Step 6. Simplify the model structure of conceptual models and estimate model structure errors based on both observation data and prediction requirements.

Step 7. Design experiments or data collection strategy. The objective of experimental design is to decrease the uncertainty of model applications, while the cost of experiment is the minimum. Usually, a compromise between the cost of the experiment and the cost of prediction (or management) risk is necessary.

Step 8. Collect new data in the field according to the designed data collection strategy.

Step 9. Re-calibrate the selected conceptual models (their structures may be simplified) with both original and new observation data. Since insufficient information may be used in the design stage (Step 7), a reliable model may not be obtained in only one circle. Generally, we have to repeat steps 3, 4, 5. If the cost of prediction (or management) risk is still too high, steps 6, 7, 8, 9 should be repeated.

It is obvious that the presented procedure is not easy to be accomplished. There are difficulties associated with each step of this procedure. To obtain a reliable simulation model for a real aquifer is still a challenging problem in groundwater modeling. We expect to see more works on the following research directions:

(1) Theoretical studies on the ill-posedness of inverse problems, especially, to find sufficient conditions for various extended identifiabilities.

(2) Directly incorporating geological information and other prior information into the formulation of inverse problem to increase the stability of inverse solutions.

(3) To find highly efficient numerical methods for solving large scale inverse problems.

(4) Theoretical studies and practical applications of computer aid experimental design in both deterministic and stochastic frameworks, especially, to develop a systematical methodology that can combine experimental design with model applications.

(5) Theoretical studies and practical applications of the calibration of three-dimensional models.

(6) The estimate of model structure errors and determination of a appropriate complexity level of model structure for aquifer simulation.

(7) Model reliability evaluations from the point of view of statistics, especially, to estimate the confidence intervals of model predictions when the model structural error is also considered.

(8) Model verification and model validation (Tsang, 1991).

(9) Some special problems. For example, the inverse problems of flow and mass transport in fractured aquifers and in unsaturated zones, the inverse problem of multiphase transport and so forth.

Mapping, Space and Norm

A.1. Mapping and Its Inverse

We have represented the subroutine of solving forward problems in a general form in section 1.2.3., i.e., $u = M(p)$, where model parameter p and model state u are both vector functions, and routine M transforms each admissible p into a u. In fact, M is a special case of *mapping* or *transformation* that has been well defined in mathematics. In what follows we will continually use the same symbols: u, M, p, but in a general sense.

Let P and U be arbitrary sets, and P_{ad} a subset of P. A "rule", M, defined on P_{ad} that associates each element (*preimage*) $p \in P_{ad}$ with a unique element (*image*) $u \in U$ is termed a *mapping from P_{ad} to U* and the corresponding relationship is denoted by $u = M(p)$. P_{ad} is named the *definition domain* of M. All images form a set U_M which is a subset of U and named the range of M. Obviously, mapping is an extension of the concept of a function. Preimages and images are no longer limited only as numbers. They may be points, vectors, functions, vector functions, data files, pictures and other things.

In the above definition, the uniqueness of a mapping is required, which means that each preimage can have only one image. The inverse statement, however, may not be true. An image may have multiple preimages. In Chapter 2 we have seen that one model parameter (preimage) only generates one model state (image), but different model parameters may be generated from a same model state.

If for each image u in U_M, there exists a unique preimage p in P_A, where P_A is a subset of P_{ad}, then a new mapping from U_M to P_A can be defined which is named an *inverse of mapping M*. According to this definition, a mapping may have more than one inverse mappings.

If the correspondence between elements of P_{ad} and U_M is *one-to-one*, then mapping M is said to be *injective*. In this case, the unique inverse of mapping M may be denoted by M^{-1}:

$$p = M^{-1}(u); \quad u \in U_M, \quad p \in P_{ad}. \tag{A.1}$$

Example A.1. A linear transformation from n-dimensional Euclidean space R^n to m-dimensional Euclidean space R^m may be represented as

$$\mathbf{y} = \mathbf{A}\mathbf{x}; \qquad \mathbf{y} \in R^m, \quad \mathbf{x} \in R^n, \tag{A.2}$$

where $\mathbf{A}$ is a $m \times n$ matrix.

When $m > n$, there does not exist any inverse for transformation $\mathbf{A}$. When $m < n$, the inverse matrix of each $m \times m$ nonsingular submatrix of $\mathbf{A}$ is an inverse of it. When $m = n$, the inverse matrix $\mathbf{A}^{-1}$ is the unique inverse of the transformation, $\mathbf{x} = \mathbf{A}^{-1}\mathbf{y}$, provided that matrix $\mathbf{A}$ is nonsingular.

A.2. Linear Vector Space and Norm

Let P be a set, in which, *addition* and *scalar multiplication* are defined ($\mathbf{p}_1 + \mathbf{p}_2 \in P$ for any $\mathbf{p}_1 \in P$ and $\mathbf{p}_2 \in P$; and $\lambda\mathbf{p} \in P$ for any real number λ and $\mathbf{p} \in P$). If the following conditions are verified:

$$\mathbf{p}_1 + \mathbf{p}_2 = \mathbf{p}_2 + \mathbf{p}_1, \tag{A.3a}$$

$$(\mathbf{p}_1 + \mathbf{p}_2) + \mathbf{p}_3 = \mathbf{p}_1 + (\mathbf{p}_2 + \mathbf{p}_3), \tag{A.3b}$$

$$\lambda(\mathbf{p}_1 + \mathbf{p}_2) = \lambda\mathbf{p}_1 + \lambda\mathbf{p}_2, \tag{A.3c}$$

$$(\lambda + \mu)\mathbf{p} = \lambda\mathbf{p} + \mu\mathbf{p}, \tag{A.3d}$$

$$1\mathbf{p} = \mathbf{p}, \qquad 0\mathbf{p} = \mathbf{0}, \tag{A.3e}$$

then P is called a *linear vector space* or a *linear space*.

Example A.2. The n-dimensional Euclidean space R^n is a linear vector space.

Example A.3. Let F be the set of all functions defined on a three-dimensional region (Ω), i.e.,

$$F = \{f(x, y, z)| \text{ all functions defined on } (\Omega)\}, \tag{A.4}$$

then F is a linear vector space. Its subspace

$$C = \{f(x, y, z)| \text{all continuous functions on } (\Omega)\} \tag{A.5}$$

is also a linear vector space. Each function is an element or a 'point' of the space.

We have known that the introducing a coordinate system and using geometric terms, such as 'point', 'line', 'distance' and so on, are very convenient in the study of functions. Similar approaches has been developed for studying transformations defined on linear vector spaces.

Let P be a linear vector space. If a mapping, $\|\mathbf{p}\|$, from P to R (the real number space) is defined and the following conditions are verified

$$\|\mathbf{p}\| > 0, \quad \text{if } \mathbf{p} \neq 0, \tag{A.6a}$$

$$\|\lambda\mathbf{p}\| = |\lambda|\,\|\mathbf{p}\|, \tag{A.6b}$$

$$\|\mathbf{p}_1 + \mathbf{p}_2\| \leq \|\mathbf{p}_1\| + \|\mathbf{p}_2\| \tag{A.6c}$$

then P is called a *normed linear space* and real number $\|\mathbf{p}\|$ the *norm* of vector $\mathbf{p}$.

Example A.4. In n-dimensional Euclidean space R^n, the "length" of a vector $\mathbf{x} = (x_1, x_2, \ldots, x_n)$,

$$\|\mathbf{x}\| = \left(\sum_{i=1}^{n} |x_i|^2\right)^{1/2} \tag{A.7}$$

is a norm.

Example A.5. For the continuous function space C, we can define

$$\|f\|_{L_2} = \left[\int_{(\Omega)} f^2(x, y, z)\,\mathrm{d}\Omega\right]^{1/2} \tag{A.8}$$

which is called the *least squares norm*.

Let P be a linear vector space, if a real number $d(\mathbf{p}_1, \mathbf{p}_2)$ is defined for any two elements $\mathbf{p}_1$ and $\mathbf{p}_2$ in the space, and the following conditions are satisfied

$$d(\mathbf{p}_1, \mathbf{p}_2) > 0 \quad \text{if} \quad \mathbf{p}_1 \neq \mathbf{p}_2, \tag{A.9a}$$

$$d(\mathbf{p}_1, \mathbf{p}_2) = d(\mathbf{p}_2, \mathbf{p}_1), \tag{A.9b}$$

$$d(\mathbf{p}_1, \mathbf{p}_3) \leq d(\mathbf{p}_1, \mathbf{p}_2) + d(\mathbf{p}_2, \mathbf{p}_3) \tag{A.9c}$$

then $d(\mathbf{p}_1, \mathbf{p}_2)$ is called the '*distance*' between points $\mathbf{p}_1$ and $\mathbf{p}_2$, and P a *metric space*.

Obviously, a normed linear space P must be a metric space, because we can define

$$d(\mathbf{p}_1, \mathbf{p}_2) = \|\mathbf{p}_1 - \mathbf{p}_2\|. \tag{A.10}$$

The concept of '*limit*' now can be extended to metric spaces. Let

$$\{\mathbf{p}_n\}: \quad \mathbf{p}_1, \mathbf{p}_2, \ldots \mathbf{p}_n, \ldots \tag{A.11}$$

be a sequence in a metric space P. If for any $\varepsilon > 0$, there exists an integer n_0, such that $n \geq n_0$ and $m \geq n_0$ imply $d(\mathbf{p}_n, \mathbf{p}_m) < \varepsilon$, then $\{\mathbf{p}_n\}$ is called a *cauchy sequence*.

If for each cauchy sequence $\{p_n\}$ in P there exists a $p \in P$, such that

$$d(p_n, p) \rightarrow 0, \quad \text{when} \quad n \rightarrow \infty \tag{A.12}$$

i.e., each cauchy sequence in P converges to a limit in P, then P is called a *complete metric space*. A complete normed linear space is named a *Banach space*.

Example A.6. The n-dimensional Euclidean space R^n is a Banach space.

Example A.7. The continuous function space C defined in (A.5) is incomplete, because a sequence of continuous functions may converge to a discontinuous function.

Suppose that P and U are both Banach spaces. A mapping M is said to be continuous at a point $p_0 \in P_{ad}$, if $M(p_n) \rightarrow M(p_0)$ in U for any sequence $p_n \rightarrow p_0$ in P_{ad}.

A.3. Inner Product Space and Orthogonal Expansions

Let P be a linear space. If a real number denoted by (p_1, p_2) is defined for any two elements p_1 and p_2 in the space, and the following conditions are satisfied

$$(p, p) > 0, \quad \text{if} \quad p \neq 0, \tag{A.13a}$$
$$(\lambda p_1, p_2) = \lambda(p_1, p_2), \tag{A.13b}$$
$$(p_1 + p_2, p_3) = (p_1, p_3) + (p_2, p_3), \tag{A.13c}$$

then (p_1, p_2) is called the inner product of elements p_1 and p_2, and P an inner product space.

Example A.8. For any two points $x = (x_1, x_2, \ldots, x_n)$, $y = (y_1, y_2, \ldots, y_n)$ in n-dimensional Euclidean space, their inner product may be defined as

$$(x, y) = \sum_{i=1}^{n} x_i y_i. \tag{A.14}$$

Example A.9. For any two functions f and g in continuous function space C, the inner product may be defined as

$$(f, g) = \int_{(\Omega)} fg \, d\Omega. \tag{A.15}$$

The norm of a vector in an inner product space can be defined as

$$\|\mathbf{p}\| = (\mathbf{p}, \mathbf{p})^{1/2}. \tag{A.16}$$

Therefore, an inner product space must be a metric space. A complete inner product space is named a *Hilbert space*.

Now let us construct 'coordinate systems' for Hilbert spaces. Vectors e_1 and e_2 in a Hilbert space P are called *orthogonal* if $(e_1, e_2) = 0$. A set of elements e_α is called an *orthonormal system* if any two members of the system are orthogonal and the norms of all members are equal to one unit, i.e.,

$$(\mathbf{e}_\alpha, \mathbf{e}_\beta) = \begin{cases} 0, & \text{if } \alpha \neq \beta, \\ 1, & \text{if } \alpha = \beta. \end{cases} \tag{A.17}$$

If there is no such an element $e \in P$ (except $e = 0$) that it is not a member of $\{e_\alpha\}$ but orthogonal to all members of $\{e_\alpha\}$, then $\{e_\alpha\}$ is called a *complete orthonormal system*.

When there is a complete orthonormal system in P, for which, the number of members is a finite number n, P is called a n-dimensional space. When the system is a sequence $\{e_i\}$, $i = 1, 2, \ldots, n, \ldots$, P is called a *separable Hilbert space* and $\{e_i\}$ an *orthonormal basis* of the space. It has been proved that any vector $\mathbf{p}$ in a separable Hilbert space can be represented as

$$\mathbf{p} = \sum_{i=1}^{\infty} \lambda_i \mathbf{e}_i, \tag{A.18}$$

where $\lambda_i = (\mathbf{p}, \mathbf{e}_i)$ is called the e_i-coordinate of $\mathbf{p}$. (A.18) is an extension of $\mathbf{p} = \sum_{i=1}^{n} \lambda_i \mathbf{e}_i$ in the n-dimensional space.

A.4. Some Frequently Used Spaces

In the practical study of groundwater modeling, we only need to deal with numerical models, where the model parameters (preimage) and state variables (image) are both in finite-dimensional spaces. However, in the theoretical and methodological studies of the inverse problem, the use of functional spaces may be more convenient. Some frequently used spaces will be discussed below.

(1) *n-dimensional vector space R^n*
Its generic element $\mathbf{p}$ is represented by a real array $(p_1, p_2, \ldots p_n)^T$. The conventional l_2-norm is defined as

$$\|\mathbf{p}\| = \left[\sum_{i=1}^{n} p_i^2 \right]^{1/2}. \tag{A.19}$$

However, in the study of inverse problems, we prefer to use the generalized l_2-norm:

$$\|\mathbf{p}\|_{l_2} = [\mathbf{p}^T \mathbf{W} \mathbf{p}]^{1/2}, \tag{A.20}$$

where $\mathbf{W}$ is an $n \times n$ positive-defined symmetric matrix. When it is a diagonal matrix

$$\mathbf{W} = \begin{bmatrix} w_1^2 & & & \\ & w_2^2 & & 0 \\ & & \ddots & \\ 0 & & & w_n^2 \end{bmatrix}, \tag{A.21}$$

(A.20) may reduce to

$$\|\mathbf{p}\|_{l_2} = \left[\sum_{i=1}^{n} w_i^2 p_i^2 \right]^{1/2}. \tag{A.22}$$

The l_p-norm $(1 \leq p \leq \infty)$ is defined as

$$\|\mathbf{p}\|_{l_p} = \left[\sum_{i=1}^{n} |w_i p_i|^p \right]^{1/p}. \tag{A.23}$$

Especially, we have l_1-norm

$$\|\mathbf{p}\|_{l_1} = \sum_{i=1}^{n} |w_i p_i| \tag{A.24}$$

and l_∞-norm

$$\|\mathbf{p}\|_{l_\infty} = \max_{1 \leq i \leq n} |w_i p_i|. \tag{A.25}$$

The inner product of two vectors $\mathbf{p}$ and $\mathbf{q}$ may be defined as

$$(\mathbf{p}, \mathbf{q}) = \sum_{i=1}^{n} (w_i p_i)(w_i q_i), \tag{A.26}$$

which is consistent with the generalized l_2-norm (A.20). It is easy to prove that an n-dimensional space furnished with inner product (A.26) must be a Hilbert space.

When inner product is defined by (A.26), the orthonormal system and corresponding coordinates of a vector $\mathbf{p}$ are given respectively by

$$\mathbf{e}_i = (0, 0, \ldots 0, \frac{1}{w_i}, 0, \ldots, 0)^T \quad \text{(only the i-th element is non-zero)} \quad \text{(A.27)}$$

$$\lambda_i = w_i p_i, \quad \text{(A.28)}$$

where $i = 1, 2, \ldots, n$.

(2) *Space l_2.*
   The generic element of this space is a real sequence $\mathbf{p} = (p_1, p_2, \ldots, p_n, \ldots)$ which is square summable with the weight $\mathbf{w} = (w_1, w_2, \ldots, w_n, \ldots)$, i.e., satisfies

$$\sum_{i=1}^{\infty} w_i^2 p_i^2 < \infty. \quad \text{(A.29)}$$

The inner product of the space is defined as

$$(\mathbf{p}, \mathbf{q}) = \sum_{i=1}^{\infty} (w_i p_i)(w_i q_i), \quad \text{(A.30)}$$

where $\mathbf{q} = (q_1, q_2, \ldots, q_n, \ldots)$ is another element of the space.
   Using the Cauchy–Schwarz inequality we can find, for any n,

$$\sum_{i=1}^{n} |w_i p_i| \, |w_i q_i| \leq \left[\sum_{i=1}^{n} w_i^2 p_i^2 \right]^{1/2} \left[\sum_{i=1}^{n} w_i^2 q_i^2 \right]^{1/2}. \quad \text{(A.31)}$$

Therefore, the series on the right-hand side of (A.30) converges absolutely and the definition of the inner product does make sense.

(3) *Space $L_2(\Omega)$.*
Space $L_2(\Omega)$ involves all functions defined on (Ω) which are square-integrable with respect to a positive weight function $W(\mathbf{x})$, that is, we have

$$\int_{(\Omega)} W^2(\mathbf{x}) f^2(\mathbf{x}) \, d\Omega < \infty \quad \text{(A.32)}$$

for any element $f(x)$ in the space. The inner product of the space is defined as

$$(f, g) = \int_{(\Omega)} [W(\mathbf{x}) f(\mathbf{x})][W(\mathbf{x}) g(\mathbf{x})] \, d\Omega, \quad \text{(A.33)}$$

where $g(\mathbf{x})$ is another element of the space. From inequality

$$2|W(\mathbf{x}) f(\mathbf{x})| \, |W(\mathbf{x}) g(\mathbf{x})| \leq |W(\mathbf{x}) f(\mathbf{x})|^2 + |W(\mathbf{x}) g(\mathbf{x})|^2 \quad \text{(A.34)}$$

we know that definition (A.33) *does* make sense. It has been proved that space $L_2(\Omega)$ is complete (for example, in de Barra, 1981). In other words, it is a *Hilbert space*. Note that space $L_2(\Omega)$ includes not only continuous functions but also some discontinuous functions.

From (A.33), we have

$$d^2(f,g) = \|f - g\|_{L_2}^2 = \int_{(\Omega)} W^2(\mathbf{x})[f(\mathbf{x}) - g(\mathbf{x})]^2 \, d\Omega. \tag{A.35}$$

Therefore, if two functions f and g are identical "*almost everywhere*", then $\|f - g\| = 0$, and we should consider them as the same element in space $L_2(\Omega)$.

Probabilities and Random Fields

This appendix lists probabilistic concepts and theorems used in this book. It is not an introduction to or summary of probability theory.

B.1. Random Variables and Probability Distributions

When we measure the water level, h_w, of a well many times (taking samples), different values may be obtained, even the flow field is stable. In the probability theory, a random variable is a real-valued function defined on the sample space of an experiment. It varies with uncontrollable factors but obeys certain distribution rules. Thus, the h_w mentioned above may be considered as a random variable.

A random variable X is described by its *probability distribution function* $F_X(x)$, which is defined as the probability of $X \leq x$, i.e.,

$$F_X(x) = P(X \leq x), \quad -\infty < x < \infty. \tag{B.1}$$

where $f_X(x)$ is called the *probability density function* *(pdf)* of X. It must satisfy conditions $f_X(x) \geq 0$ for any x and $\int_{-\infty}^{\infty} f_X(x)\,dx = 1$. From this definition, the probability of the values of the random variable that falls in an interval $[a, b]$ is given by

$$P(a < X \leq b) = \int_a^b f_X(x)\,dx = F_X(b) - F_X(a). \tag{B.2}$$

The joint distribution of two or more random variables defined on the same sample space is a multivariate distribution. An example is the distribution connected with simultaneous measurements of water level, water quality and water temperature of a well. Let X and Y be two random variables. Their *joint distribution* is defined as

$$F_{X,Y}(x,y) = P(X \leq x \text{ and } Y \leq y)$$
$$= \int_{-\infty}^{x} \int_{-\infty}^{y} f_{X,Y}(t_1, t_2)\,dt_1\,dt_2, \tag{B.3}$$

where $f_{X,Y}(x,y)$ is the *joint probability density function* which must satisfy conditions $f_{X,Y}(x,y) \geq 0$ for any x and y, and $\int_{-\infty}^{\infty} \int_{-\infty}^{\infty} f_{X,Y}(x,y) \, dx \, dy = 1$. From the joint *pdf*, $f_{X,Y}(x,y)$, we can find

$$f_X(x) = \int_{-\infty}^{\infty} f_{X,Y}(x,y) \, dy \tag{B.4}$$

and

$$f_Y(y) = \int_{-\infty}^{\infty} f_{X,Y}(x,y) \, dx. \tag{B.5}$$

The so defined $f_X(x)$ and $f_Y(y)$ are called *marginal density functions*. Two random variables X and Y are *independent* if and only if

$$F_{X,Y}(x,y) = F_X(x)F_Y(y) \tag{B.6}$$

or

$$f_{X,Y}(x,y) = f_X(x)f_Y(y). \tag{B.7}$$

The joint distribution $F_{\mathbf{X}}(\mathbf{x})$ and its *pdf* $f_{\mathbf{X}}(\mathbf{x})$ of a random vector $\mathbf{X}$ are related by

$$F_{\mathbf{X}}(\mathbf{x}) = \int_{-\infty}^{\mathbf{x}} f_{\mathbf{x}}(\mathbf{t}) \, dt, \tag{B.8}$$

where the multi-integral is defined in the n-dimensional space.

Hereafter, the subscript of a *pdf* is often ignored. This simplification will not cause any confusion.

B.2. Conditional Probabilities and Bayes's Theorem

Random variables may be dependent on each other. For example, the head in an irrigation well and the pumping rate of the well are closely related, and both of them depend on precipitation. Their relationships may be described by *conditional probability density functions* when they are considered as random variables.

Let X and Y be two random variables. When $f(y) \neq 0$, the conditional *pdf* of X given Y is defined as

$$f(x|y) = \frac{f(x,y)}{f(y)}. \tag{B.9}$$

If $f(x) \neq 0$ is also assumed, we then have

$$f(y|x) = \frac{f(x,y)}{f(x)}. \tag{B.10}$$

The combination of (B.9) and (B.10) yields

$$f(x|y) = \frac{f(x)f(y|x)}{f(y)},\tag{B.11}$$

which is known as the Bayes's theorem. Further, assume that X is the sum of m independent random variables $Z_1, Z_2, \ldots, Z_m$. Since $f(y) = \sum_{i=1}^{m} f(z_i)f(y|z_i)$, Bayes's theorem (B.11) now is expressed by

$$f(z_j|y) = \frac{f(z_j)f(y|z_j)}{\sum_{i=1}^{m} f(z_i)f(y|z_i)}.\tag{B.12}$$

Similarly, we can define conditional *pdf* for two vectors $\mathbf{X}$ and $\mathbf{Y}$. The Bayes's theorem for this case is

$$f(\mathbf{x}|\mathbf{y}) = \frac{f(\mathbf{x})f(\mathbf{y}|\mathbf{x})}{f(\mathbf{y})}.\tag{B.13}$$

The continuous form of Bayes's theorem (B.12) is given by

$$f(\mathbf{x}|\mathbf{y}) = \frac{f(\mathbf{x})f(\mathbf{y}|\mathbf{x})}{\int_{-\infty}^{\infty} f(\mathbf{t})f(\mathbf{y}|\mathbf{t})\, dt}.\tag{B.14}$$

For the case of parameter identification, we have observation vector $\mathbf{Z}_D$ and parameter vector $\boldsymbol{\theta}$. *pdf* $f(\boldsymbol{\theta})$ is regarded as prior distribution $p_0(\boldsymbol{\theta})$, while conditional probability density functions $f(\boldsymbol{\theta}|\mathbf{Z}_D)$ and $f(\mathbf{Z}_D|\boldsymbol{\theta})$ are posterior distribution $p_*(\boldsymbol{\theta})$ and likelihood function $L(\boldsymbol{\theta})$, respectively. Thus, (B.14) has the following form

$$p_*(\boldsymbol{\theta}) = \frac{p_0(\boldsymbol{\theta})L(\boldsymbol{\theta})}{\int_{(\Omega)} p_0(\boldsymbol{\theta})L(\boldsymbol{\theta})\, d\boldsymbol{\theta}},\tag{B.15}$$

where (Ω) is the admissible region of $\boldsymbol{\theta}$.

B.3. Expected Value, Variance and Covariance

If X is a continuous random variable, the *mean*, or *expected value* of X is defined by

$$E(X) \equiv \int_{-\infty}^{\infty} x f(x)\, dx,\tag{B.16}$$

where $f(x)$ is the *pdf* of X. Besides $E(X)$, another common notation for expected value of X is μ_X. Furthermore, if random variable Y is a function of X, $Y = g(X)$, then the expected value of Y is given by

$$E(Y) = \int_{-\infty}^{\infty} g(x)f(x)\, dx.\tag{B.17}$$

The expected value has the following properties:

· $E(aX + bY) = aE(X) + bE(Y)$,, where a and b are constant.

· $E(XY) = E(X)E(Y)$, if X and Y are independent.

The *variance* of a random variable X is defined by

$$V(X) = \int_{-\infty}^{\infty} (x - \mu_X)^2 f(x)\, dx. \tag{B.18}$$

Variance $V(X)$ is often denoted by σ_X^2, where σ_X is called the *standard deviation*. Using the properties of expected value given above, we can reduce (B.18) to

$$V(X) = E(X^2) - [E(X)]^2. \tag{B.19}$$

Relating two random variables X and Y defined on the same sample space, we have the covariance

$$\text{Cov}(X, Y) \equiv E[(X - \mu_X)(Y - \mu_Y)] \tag{B.20a}$$

or equivalently,

$$\text{Cov}(X, Y) = E(XY) - \mu_X \mu_Y. \tag{B.20b}$$

The following properties of variance and covariance are easily obtained from their definitions

· $V(aX) = a^2 V(X)$,

· $V(aX + c) = a^2 V(X)$,

· $V(aX + bY) = a^2 V(X) + b^2 V(Y) + 2ab\text{Cov}(X, Y)$,

where a, b, c are constant.

Let $\mathbf{X} = (X_1, X_2, \ldots, X_n)$ be a random vector. Its expected value is

$$E(\mathbf{X}) = \int_{(R^n)} \mathbf{x} f(\mathbf{x})\, d\mathbf{x}, \tag{B.21}$$

where R^n is the n-dimensional space. The *covariance matrix* of X is defined by

$$\text{Cov}(\mathbf{X}) \equiv \begin{bmatrix} \text{Var}(X_1) & \text{Cov}(X_1, X_2) & \cdots & \text{Cov}(X_1, X_n) \\ \text{Cov}(X_1, X_2) & \text{Var}(X_2) & \cdots & \text{Cov}(X_2, X_n) \\ \vdots & \vdots & \ddots & \vdots \\ \text{Cov}(X_1, X_n) & \text{Cov}(X_2, X_n) & \cdots & \text{Var}(X_n) \end{bmatrix}. \tag{B.22}$$

When all $X_1, X_2, \ldots, X_n$ are independent, $\text{Cov}(\mathbf{X})$ reduces to a diagonal matrix. Furthermore, if $\text{Var}(X_i) = \sigma^2$ for all $i = 1, 2, \ldots, n$, we will have $\text{Cov}(\mathbf{X}) = \sigma^2 \mathbf{I}$, where $\mathbf{I}$ is the unit matrix. Sometimes, we write $\text{Cov}(\mathbf{X})$ in a shorthand notation as $\mathbf{V_X}$.

B.4. Random Fields, Autocovariance and Cross-Covariance Functions

A *Random field* (or *stochastic function*, stochastic process) is a set of random variables that depend on time and/or space variables. For example, the water level of an observation well, $h_w(t)$, in a transient flow field is a stochastic function with respect to time t. Another example is the conductivity distribution in an aquifer, which may be considered as a stochastic function with respect to space variables. In what follows, we use $X(t), t \in T$ to denote a random field, where t represents independent (time and/or space) variables and T is the definition region of the random field.

For a random field $X(t)$, $t \in T$, let $X(t_1)$ and $X(t_2)$ be two random variables corresponding to $t_1 \in T$ and $t_2 \in T$, respectively. The covariance of $X(t_1)$ and $X(t_2)$, $Cov(X(t_1), X(t_2))$, now is called the *autocovariance*. Specifically, when $t_1 = t_2$, function

$$Var(X(t)) = Cov(X(t), X(t)) \tag{B.23}$$

is called the *variance function* of the random field.

Furthermore, let $X(t)$ and $Y(t)$ be two random field defined on the same set T. Then, for any $t_1 \in T$ and $t_2 \in T$, $X(t_1)$ and $Y(t_2)$ are random variables. Their covariance $Cov(X(t_1), Y(t_2))$ is called cross-covariance. For example, we can define cross-covariances between head observations and hydraulic conductivity measurements.

A random field $X(t)$ is called *stationary*, if the distribution of $X(t)$ does not depend on t and the joint distribution of $X(t_1)$ and $X(t_2)$ depends only on the distance between t_1 and t_2. These requirements are difficult to be satisfied in practice. In Section 7.3, we have defined weakly stationary fields, increment stationary fields and general unstationary fields.

A FORTRAN Program

The program given here is able to solve both forward and inverse problems of coupled flow and mass transport in confined or unconfined aquifers. It is useful not only for teaching and learning purposes, but also for solving practical problems. By changing a few control numbers, the program can be switched to different applications. Five kinds of parameters: hydraulic conductivity, specific storage coefficient, porosity, longitudinal dispersivity and transverse dispersivity, can be identified individually or simultaneously, provided that observation data are sensitive enough to the unknown parameters. From this program, readers can learn how to link a simulation routine with an optimization routine for inverse solution, and how to incorporate prior information and a parameterization method into the parameter identification procedure. Many exercises of this book can be completed by revising this program.

C.1. Functions, Methods and Assumptions

There are three switch numbers: **INV**, **IHC** and **NST**. Setting **INV**=0, the programs only solves forward problems. When setting **INV**=1, the program solves inverse problems. Setting **IHC**=1 for solving groundwater flow problems, while setting **IHC**=2 for solving coupled flow and mass transport problems. When the flow field tends to steady state, setting **NST**=1, otherwise, setting **NST**=0. Thus, we may have six different combinations. For example, if we set **INV**=1, **IHC**=2, **NST**=0, the program solves inverse problems of mass transport in transient flow fields.

The multiple cell balance (MCB) method (Sun and Yeh, 1983), a variety of the finite element method, is used to solve forward problems. Inverse problems are solved by an indirect method, in which the Gauss-Newton-Levenberg-Marquardt method is used to minimize the generalized least squares criterion.

The program is designed to solve coupled groundwater flow-mass transport problems described in Example 1.1.6. In equations (1.1.8)-(1.1.11), we have assumed that the flow field is basically horizontal, the aquifer is isotropic but may

265

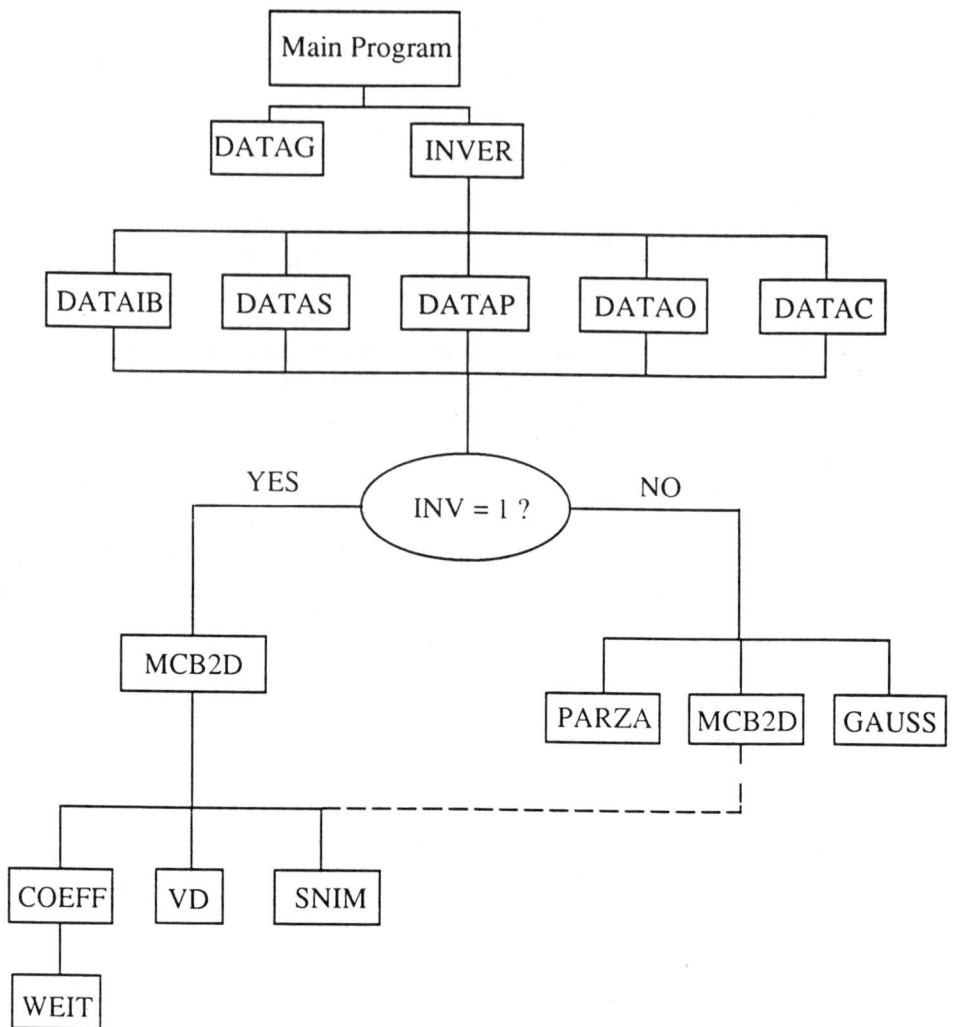

Fig. C.1   Program Structure

be inhomogeneous. For unconfined aquifers, Dupuit's assumptions must be approximately applicable.

C.2 Program Structure

A flow chart shown in Figure C.1 describes the structure of the program. The whole program consists of a main program and 14 subroutines.

Main Program	reads control integers; allocates storage spaces to all arrays; calls subroutines **DATAG** and **INVER.**
Subroutine **DATAG**	reads geometric and geologic information which includes the shape of the considered aquifer, inhomogeneous zones and so forth; forms element and node systems.
Subroutine **DATAIB**	reads initial and boundary conditions.
Subroutine **DATAS**	reads extraction and recharge data, then forms distributed sink/source terms.
Subroutine **DATAP**	reads aquifer parameters associated with each zone, then forms distributed parameters. When solving inverse problems, the values of unknown parameters are not used. Thus, it is allowed to assign any values for them.
Subroutine **DATAO**	reads locations of observation wells and observation times. When solving forward problems, we can input those locations and times, for which the values of head and/or concentration are of interest. When solving inverse problems, the subroutine also reads the observed heads and observed concentrations (when **IHC**=2).
Subroutine **DATAC**	reads computation control numbers for forward and inverse solutions, such as iteration factors, stop criteria, weighting coefficients, initial time step size and so forth.
Subroutine **PARZA**	generates distributed parameters for the simulation model, when we want to calculate the value of least squares corresponding to a set of gien values of unknown parameters.
Subroutine **GAUSS**	solves systems of linear equations.
Subroutine **VD**	calculates velocity distributions according to Darcy's law and head distributions, then calculates dispersion coefficients.
Subroutine **COEFF**	forms coefficient matrices of MCB discretization equations.
Subroutine **WEIT**	generates upstream weights for eliminating oscilations around sharp concentration fronts.
Subroutine **SNIM**	solves MCB discretization equations.

Subroutine **MCB2D** solves forward problems. When solving inverse problems, it returns calculated heads and concentrations corresponding to observed heads and concentrations. It also returns the least suares error.

Subroutine **INVER** solves inverse problems.

C.3. Variables and Arrays

(1) Integer numbers

NP	number of nodes;
NE	number of elements;
NB	maximum number of neighboring nodes around a node. Usually, setting **NB**=9 is enough.
IHC	switch number as defined in C.1.
INV	switch number as defined in C.1.
NST	switch number as defined in C.1.
IPR	print control number. Setting **IPR**=6, only parameter identification results are printed. Setting **IPR**=5, the comparison between calculated and observed heads (concentrations) are also printed. Setting **IPR**=4, all middle results of inverse solution are printed for testing. When solving forward problems, we may set **IPR**=0, 1, 2, 3. Setting **IPR**=3, the program only prints heads (concentrations) at designated locations and times. Setting **IPR**=2, head (concentration) distributions on the whole region are printed at the designated times. Setting **IPR**=1, all middle results of forward solution are printed for testing. Setting **IPR**=0, the program prints all input data for recording.
NW	total number of extraction and recharge wells.
NHB	total number of given head boundary nodes.
NCB	total number of given concentration boundary nodes.
NOB	total number of observation wells.
NCR	total number of observation times.
NZON	total number of inhomogeneous zones.
NID	total number of the identified parameters.
ITRMAX	maximal number of Gauss-Newton iterations.
MRQMAX	maximal number of Levenberg-Marquardt modifications.

(2) Real numbers

PT	upstream weighting coefficient. When there is no sharp concentration fronts, set **PT**=0.0; Otherwise, set **PT**=0.001.
ALF	relax factor for the solution of MCB equations. Its best value may be in the range of 1.1. to 1.6.
EPH	accuracy requirement for head solution.
EPC	accuracy requirement for concentration solution.
DT	initial size of the time step.
WTH	weighting coefficient of the head least squares error.
WTC	weighting coefficient of the concentration least squares error.
ALMDA	modification factor in the Levenberg-Marquardt method.
HLSQ	head least squares error.
CLSQ	concentration least squares error.
TLSQ	total least square error: **TLSQ = WTH*HLSQ + WTC*CLSQ.**
AHLSQ	average least squares error for each head observation.
ACLSQ	average least squares error for each concentration observation.

(3) Integer arrays

IJM	node numbers of each element.
MA	number of neighboring nodes of each node.
MCPN	node numbers of neighboring nodes of each node.
NZONP	zonation number of each node.
NHBP	node numbers of given head boundary nodes.
NCBP	node numbers of given concentration boundary nodes.
NWP	node numbers of extraction and recharge wells.
NOBP	node numbers of observation wells.
KGUP	group number of each identified parameter. **KGUP**(i)=1 means i-th unknown parameter is a hydraulic conductivity. **KGUP**(i)=2, 3, 4, 5 correspond to specific storage coefficient, porosity, longitudinal dispersivity and transverse dispersivity, respectively.
KZON	zonation number of each identified parameter.

All integer arrays are storaged in a one-dimensional array N.

(4) Real arrays

X	x-coordinates of all nodes.
Y	y-coordinates of all nodes.
TR	elevation of top impermeable layer or ground surface.
BR	elevation of bed rock.
TH	saturated thickness.
PS	area of exclusive subdomain of each node.
H00	initial head distribution.
C00	initial concentration distribution.
HB1	given head values of given head boundary nodes.
CB1	given concentration values of given concentration boundary nodes.
WR	extraction or recharge rates. It is positive for extraction and negative for recharge.
WC	concentration of recharge water at each recharge well. Set any number for an extraction well.
DWR	the distribution of WR.
DWC	the distribution of WC.

ZTK, ZSE, ZPR, ZAL, ZAT ---

values of hydraulic conductivity, specific storage coefficient, porosity, longitudinal dispersivity and transverse dispersivity for each zone, respectively.

DTK, DSE, DPR, DAL, DAT ---

distributions of **ZTK, ZSE, ZPR, ZAL** and **ZAT**, respectively.

DXX, DXY, DYY ---

elements of dispersion coefficient tensor.

VX, VY	velocity components.
CFA, CFB	coefficient matrices of MCB discretization equations.
HOB	head observations.
HSDV	standard deviation of each head observation.
COB	concentration observations.
CSDV	standard deviation of each concentration observation.
HOBC	model output corresponding to **HOB**.
COBC	model output corresponding to **COB**.

TN	observation times.
PARAI	initial estimates of the identified parameters.
PARAL	lower bounds of the identified parameters.
PARAV	upper bounds of the identified parameters.
GPARA	updated values of the identified parameters.
DHDP	partial derivatives of head with respect to the identified parameters.
DCDP	partial derivatives of concentration with respect to the identified parameters.
AA	coefficient matrix of Gauss-Newton equations.
BB	right-hand side vector of Gauss-Newton equations.

All real arrays are storaged in a one-dimensional array A.

C.4. Input and Output Files

All input data of this program are separately storaged in seven data files:

DATAN	storages all control integer numbers (**NP, NE, NB, IHC, INV, NST, IPR, NW, NHB, NOB, NCR, NZON, NID**).
DATAG	storages geometric and geologic information (**X, Y, NZONP, TR, BR, IJM**).
DATAB	storages initial and boundary conditions (**H00, HB1, NHBP**; and **C00, CB1, NCBP** when **IHC** = 2).
DATAS	storages extration and recharge data (**NWP, WR**; and **WC** when **IHC** = 2).
DATAP	storages parameters for each zone (**ZTK, ZSE, ZPR, ZAL, ZAT**).
DATAO	storages observation data (**NOBP, TN**; and **HOB, HSDV** when **INV** = 1; and **COB, CSDV**, when **INV** = 1 and **IHC** = 2).
DATAC	storages computation control numbers (**PT, ALF, EPH, EPC, DT, WTH, WTC**; and **ITRMAX, MAQMAX** when **INV** = 1).

When the user prepares the above data files for him/her problems, he or she must exactly follow the formats given in the following example. All printed results can be found in the file **RESUL**. The content of printed results depends on the value of control number **IPR** as defined before.

C.5. Source Program

```
C
C****************************************************************
C
C          Main program:  INVERSE.FOR
C
C****************************************************************
C
      DIMENSION A(80000), N(10000), CHA(70)
      MA=80000
      MN=10000
C
      OPEN(5,FILE='PARAM')
      OPEN(6,FILE='RESUL')
      OPEN(1,FILE='DATAN')
      OPEN(2,FILE='DATAG')
      OPEN(3,FILE='DATAB')
      OPEN(4,FILE='DATAS')
      OPEN(7,FILE='DATAP')
      OPEN(8,FILE='DATAO')
      OPEN(9,FILE='DATAC')
C
C Reading control integers
C
      READ (1, 2) CHA
      READ (1, 2) CHA
      READ (1, 2) CHA
    2 FORMAT (70A1)
      READ (1,10)  NP, NE, NB, IHC, INV, NST, IPR
   10 FORMAT (7I10)
      IF ( IPR .LT. 1) THEN
       WRITE (6, 2) CHA
       WRITE (6, 10) NP,NE,NB,IHC,INV,NST,IPR
      ENDIF
      READ (1, 2) CHA
      READ (1, 2) CHA
      READ (1, 2) CHA
      READ (1,10)  NW, NHB, NCB, NOB, NCR, NZON, NID
      IF ( IPR .LT. 1) THEN
       WRITE (6, 2) CHA
       WRITE (6, 10) NW,NHB,NCB,NOB,NCR,NZON,NID
```

```
      ENDIF
C
C Allocating storage spaces
C
      NP1=NP+1
      IG1=1
      JG1=1
C
C X, Y, TR, BR, TH, PS, IJM, MA, MCPN, NZONP
C
      IG2=IG1+NP
      IG3=IG2+NP
      IG4=IG3+NP
      IG5=IG4+NP
      IG6=IG5+NP
      IB1=IG6+NP
      JG2=JG1+3*NE
      JG3=JG2+NP1
      JG4=JG3+NB*NP
      JB1=JG4+NP
C
C Reading and processing geometric and geologic information
C
      CALL DATAG (A(IG1),A(IG2),A(IG3),A(IG4),A(IG5),A(IG6),N(JG1),
     *      N(JG2),N(JG3),N(JG4),NP,NP1,NE,NB,NZON,IH,IPR)
C
C Continuously allocating storage spaces
C
C
C H00, C00, HB1, CB1, NHBP, NCBP
C
      IB2=IB1+NP
      IB3=IB2+NP
      IB4=IB3+NHB
      IS1=IB4+NCB
      JB2=JB1+NHB
      JS1=JB2+NCB
C
C DWR, DWC, WR, WC, NWP
C
      IS2=IS1+NP
      IS3=IS2+NP
```

```
   IS4=IS3+NW
   IP1=IS4+NW
   JE1=JS1+NW
C
C ZTK, ZSE, ZPR, ZAL, ZAT, DTK, DSE, DPR, DAL, DAT
C
   IP2=IP1+NZON
   IP3=IP2+NZON
   IP4=IP3+NZON
   IP5=IP4+NZON
   IP6=IP5+NZON
   IP7=IP6+NP
   IP8=IP7+NP
   IP9=IP8+NP
   IP10=IP9+NP
   IH1=IP10+NP
C
C H0, H1, H2, C0, C1, C2, HF00, HF0, CF00, CF0
C
   IH2=IH1+NP
   IH3=IH2+NP
   IH4=IH3+NP
   IH5=IH4+NP
   IH6=IH5+NP
   IH7=IH6+NP
   IH8=IH7+NP
   IH9=IH8+NP
   IH10=IH9+NP
   IE1=IH10+NP
C
C DXX, DXY, DYY, VX, VY, CFA, CFB, IE, IEE
C
   IE2=IE1+NE
   IE3=IE2+NE
   IE4=IE3+NE
   IE5=IE4+NE
   IE6=IE5+NE
   IE7=IE6+IH
   IO1=IE7+IH
   JE2=JE1+NP
   JO1=JE2+NP
C
```

```
C HOB, HSDV, COB, CSDV, HOBC, HOBD, COBC, COBD, TN, NOBP
C
    IO2=IO1+NOB*NCR
    IO3=IO2+NOB*NCR
    IO4=IO3+NOB*NCR
    IO5=IO4+NOB*NCR
    IO6=IO5+NOB*NCR
    IO7=IO6+NOB*NCR
    IO8=IO7+NOB*NCR
    IO9=IO8+NOB*NCR
    ID1=IO9+NCR
    JD1=JO1+NOB
C
C PARA, PARAI, PARAL, PARAU, GPARA, KGUP. KZON
C
    ID2=ID1+NID
    ID3=ID2+NID
    ID4=ID3+NID
    ID5=ID4+NID
    IC1=ID5+NID
    JD2=JD1+NID
    JD3=JD2+NID
C
C DHDP, DCDP, AA, WAA, BB, WBB
C
    IC2=IC1+NOB*NCR*NID
    IC3=IC2+NOB*NCR*NID
    IC4=IC3+NID*NID
    IC5=IC4+NID*NID
    IC6=IC5+NID
    IC7=IC6+NID
C
C Calculating the required total storage space
C
    NTOTAL=IC7+JD3
    IF (IPR .LT. 1) THEN
       WRITE  (6,20) NTOTAL
 20 FORMAT (//,1X,'THE REQUIRED TOTAL STORAGE SPACE IS:',I15,//)
    ENDIF
    IF(IC7 .GT. MA .OR. JD3 .GT. MN) GO TO 30
    WRITE (6, 50)
 50 FORMAT (/,15X,'***** Program runs ******')
```

```
      WRITE (6, 52) INV, IHC, NST
   52 FORMAT (/,10X,'INV=',I2,10X,'IHC=',I2,10X,'NST=',I2)
C
C Solving either forward or inverse problems
C
      CALL INVER(A(IG1),A(IG2),A(IG3),A(IG4),A(IG5),A(IG6),N(JG1),
     1      N(JG2),N(JG3),N(JG4),
     2      A(IB1),A(IB2),A(IB3),A(IB4),N(JB1),N(JB2),
     3      A(IS1),A(IS2),A(IS3),A(IS4),N(JS1),
     4      A(IP1),A(IP2),A(IP3),A(IP4),A(IP5),
     5      A(IP6),A(IP7),A(IP8),A(IP9),A(IP10),
     6      A(IH1),A(IH2),A(IH3),A(IH4),A(IH5),
     7      A(IH6),A(IH7),A(IH8),A(IH9),A(IH10),
     8      A(IE1),A(IE2),A(IE3),A(IE4),A(IE5),A(IE6),
     8      A(IE7),N(JE1),N(JE2),
     1      A(IO1),A(IO2),A(IO3),A(IO4),A(IO5),A(IO6),
     2      A(IO7),A(IO8),A(IO9),N(JO1),
     3      A(ID1),A(ID2),A(ID3),A(ID4),A(ID5),N(JD1),N(JD2),
     4      A(IC1),A(IC2),A(IC3),A(IC4),A(IC5),A(IC6),
     5      NP,NP1,NE,NB,IHC,INV,NID,NST,
     6      NW,NHB,NCB,NOB,NCR,NZON,IH,IPR)
C
C
C
   30 STOP
C
      CLOSE(5)
      CLOSE(6)
      CLOSE(1)
      CLOSE(2)
      CLOSE(3)
      CLOSE(4)
      CLOSE(7)
      CLOSE(8)
      CLOSE(9)
      END
C
C**************************************************************
C
C     Subroutine INVER solves either forward or inverse problems
C
C**************************************************************
```

```
C
      SUBROUTINE INVER(X,Y,TR,BR,TH,PS,IJM,MA,MCPN,NZONP,
     1        H00,C00,HB1,CB1,NHBP,NCBP,
     2        DWR,DWC,WR,WC,NWP,
     3        ZTK,ZSE,ZPR,ZAL,ZAT,DTK,DSE,DPR,DAL,DAT,
     4        H0,H1,H2,C0,C1,C2,HF00,HF0,CF00,CF0,
     5        DXX,DXY,DYY,VX,VY,CFA,CFB,IE,IEE,
     6        HOB,HSDV,COB,CSDV,HOBC,HOBD,COBC,COBD,TN,NOBP,
     7        PARA,PARAI,PARAL,PARAU,GPARA,KGUP,KZON,
     8        DHDP,DCDP,AA,WAA,BB,WBB,
     9      NP,NP1,NE,NB,IHC,INV,NID,NST,
     1      NW,NHB,NCB,NOB,NCR,NZON,IH,IPR)
C
C
C
      DIMENSION  X(NP),Y(NP),TR(NP),BR(NP),TH(NP),PS(NP),IJM(3,NE),
     1      MA(NP1),MCPN(NB,NP),NZONP(NP),
     2      H00(NP),C00(NP),HB1(NHB),CB1(NCB),NHBP(NHB),NCBP(NCB),
     3      DWR(NP),DWC(NP),WR(NW),WC(NW),NWP(NW),
     4      ZTK(NZON),ZSE(NZON),ZPR(NZON),ZAL(NZON),ZAT(NZON),
     5      DTK(NP),DSE(NP),DPR(NP),DAL(NP),DAT(NP),
     6      H0(NP),H1(NP),H2(NP),C0(NP),C1(NP),C2(NP),
     7      HF00(NP),HF0(NP),CF00(NP),CF0(NP),CFA(IH),CFB(IH),
     8      VX(NE),VY(NE),DXX(NE),DXY(NE),DYY(NE),IE(NP),IEE(NP),
     9      HOB(NOB,NCR),COB(NOB,NCR),HSDV(NOB,NCR),CSDV(NOB,NCR),
     1      HOBC(NOB,NCR),COBC(NOB,NCR),HOBD(NOB,NCR),
     2      COBD(NOB,NCR),TN(NCR),NOBP(NOB),
     3      PARA(NID),PARAI(NID),PARAL(NID),PARAU(NID),GPARA(NID),
     4      KGUP(NID),KZON(NID),DHDP(NOB,NCR,NID),DCDP(NOB,NCR,NID),
     5      AA(NID,NID),WAA(NID,NID),BB(NID),WBB(NID),
     6      EPS(2),ME(3),CHA(70)
C
C Reading initial and boundary conditions, sink/source terms,
C parameters, observations and computation control parameters.
C
      CALL DATAIB (H00,C00,HB1,CB1,NHBP,NCBP,NP,NHB,NCB,IHC)
      CALL DATAS  (DWR,DWC,WR,WC,NWP,NP,NW,IHC)
      CALL DATAP  (ZTK,ZSE,ZPR,ZAL,ZAT,DTK,DSE,DPR,DAL,DAT,
     *        NZONP,NP,NZON)
      CALL DATAO  (HOB,COB,HSDV,CSDV,NOBP,TN,NOB,NCR,IHC,INV)
      CALL DATAC  (PT,ALF,EPS(1),EPS(2),DT,WTH,WTC,itrmax,maqmax)
C
```

```
C Printing all input data when setting IPR=0
C
    IF (IPR .LT. 1) THEN
      WRITE (6, 2)
  2 FORMAT (//,6X,'NODE', 7X,'DWR',7X,'DWC',7X,'H00',7X,'C00')
      DO I=1,NP
        WRITE(6, 4) I, DWR(I),DWC(I),H00(I),C00(I)
  4 FORMAT(I10, 4F10.2)
      END DO
C
      WRITE (6, 6)
  6 FORMAT (//,6X,'NODE', 7X,'DTK',7X,'DSE',7X,'DPR',7X,'DAL',
    *     7X, 'DAT')
      DO I=1,NP
        WRITE(6, 8) I, DTK(I),DSE(I),DPR(I),DAL(I),DAT(I)
  8 FORMAT(I10, 5F10.2)
      END DO
C
      WRITE (6, 12)
 12 FORMAT (//,5X,'NUMBER',6X,'NODE',7X, 'HB1')
      DO I=1,NHB
        WRITE(6, 14) I, NHBP(I), HB1(I)
 14 FORMAT(I10, I10, F10.2)
      END DO
      IF (IHC .EQ. 2) THEN
        WRITE (6, 16)
 16    FORMAT (//,5X,'NUMBER',6X,'NODE',7X, 'CB1')
        DO I=1,NHB
          WRITE(6, 14) I, NHBP(I), CB1(I)
        END DO
      ENDIF
C
      IF (INV .EQ. 1) THEN
        WRITE (6,18)
 18 FORMAT (/, 5X, 'HEAD OBSERVATIONS')
        DO I=1,NOB
          WRITE (6, 20) I
 20 FORMAT (/,10X,'OBSERVATION WELL NO.',I5 )
          WRITE (6, 22)
 22 FORMAT (5X, 'TIME', 14X, 'HEAD.', 15X, 'HSDV.')
          DO J=1,NCR
            WRITE (6, 24) J, HOB(I,J), HSDV(I,J)
```

```
 24  FORMAT(I10,2F20.2)
         END DO
         END DO
C
         IF (IHC .EQ. 2) THEN
            WRITE (6,26)
 26  FORMAT (/, 5X, 'CONCENTRATION  OBSERVATIONS')
         DO I=1,NOB
            WRITE (6, 28) I
 28  FORMAT (/,10X,'OBSERVATION WELL NO.', I5 )
            WRITE (6, 30)
 30  FORMAT (5X, 'TIME', 14X, 'CONC.', 15X, 'CSDV.')
            DO J=1,NCR
               WRITE (6, 32) J, COB(I,J), CSDV(I,J)
 32  FORMAT(I10,2F20.2)
            END DO
            END DO
         ENDIF
C
C
      ENDIF
C
    ENDIF
C
C Solving forward problems when setting INV=0
C
    IF (INV .EQ. 0) THEN
    WRITE (6, 33)
 33  FORMAT (/,10X,'*** SOLVE THE FORWARD PROBLEM ONLY ***',/)
C
    CALL MCB2D (X,Y,TR,BR,TH,PS,IJM,MA,MCPN,NZONP,
   1         H00,C00,HB1,CB1,NHBP,NCBP,
   2         DWR,DWC,WR,WC,NWP,
   3         DTK,DSE,DPR,DAL,DAT,
   4         H0,H1,H2,C0,C1,C2,HF00,HF0,CF00,CF0,
   5         DXX,DXY,DYY,VX,VY,CFA,CFB,IE,IEE,
   6         HOB,HSDV,COB,CSDV,HOBC,COBC,TN,NOBP,
   7         EPS,PT,ALF,DT,WTH,WTC,TLSQ,
   8         NP,NP1,NE,NB,IHC,INV,NID,NST,
   9         NW,NHB,NCB,NOB,NCR,NZON,IH,IPR)
C
C The forward solution is completed
```

```
C
      RETURN
      ENDIF
C
C Reading initial estimates and bounds of the unknown parameters
C
      READ (5, 34) CHA
      READ (5, 34) CHA
      READ (5, 34) CHA
   34 FORMAT (70A1)
      DO I=1,NID
         READ (5,36) II,KGUP(I),KZON(I),PARAL(I),PARAI(I),PARAU(I)
   36 FORMAT (3I10, 3F10.4)
      END DO
C
      WRITE (6, 38)
   38 FORMAT (/,10X,'*** BEGIN TO SOLVE THE INVERSE PROBLEM ***',/)
C
C Calculating least squares error associated with initial estimates
C
      CALL PARZA (PARAI,KGUP,KZON,NZONP,ZTK,ZSE,ZPR,ZAL,ZAT,
     *            DTK,DSE,DPR,DAL,DAT,NP,NZON,NID)
C
      CALL MCB2D (X,Y,TR,BR,TH,PS,IJM,MA,MCPN,NZONP,
     1            H00,C00,HB1,CB1,NHBP,NCBP,
     2            DWR,DWC,WR,WC,NWP,
     3            DTK,DSE,DPR,DAL,DAT,
     4            H0,H1,H2,C0,C1,C2,HF00,HF0,CF00,CF0,
     5            DXX,DXY,DYY,VX,VY,CFA,CFB,IE,IEE,
     6            HOB,HSDV,COB,CSDV,HOBC,COBC,TN,NOBP,
     7            EPS,PT,ALF,DT,WTH,WTC,TLSQI,
     8            NP,NP1,NE,NB,IHC,INV,NID,NST,
     9            NW,NHB,NCB,NOB,NCR,NZON,IH,IPR)
C
C
C
      WRITE (6, 40)
   40 FORMAT (/,10X, '---- Write the initial parameters ----',/)
      WRITE (6, 42)
   42 FORMAT (7X,'NO.', 6X,'KGUP', 6X, 'KZON', 11X, 'PARAMETER')
      DO I=1, NID
        GPARA(I)=PARAI(I)
```

```
      WRITE (6, 44) I, KGUP(I), KZON(I), GPARA(I)
   44 FORMAT (3I10, F20.5)
      END DO
      WRITE (6, 46) TLSQI
   46 FORMAT (/,10x,'---- The value of initial LS error is:', F15.4)
C
C Entering the iteration procedure of inverse solution
C
      ITR=0
   50 ITR=ITR+1
      IF (ITR .GT. itrmax) GOTO 100
C
C Calculating sensitivity coefficients of model output to unknown
C parameters
C
      DO L=1, NID
        DO K=1, NID
          PARA(K)=GPARA(K)
        END DO
        DPL=0.05*GPARA(L)
        PARA(L)=GPARA(L) + DPL
C
        CALL PARZA (PARA,KGUP,KZON,NZONP,ZTK,ZSE,ZPR,ZAL,ZAT,
     *         DTK,DSE,DPR,DAL,DAT,NP,NZON,NID)
        CALL MCB2D (X,Y,TR,BR,TH,PS,IJM,MA,MCPN,NZONP,
     1         H00,C00,HB1,CB1,NHBP,NCBP,
     2         DWR,DWC,WR,WC,NWP,
     3         DTK,DSE,DPR,DAL,DAT,
     4         H0,H1,H2,C0,C1,C2,HF00,HF0,CF00,CF0,
     5         DXX,DXY,DYY,VX,VY,CFA,CFB,IE,IEE,
     6         HOB,HSDV,COB,CSDV,HOBD,COBD,TN,NOBP,
     7         EPS,PT,ALF,DT,WTH,WTC,TLSQ,
     8         NP,NP1,NE,NB,IHC,INV,NID,NST,
     9         NW,NHB,NCB,NOB,NCR,NZON,IH,IPR)
        DO I=1,NOB
          DO J=1,NCR
            DHDP(I,J,L)=(HOBD(I,J)-HOBC(I,J))/DPL
            IF (IHC .EQ. 2) THEN
              DCDP(I,J,L)=(COBD(I,J)-COBC(I,J))/DPL
            ENDIF
          END DO
        END DO
```

```
      END DO
C
C Generating the Gauss-Newton equations
C
      DO L=1, NID
        DO K=1, NID
         AA(L,K)=0.
        END DO
        BB(L)=0.
      END DO
C
      DO L=1, NID
        DO K=1, NID
         DO I=1, NOB
          DO J=1, NCR
      AA(L,K)=AA(L,K) + WTH*DHDP(I,J,L)*DHDP(I,J,K)
         IF ( IHC .EQ. 2) THEN
      AA(L,K)=AA(L,K) + WTC*DCDP(I,J,L)*DCDP(I,J,K)
          ENDIF
          END DO
         END DO
        END DO
      END DO
C
      DO L=1, NID
        DO I=1, NOB
         DO J=1, NCR
          BB(L)=BB(L) + WTH*(HOB(I,J)-HOBC(I,J))*DHDP(I,J,L)
          IF (IHC .EQ. 2) THEN
          BB(L)=BB(L) + WTC*(COB(I,J)-COBC(I,J))*DCDP(I,J,L)
          ENDIF
          END DO
         END DO
      END DO
C
C Using the Levenberg-Marquardt method
C
      ALMDA=0.
      MARQ=0
   54 DO I=1, NID
        DO J=1, NID
         WAA(I,J)=AA(I,J)
```

```
      END DO
      WAA(I,I)=AA(I,I) * (1.+ALMDA)
      WBB(I)=BB(I)
      END DO
C
C Solving Gauss-Newton-Levenberg-Marquardt equations
C
      CALL GAUSS (NID, WAA, WBB, MCH)
      IF ( MCH .EQ. 0) THEN
      WRITE (6, 56)
   56 FORMAT (/, 10X,'!!!!!  SINGULAR MATRIX   !!!!!', /)
      GOTO 100
      ENDIF
C
C Obtaining a set of new parameters
C
      DO I=1, NID
       PARA(I)=GPARA(I) + WBB(I)
      END DO
C
C Using the lower and upper bound constraints
C
      DO I=1, NID
       IF (PARA(I) .LT. PARAL(I)) PARA(I)=PARAL(I)
       IF (PARA(I) .GT. PARAU(I)) PARA(I)=PARAU(I)
      END DO
C
C Calculating the least squares error for the new parameters
C
      CALL PARZA (PARA,KGUP,KZON,NZONP,ZTK,ZSE,ZPR,ZAL,ZAT,
     *       DTK,DSE,DPR,DAL,DAT,NP,NZON,NID)
C
      CALL MCB2D (X,Y,TR,BR,TH,PS,IJM,MA,MCPN,NZONP,
     1       H00,C00,HB1,CB1,NHBP,NCBP,
     2       DWR,DWC,WR,WC,NWP,
     3       DTK,DSE,DPR,DAL,DAT,
     4       H0,H1,H2,C0,C1,C2,HF00,HF0,CF00,CF0,
     5       DXX,DXY,DYY,VX,VY,CFA,CFB,IE,IEE,
     6       HOB,HSDV,COB,CSDV,HOBC,COBC,TN,NOBP,
     7       EPS,PT,ALF,DT,WTH,WTC,TLSQM,
     8       NP,NP1,NE,NB,IHC,INV,NID,NST,
     9       NW,NHB,NCB,NOB,NCR,NZON,IH,IPR)
```

```
C
C Testing whether the new parameters are acceptable
C
      IF (TLSQM .LT. TLSQI) THEN
C
C Printing the modified parameters and associated least squares error
C
      WRITE (6, 60) MARQ
  60  FORMAT (/,10X,'---- The modefied parameters at MARQ ----',I3,/)
      WRITE (6, 42)
      DO I=1, NID
       GPARA(I)=PARA(I)
       WRITE (6, 44) I, KGUP(I), KZON(I), GPARA(I)
      END DO
C
      WRITE (6, 62) TLSQM
  62  FORMAT (/,10X, '---The value of modified LS error is:', F15.4)
C
C Printing the comparison between calculated and observed heads C
C
      IF (IPR .LT. 6) THEN
         WRITE (6, 64)
  64  FORMAT (/,5X,'COMPARING CALCULATED HEADS WITH OBSERVED
HEADS')
         DO I=1,NOB
           WRITE (6, 66) I
  66  FORMAT (/, 15X, 'FOR OBSERVATION WELL NO.', I3)
           WRITE (6, 68)
  68  FORMAT (7X,'NO.',6X,'TIME',11X,'CAL. HEAD',11X,'OBS. HEAD')
           DO K=1,NCR
             WRITE (6, 70) K, TN(K), HOBC(I,K), HOB(I,K)
  70  FORMAT (I10, F10.2, 2F20.5)
           END DO
         END DO
C
         IF (IHC .EQ. 2) THEN
           WRITE (6, 72)
  72  FORMAT (/,5X,'COMPARING CALCULATED CONC. WITH OBSERVED
CONC.')
           DO I=1,NOB
             WRITE (6, 66) I
             WRITE (6, 74)
```

```
 74  FORMAT (7X,'NO.',6X,'TIME',10X,'CAL. CONC.',11X,'OBS. CONC.')
         DO K=1,NCR
             WRITE (6, 70) K, TN(K), COBC(I,K), COB(I,K)
             END DO
         END DO
       ENDIF
C
     ENDIF
C
C Returning to the iteration procedure
C
     TLSQI=TLSQM
     GOTO 50
C
     ELSE
C
C Increasing the factor of Levenberg-Marquardt technique when the
C solution is not acceptable
C
     ALMDA=MARQ*10.*ALMDA + 0.001
     MARQ=MARQ + 1
     IF (MARQ .LT. maqmax) GOTO 54
     WRITE (6, 80) ITR, MARQ
 80  FORMAT (5X, '*** GAUSS-NEWTON MARQUARDT METHOD STOPS AT
****',
     *      /, 'ITR=',I3, 15X, 'MARQ=', I3)
C
     ENDIF
C
100  RETURN
     END
C
C***************************************************************
C
C      Subroutine GAUSS sovles linear equations
C
C***************************************************************
C
C
     SUBROUTINE GAUSS (N, A, B, MCH)
CC   IMPLICIT REAL*8 (A-H, O-Z)
     DIMENSION A(N,N), B(N)
```

```
C
      DO 1 J=1,N
      IF (A(J,J) .LT. 1.E-20) GOTO 25
  1   CONTINUE
      L=N-1
      DO 10 K=1,L
      C=0.
      DO 2 I=K,N
      IF ( A(I,K) .LE. ABS(C)) GOTO 2
      C=A(I,K)
      M=I
  2   CONTINUE
      IF (M .EQ. K) GOTO 6
      DO 4 J=K, N
      T=A(K,J)
      A(K,J)=A(M,J)
  4   A(M,J)=T
      T=B(K)
      B(K)=B(M)
      B(M)=T
  6   KL=K+1
      C=1./C
      B(K)=B(K)*C
      DO 10 J=KL,N
      A(K,J)=A(K,J)*C
      DO 8 I=KL,N
  8   A(I,J)=A(I,J)-A(I,K)*A(K,J)
 10   B(J)=B(J)-A(J,K)*B(K)
      B(N)=B(N)/A(N,N)
      DO 40 K=1,L
      I=N-K
      C=0.
      LP=I+1
      DO 50 J=LP, N
 50   C=C+A(I,J)*B(J)
 40   B(I)=B(I)-C
      MCH=1
      GOTO 26
 25   MCH=0
 26   RETURN
      END
C
```

```
C
C****************************************************************
C
C       Subroutine DATAG reads and processes geometric and
C       geologic information
C
C****************************************************************
C
      SUBROUTINE DATAG (X,Y,TR,BR,TH,PS,IJM,MA,MCPN,NZONP,
     1        NP,NP1,NE,NB,NZON,IH,IPR)
      DIMENSION  X(NP),Y(NP),TR(NP),BR(NP),TH(NP),PS(NP),
     1        IJM(3,NE),MA(NP1),MCPN(NB,NP),NZONP(NP),
     2        ME(3),CHA(70),CHB(70)
C
C reading nodal coordinates
C
      READ (2, 1) CHA
      READ (2, 1) CHA
      READ (2, 1) CHB
    1 FORMAT (70A1)
C
      DO  N=1, NP
       READ (2, 2) NN, X(N), Y(N), NZONP(N),TR(N), BR(N)
    2 FORMAT (I10, 2F15.2, I10, 2F10.2)
      END DO
      IF (IPR .LT. 1) THEN
       WRITE (6, 4) CHB
    4 FORMAT (//, 70A1, /)
       DO  N=1, NP
         WRITE(6, 2) N, X(N), Y(N), NZONP(N), TR(N), BR(N)
       END DO
      ENDIF
C
C reading node numbers of each element
C
      READ (2, 1) CHA
      READ (2, 1) CHA
      READ (2, 1) CHB
C
      DO  M=1, NE
       READ (2, 12) MM, IJM(1,M), IJM(2,M), IJM(3,M)
   12 FORMAT(4I10)
```

```
      END DO
      IF (IPR .LT. 1) THEN
        WRITE (6, 14) CHB
 14   FORMAT (/, 70A1,/)
        DO M=1, NE
          WRITE (6, 12) M, IJM(1,M), IJM(2,M), IJM(3,M)
        END DO
      ENDIF
C
C calculating arrays MCPN, MA and integer IH
C
      DO I=1, NP
       MCPN(1,I)=I
       DO J=2, NB
       MCPN(J,I)=0
       END DO
      END DO
      DO I=1,NP1
       MA(I)=1
      END DO
      DO N=1,NE
       DO J=1,3
        ME(J)=IJM(J,N)
       END DO
       DO L=1,3
        DO K=2,3
         IBR=0
         M1=ME(1)
         NG=MA(M1+1)
         DO I=1, NG
           IF(ME(K).EQ.MCPN(I+1,M1)) IBR=1
         END DO
          IF(IBR.EQ.0) THEN
          MA(M1+1)=MA(M1+1)+1
          NH=MA(M1+1)
          MCPN(NH,M1)=ME(K)
          ENDIF
        END DO
        J=ME(1)
        ME(1)=ME(2)
        ME(2)=ME(3)
        ME(3)=J
```

```
      END DO
      END DO
C
C writing arrays MCPN and MA
C
      IF (IPR .LT. 2) THEN
        WRITE (6, 40)
 40   FORMAT (1X, 'MCPN'/)
        DO J=1,NP
          WRITE (6, 42) (MCPN(I,J), I=1,NB)
 42   FORMAT (1X, 9I8)
        END DO
        WRITE (6, 44) MA
 44   FORMAT (1X,'MA'/,(1X,10I5))
      ENDIF
C
C modifying array MA
C
      MA(1)=1
      DO N=1,NP
       MA(N+1)=MA(N)+MA(N+1)
      END DO
      IH=MA(NP+1)
      IF (IPR .LT. 2) THEN
        WRITE (6, 46) MA
 46   FORMAT (1X,'THE MODIFIED MA IS:'/,(1X,10I5))
        WRITE (6, 48) IH
 48   FORMAT (1X,'THE LENTH OF COEFFICIENT MATRIX IS',I6)
      ENDIF
C
C calculating the area of each exclusive subdomain
C
      DO I=1,NP
       PS(I)=0.
      END DO
      DO N=1,NE
       DO J=1, 3
         ME(J)=IJM(J,N)
       END DO
      MI=ME(1)
      MJ=ME(2)
      MK=ME(3)
```

```
      BI=Y(MJ)-Y(MK)
      BJ=Y(MK)-Y(MI)
      BK=Y(MI)-Y(MJ)
      CI=X(MK)-X(MJ)
      CJ=X(MI)-X(MK)
      CK=X(MJ)-X(MI)
      TRS=ABS(CJ*BI-BJ*CI)/6.
      PS(MI)=PS(MI)+TRS
      PS(MJ)=PS(MJ)+TRS
      PS(MK)=PS(MK)+TRS
      END DO
      IF (IPR .LT. 2) THEN
        WRITE (6, 56)
   56 FORMAT (1X,'THE AREA OF EXCLUSIVE SUBDOMAIN OF EACH NODE:',/)
        DO N=1, NP
         WRITE (6, 58) N, PS(N)
   58 FORMAT (I10, F15.2)
        END DO
      ENDIF
  100 RETURN
      END
C
C*****************************************************************
C
C     Subroutine DATAIB reads initial and boundary conditions
C
C*****************************************************************
C
      SUBROUTINE DATAIB (H00,C00,HB1,CB1,NHBP,NCBP,NP,NHB,NCB,IHC)
      DIMENSION  H00(NP),C00(NP),HB1(NHB),CB1(NCB),NHBP(NHB),
     1           NCBP(NCB),CHA(70)
C
C Reading initial and boundary conditions of the flow problem
C
      READ (3, 2) CHA
      READ (3, 2) CHA
      READ (3, 2) CHA
    2 FORMAT (70A1)
C
      DO I=1, NP
      READ (3, 10) II, H00(I)
   10 FORMAT (I10, F20.5)
```

```
      END DO
C
      READ (3, 2) CHA
      READ (3, 2) CHA
      READ (3, 2) CHA
C
      DO I=1, NHB
      READ (3, 20) II, NHBP(I), HB1(I)
  20  FORMAT (2I10, F20.5)
      END DO
C
C Reading initial and boundary conditions of the dispersion problem
C
      IF (IHC .EQ. 2) THEN
      READ (3, 2) CHA
      READ (3, 2) CHA
      READ (3, 2) CHA
C
      DO I=1, NP
      READ (3, 10) II, C00(I)
      END DO
C
      READ (3, 2) CHA
      READ (3, 2) CHA
      READ (3, 2) CHA
C
      DO I=1, NCB
      READ (3, 20) II, NCBP(I), CB1(I)
      END DO
      ENDIF
C
      RETURN
      END
C
C****************************************************************
C
C      Subroutine DATAS readis sink/source terms
C
C****************************************************************
C
      SUBROUTINE DATAS (DWR,DWC,WR,WC,NWP,NP,NW,IHC)
      DIMENSION  DWR(NP),DWC(NP),WR(NW),WC(NW),NWP(NW),CHA(70)
```

```
C
C Reading extration and injection data
C
      READ (4, 2) CHA
      READ (4, 2) CHA
      READ (4, 2) CHA
    2 FORMAT (70A1)
C
      DO I=1,NW
      READ (4,10) II, NWP(I), WR(I)
   10 FORMAT (I10, I10, F20.2)
      END DO
C
C Generating distributed sink/source terms for the flow equation
C
      DO I=1,NP
      DWR(I)=0.
      DO J=1,NW
      K=NWP(J)
      IF (K .EQ. I) DWR(I)=WR(J)
      END DO
      END DO
C
C Reading water quality data of the injected water
C
      IF (IHC .EQ. 2) THEN
      READ (4, 2) CHA
      READ (4, 2) CHA
      READ (4, 2) CHA
C
      DO I=1,NW
      READ (4,10) II, NWP(I), WC(I)
      END DO
C
C Generating distributed sink/souece terms for the dispersion equation
C
      DO I=1,NP
      DWC(I)=0.
      DO J=1,NW
      K=NWP(J)
      IF (K .EQ. I) DWC(I)=WC(J)
      END DO
```

```
      END DO
C
      ENDIF
C
      RETURN
      END
C
C*********************************************************************
C
C     Subroutine DATAP reads known parameters and the initial
C     estimates of unknown parameters
C
C*********************************************************************
C
C
      SUBROUTINE DATAP  (ZTK,ZSE,ZPR,ZAL,ZAT,DTK,DSE,DPR,DAL,DAT,
     *            NZONP,NP,NZON)
      DIMENSION  ZTK(NZON),ZSE(NZON),ZPR(NZON),ZAL(NZON),ZAT(NZON),
     *      DTK(NP),DSE(NP),DPR(NP),DAL(NP),DAT(NP),NZONP(NP),
     *      CHA(70)
C
C Reading parameters associated with each zone
C
      READ (7, 2) CHA
      READ (7, 2) CHA
      READ (7, 2) CHA
    2 FORMAT (70A1)
C
      DO I=1,NZON
      READ (7, 10) II, ZTK(I),ZSE(I),ZPR(I),ZAL(I),ZAT(I)
   10 FORMAT (I5, F15.5, 4F10.5)
      END DO
C
C Generating distributed parameters
C
      DO N=1, NP
      DTK(N)=ZTK(NZONP(N))
      DSE(N)=ZSE(NZONP(N))
      DPR(N)=ZPR(NZONP(N))
      DAL(N)=ZAL(NZONP(N))
      DAT(N)=ZAT(NZONP(N))
      END DO
```

```
C
      RETURN
      END
C
C
C******************************************************************
C
C        Subroutine DATAO reads observation data
C
C******************************************************************
C
C
      SUBROUTINE DATAO (HOB,COB,HSDV,CSDV,NOBP,TN,NOB,NCR,IHC,INV)
      DIMENSION
HOB(NOB,NCR),COB(NOB,NCR),HSDV(NOB,NCR),CSDV(NOB,NCR),
     *      NOBP(NOB),TN(NCR),CHA(70)
C
C Reading locations of observation wells
C
      READ (8, 2) CHA
      READ (8, 2) CHA
      READ (8, 2) CHA
   2 FORMAT (70A1)
C
      DO I=1, NOB
      READ (8,10) II, NOBP(I)
  10 FORMAT (I10, I10)
      END DO
C
C Reading observation times
C
      READ (8, 2) CHA
      READ (8, 2) CHA
      READ (8, 2) CHA
C
      DO  I=1, NCR
      READ (8, 20) II, TN(I)
  20 FORMAT (I10, F15.5)
      END DO
C
      IF (INV .EQ. 0) GOTO 100
C
```

```
C Reading head observations
C
    READ (8, 2) CHA
    READ (8, 2) CHA
C
    DO  I=1,NOB
    READ (8, 2) CHA
    READ (8, 2) CHA
    READ (8, 2) CHA
C
    DO  J=1,NCR
    READ (8, 30) JJ, HOB(I,J), HSDV(I,J)
 30 FORMAT (I10, F20.2, F20.2)
C   HSDV(I,J)=1.
    END DO
    END DO
C
C Reading concentration observations
C
    IF (IHC .EQ. 2) THEN
C
    READ (8, 2) CHA
    READ (8, 2) CHA
C
    DO  I=1,NOB
    READ (8, 2) CHA
    READ (8, 2) CHA
    READ (8, 2) CHA
C
    DO  J=1,NCR
    READ (8, 30) JJ, COB(I,J), CSDV(I,J)
C   CSDV(I,J)=1.
    END DO
    END DO
C
    ENDIF
C
 100 RETURN
    END
C
C
C**************************************************************
```

```
C
C        Subroutine DATAC reads computation control numbers
C
C******************************************************************
C
C
       SUBROUTINE DATAC (PT,ALF,EPH,EPC,DT,WTH,WTC,ITRMAX,MAQMAX)
       DIMENSION CHA(70)
C
C Reading computation control numbers of forward solution
C
       READ (9, 2) CHA
       READ (9, 2) CHA
       READ (9, 2) CHA
     2 FORMAT (70A1)
C
       READ (9, 10) PT, ALF, EPH, EPC, DT, WTH, WTC
    10 FORMAT (7F10.5)
C
C Reading computation control numbers of inverse solution
C
       READ (9, 2) CHA
       READ (9, 2) CHA
       READ (9, 2) CHA
C
       READ (9,12) ITRMAX, MAQMAX
    12 FORMAT (2I10)
C
       RETURN
       END
C
C
C******************************************************************
C
C        Subroutine PARZA generates distributed parameters when
C        the values of unknown parameters are changed
C
C******************************************************************
C
C
       SUBROUTINE PARZA (PARA,KGUP,KZON,NZONP,ZTK,ZSE,ZPR,ZAL,ZAT,
      *          DTK,DSE,DPR,DAL,DAT,NP,NZON,NID)
```

```
C
      DIMENSION  ZTK(NZON),ZSE(NZON),ZPR(NZON),ZAL(NZON),ZAT(NZON),
     *        DTK(NP),DSE(NP),DPR(NP),DAL(NP),DAT(NP),
     *        PARA(NID),KGUP(NID),KZON(NID),NZONP(NP)
C
C Generating parameters associated with zones
C
      DO I=1,NID
      J=KGUP(I)
      K=KZON(I)
      IF (J .EQ. 1) ZTK(K)=PARA(I)
      IF (J .EQ. 2) ZSE(K)=PARA(I)
      IF (J .EQ. 3) ZPR(K)=PARA(I)
      IF (J .EQ. 4) ZAL(K)=PARA(I)
      IF (J .EQ. 5) ZAT(K)=PARA(I)
      END DO
C
C generating distributed parameters
C
      DO N=1, NP
      DTK(N)=ZTK(NZONP(N))
      DSE(N)=ZSE(NZONP(N))
      DPR(N)=ZPR(NZONP(N))
      DAL(N)=ZAL(NZONP(N))
      DAT(N)=ZAT(NZONP(N))
      END DO
C
      RETURN
      END
C
C******************************************************************
C
C  Subroutine VD generates velocities and dispersion coefficients.
C
C******************************************************************
C
      SUBROUTINE VD (X,Y,IJM,TK,PR,ALD,ATD,DXX,DXY,DYY,VX,VY,H0,
     *        NP,NE)
      DIMENSION  X(NP),Y(NP),IJM(3,NE),TK(NP),PR(NP),
     1        ALD(NP),ATD(NP),DXX(NE),DXY(NE),DYY(NE),
     2        VX(NE),VY(NE),H0(NP)
C
```

```
C
C
    DO N=1, NE
      MI=IJM(1,N)
      MJ=IJM(2,N)
      MK=IJM(3,N)
      BI=Y(MJ)-Y(MK)
      BJ=Y(MK)-Y(MI)
      BK=Y(MI)-Y(MJ)
      CI=X(MK)-X(MJ)
      CJ=X(MI)-X(MK)
      CK=X(MJ)-X(MI)
      TRS=ABS(CJ*BI-BJ*CI)
C
C Calculating average values of parameters in each element
C
      TKN=(TK(MI)+TK(MJ)+TK(MK))/3.
      PRN=(PR(MI)+PR(MJ)+PR(MK))/3.
      A=-TKN/PRN/TRS
      ALL=(ALD(MI)+ALD(MJ)+ALD(MK))/3.
      ATT=(ATD(MI)+ATD(MJ)+ATD(MK))/3.
C
C Calculating the velocity distribution
C
      VX(N)=A*(BI*H0(MI)+BJ*H0(MJ)+BK*H0(MK))
      VY(N)=A*(CI*H0(MI)+CJ*H0(MJ)+CK*H0(MK))
      V=SQRT(VX(N)*VX(N)+VY(N)*VY(N))
      IF (V .LT. 1.E-10) V=1.E-10
C
C Calculating dispersion coefficients
C
      DXX(N)=(ALL*VX(N)*VX(N)+ATT*VY(N)*VY(N))/V
      DYY(N)=(ATT*VX(N)*VX(N)+ALL*VY(N)*VY(N))/V
      DXY(N)=(ALL-ATT)*(VX(N)*VY(N))/V
    END DO
C   WRITE (6, 20)
C 20 FORMAT(//,3X,'ELEMEN',8X,'VX',8X,'VY',7X,'DXX',7X,'DXY',7X,'DYY')
C   DO M=1, NE
C   WRITE (6, 25) M, VX(M), VY(M), DXX(M), DXY(M), DYY(M)
C 25  FORMAT (I10, 5E10.5)
C   END DO
C
```

```
      RETURN
      END
C
C
C*******************************************************************
C
C   Subroutine COEFF forms coefficient matrices of MCB equations
C
C*******************************************************************
C
C
      SUBROUTINE COEFF
     (X,Y,IJM,MA,MCPN,PS,VX,VY,DXX,DXY,DYY,TR,BR,TH,
     1        H0,TK,SE,PR,WP,CFA,CFB,PT,NP,NP1,NE,NB,IH,IHC)
      DIMENSION  X(NP),Y(NP),IJM(3,NE),MA(NP1),MCPN(NB,NP),PS(NP),
     1        VX(NE),VY(NE),DXX(NE),DXY(NE),DYY(NE),TR(NP),BR(NP),
     2        TH(NP),TK(NP),SE(NP),PR(NP),WP(NP),H0(NP),
     3        CFA(IH),CFB(IH),ME(3)
      DO  K=1, IH
        CFA(K)=0.
        CFB(K)=0.
      END DO
C
C
C
      DO  I=1, NP
        IF (H0(I) .GT. TR(I)) THEN
         TH(I)=TR(I)-BR(I)
        ELSE
         TH(I)=H0(I)-BR(I)
        ENDIF
      END DO
C
C
C
      DO N=1, NE
        MI=IJM(1,N)
        MJ=IJM(2,N)
        MK=IJM(3,N)
C
        IF (IHC .EQ. 1) THEN
         AVH0=(H0(MI)+H0(MJ)+H0(MK))/3.
```

```
                AVTR=(TR(MI)+TR(MJ)+TR(MK))/3.
                AVBR=(BR(MI)+BR(MJ)+BR(MK))/3.
                AVTK=(TK(MI)+TK(MJ)+TK(MK))/3.
                AVSE=(SE(MI)+SE(MJ)+SE(MK))/3.
                AVPR=(PR(MI)+PR(MJ)+PR(MK))/3.
                  IF (AVH0 .GT. AVTR) THEN
                   AVTH=AVTR-AVBR
                   SS=AVSE
                   TT=AVTK*AVTH
                  ELSE
                   AVTH=AVH0-AVBR
                   SS=AVPR
                   TT=AVTK*AVTH
                  ENDIF
                D11=TT
                D22=TT
                D12=0.
              ELSE
                SS=1.
                D11=DXX(N)
                D12=DXY(N)
                D22=DYY(N)
              ENDIF
C
C
C
        DO  J=1, 3
          ME(J)=IJM(J,N)
        END DO
        DO L=1, 3
          MI=ME(1)
          MJ=ME(2)
          MK=ME(3)
          XM=(X(MI)+X(MJ)+X(MK))/3
          YM=(Y(MI)+Y(MJ)+Y(MK))/3
          BI=Y(MJ)-Y(MK)
          BJ=Y(MK)-Y(MI)
          BK=Y(MI)-Y(MJ)
          CI=X(MK)-X(MJ)
          CJ=X(MI)-X(MK)
          CK=X(MJ)-X(MI)
          BJI=Y(MJ)-YM
```

```
        BJJ=YM-Y(MI)
        BJM=Y(MI)-Y(MJ)
        CJI=XM-X(MJ)
        CJJ=X(MI)-XM
        CJM=X(MJ)-X(MI)
        BKI=YM-Y(MK)
        BKM=Y(MK)-Y(MI)
        BKK=Y(MI)-YM
        CKI=X(MK)-XM
        CKM=X(MI)-X(MK)
        CKK=XM-X(MI)
        TRS=ABS(CJ*BI-BJ*CI)
        IF (IHC .EQ. 1) THEN
C
C introducing upstream weights
C
        WI=1./3.
        WJ=1./3.
        WK=1./3.
        ELSE
       CALL WEIT(BI,BJ,BK,CI,CJ,CK,DXX(N),VX(N),VY(N),WI,WJ,WK,PT)
        ENDIF
        BBJI=BJI+WI*BJM
        BBJJ=BJJ+WJ*BJM
        BBJK=WK*BJM
        CCJI=CJI+WI*CJM
        CCJJ=CJJ+WJ*CJM
        CCJK=WK*CJM
        BBKI=BKI+WI*BKM
        BBKJ=WJ*BKM
        BBKK=BKK+WK*BKM
        CCKI=CKI+WI*CKM
        CCKJ=WJ*CKM
        CCKK=CKK+WK*CKM
C
C generating coefficient matrices of discretization equations
C
        NI=MA(MI)
        CFA(NI+1)=CFA(NI+1)
      1     +((D11*BBJI+D12*CCJI)*(BI-BJ)
      2     + (D12*BBJI+D22*CCJI)*(CI-CJ)
      3     + (D11*BBKI+D12*CCKI)*(BI-BK)
```

```
4        + (D12*BBKI+D22*CCKI)*(CI-CK))/(2.*TRS)
         IF (IHC .EQ. 2) THEN
           VV=(VX(N)*(BBJI+BBKI)+VY(N)*(CCJI+CCKI))/4.
           GA=WP(MI)*TRS/6./PS(MI)/PR(MI)/TH(MI)
           CFA(NI+1)=CFA(NI+1)+VV-GA
         ENDIF
         CFB(NI+1)=CFB(NI+1)+TRS*SS/6.
         NS=MA(MI+1)-MA(MI)
         DO I=1,NS
           IF(MCPN(I,MI) .EQ. MJ) THEN
             CFA(NI+I)=CFA(NI+I)
1              +((D11*BBJJ+D12*CCJJ)*(BI-BJ)
2              + (D12*BBJJ+D22*CCJJ)*(CI-CJ)
3              + (D11*BBKJ+D12*CCKJ)*(BI-BK)
4              + (D12*BBKJ+D22*CCKJ)*(CI-CK))/(2.*TRS)
             IF (IHC .EQ. 2) THEN
               VV=(VX(N)*(BBJJ+BBKJ)+VY(N)*(CCJJ+CCKJ))/4.
               CFA(NI+I)=CFA(NI+I)+VV
             ENDIF
             CFB(NI+I)=CFB(NI+I)
           ENDIF
         END DO
         DO I=1,NS
           IF(MCPN(I,MI) .EQ. MK) THEN
             CFA(NI+I)=CFA(NI+I)
1              +((D11*BBJK+D12*CCJK)*(BI-BJ)
2              + (D12*BBJK+D22*CCJK)*(CI-CJ)
3              + (D11*BBKK+D12*CCKK)*(BI-BK)
4              + (D12*BBKK+D22*CCKK)*(CI-CK))/(2.*TRS)
             IF (IHC .EQ. 2) THEN
               VV=(VX(N)*(BBJK+BBKK)+VY(N)*(CCJK+CCKK))/4.
               CFA(NI+I)=CFA(NI+I)+VV
             ENDIF
             CFB(NI+I)=CFB(NI+I)
           ENDIF
         END DO
         J=ME(1)
         ME(1)=ME(2)
         ME(2)=ME(3)
         ME(3)=J
       END DO
     END DO
```

```
C
C
      RETURN
      END
C
C*******************************************************************
C
C    Subroutine  WEIT  generates adaptive upstream weights.
C
C*******************************************************************
C
      SUBROUTINE WEIT (BI,BJ,BK,CI,CJ,CK,DXX,VX,VY,WI,WJ,WK,PT)
      VIJ=CK*VX-BK*VY
      VIK=BJ*VY-CJ*VX
      VKJ=BI*VY-CI*VX
      IF (DXX .GT. 1.E-20) THEN
        WI=( VIJ+VIK)/DXX
        WJ=(-VIJ-VKJ)/DXX
        WK=(-VIK+VKJ)/DXX
      ELSE
        WI=0.
        WJ=0.
        WK=0.
      ENDIF
      WI=1./3.+PT*WI
      WJ=1./3.+PT*WJ
      WK=1./3.+PT*WK
C
      RETURN
      END
C
C*******************************************************************
C
C    Subroutine  SNIM  solves discretization equations.
C
C*******************************************************************
C
      SUBROUTINE SNIM (MA,MCPN,PS,IE,IEE,CFA,CFB,NHBP,H0,H1,H2,
     1          DF00,DF0,HB1,WP,WC,TK,TM,SE,PR,EPS,ALF,
     2          DT00,DT0,DT1,NP,NP1,NE,NB,NHB,IH,IHC)
      DIMENSION  MA(NP1),MCPN(NB,NP),PS(NP),IE(NP),IEE(NP),CFA(IH),
     1      CFB(IH),NHBP(NHB),H0(NP),H1(NP),H2(NP),
```

```
    2        DF00(NP),DF0(NP),HB1(NHB),WP(NP),WC(NP),
    3        TK(NP),TM(NP),SE(NP),PR(NP),EPS(2)
C
C Identifying given head or given concentration boundary nodes
C
      DO I=1, NP
        IE(I)=0
        DO  J=1, NHB
        IF (I .EQ. NHBP(J)) THEN
          H1(I)=HB1(J)
          H0(I)=HB1(J)
          IE(I)=2
        ENDIF
        END DO
      END DO
C
C extrapolation
C
      DO N=1, NP
        IF(IE(N).LT.2) THEN
          IF(ABS(DF00(N)).GE.1.E-5) THEN
          DD=(DT00/DT0)*(DF0(N)/DF00(N))
          ELSE
           DD=1.
          ENDIF
          IF(DD. GT. 5.) DD=5.
          IF(DD. LT. 0.) DD=0.
          H1(N)=H0(N)-(DT1*DD*DF0(N))/DT0
        ENDIF
      END DO
C
C forming iteration equations
C
   40 DO N=1, NP
        IF(IE(N) .EQ. 0) THEN
          NA=MA(N+1)
          NC=MA(N)
          TRS=CFA(NC+1)+CFB(NC+1)/DT1
          DD=CFB(NC+1)/DT1*H0(N)
          ND=NA-NC
          DO K=2, ND
            MC=MCPN(K,N)
```

```
      DD=DD+CFB(NC+K)/DT1*(H0(MC)-H1(MC))-CFA(NC+K)*H1(MC)
      END DO
C
C incorporating sink/source terms
C
      IF (IHC .EQ. 1) THEN
        DD=DD-WP(N)
      ELSE
       IF (WP(N) .GT. 0.) DD=DD-WP(N)*H1(N)/TM(N)/PR(N)
       IF (WP(N) .LT. 0.) DD=DD-WP(N)*WC(N)/TM(N)/PR(N)
      ENDIF
C
C The selecting node iteration procedure
C
      H2(N)=(1.-ALF)*H1(N)+ALF*DD/TRS
      IF ( ABS(H2(N)-H1(N)) .LT. EPS(IHC)) IE(N)=1
      H1(N)=H2(N)
C
    ENDIF
    END DO
C
C
C
    DO  N=1, NP
      IEE(N)=1
    END DO
    DO N=1,NP
      IF(IE(N) .EQ. 0) THEN
        IEE(N)=0
        NA=MA(N+1)
        NC=MA(N)
        ND=NA-NC
        DO K=2,ND
          MC=MCPN(K,N)
          IF(IE(MC). LT. 2) IEE(MC)=0
        END DO
      ENDIF
    END DO
    DO  N=1, NP
      IF(IE(N) .LT. 2) IE(N)=IEE(N)
    END DO
C
```

```
      DO N=1 ,NP
        IF(IE(N) .EQ. 0) GO TO 40
      END DO
C
C The solution is obtained for this time step.
C
      DT00=DT0
      DT0=DT1
      DO N=1,NP
        IF(IE(N).LT.2) THEN
          DF00(N)=DF0(N)
          DF0(N)=H0(N)-H1(N)
          H0(N)=H1(N)
        ENDIF
      END DO
C
      RETURN
      END
C
C*****************************************************************
C
C      Subroutine MCB2D is the simulation program
C
C*****************************************************************
C
      SUBROUTINE MCB2D (X,Y,TR,BR,TH,PS,IJM,MA,MCPN,NZONP,
     1          H00,C00,HB1,CB1,NHBP,NCBP,
     2          DWR,DWC,WR,WC,NWP,
     3          DTK,DSE,DPR,DAL,DAT,
     4          H0,H1,H2,C0,C1,C2,HF00,HF0,CF00,CF0,
     5          DXX,DXY,DYY,VX,VY,CFA,CFB,IE,IEE,
     6          HOB,HSDV,COB,CSDV,HOBC,COBC,TN,NOBP,
     7          EPS,PT,ALF,DT,WTH,WTC,TLSQ,
     8       NP,NP1,NE,NB,IHC,INV,NID,NST,
     9       NW,NHB,NCB,NOB,NCR,NZON,IH,IPR)
C
      DIMENSION  X(NP),Y(NP),TR(NP),BR(NP),TH(NP),PS(NP),IJM(3,NE),
     1       MA(NP1),MCPN(NB,NP),NZONP(NP),
     2       H00(NP),C00(NP),HB1(NHB),CB1(NCB),NHBP(NHB),NCBP(NCB),
     3       DWR(NP),DWC(NP),WR(NW),WC(NW),NWP(NW),
     4       DTK(NP),DSE(NP),DPR(NP),DAL(NP),DAT(NP),
     5       H0(NP),H1(NP),H2(NP),C0(NP),C1(NP),C2(NP),
```

```
      6       HF00(NP),HF0(NP),CF00(NP),CF0(NP),CFA(IH),CFB(IH),
      7       VX(NE),VY(NE),DXX(NE),DXY(NE),DYY(NE),IE(NP),IEE(NP),
      8       HOB(NOB,NCR),COB(NOB,NCR),HSDV(NOB,NCR),CSDV(NOB,NCR),
      9       HOBC(NOB,NCR),COBC(NOB,NCR),TN(NCR),NOBP(NOB),
      1       EPS(2),ME(3)
C
C Setting initial conditions
C
      TSUM=0.
      DT00=DT
      DT0=DT
      DT1=DT
      DO N=1, NP
        HF00(N)=0.
        HF0(N)=0.
        CF00(N)=0.
        CF0(N)=0.
        H0(N)=H00(N)
        C0(N)=C00(N)
      END DO
C
C  calculating mass transport in transient flow field
C
      L=1
      IF (NST .EQ. 1) GOTO 45
   10 TSUM=TSUM + DT1
      CALL COEFF (X,Y,IJM,MA,MCPN,PS,VX,VY,DXX,DXY,DYY,TR,BR,TH,
      1           H0,DTK,DSE,DPR,DWR,CFA,CFB,PT,NP,NP1,NE,NB,IH,1)
      CALL SNIM  (MA,MCPN,PS,IE,IEE,CFA,CFB,NHBP,H0,H1,H2,
      1           HF00,HF0,HB1,DWR,DWC,TH,DTK,DSE,DPR,EPS,ALF,
      2           DT00,DT0,DT1,NP,NP1,NE,NB,NHB,IH,1)
C
C
C
      IF (IHC .EQ. 2) THEN
C
      CALL VD    (X,Y,IJM,DTK,DPR,DAL,DAT,DXX,DXY,DYY,VX,VY,H0,
      1           NP,NE)
      CALL COEFF (X,Y,IJM,MA,MCPN,PS,VX,VY,DXX,DXY,DYY,TR,BR,TH,
      1           H0,DTK,DSE,DPR,DWR,CFA,CFB,PT,NP,NP1,NE,NB,IH,2)
      CALL SNIM  (MA,MCPN,PS,IE,IEE,CFA,CFB,NCBP,C0,H1,H2,
      1           CF00,CF0,CB1,DWR,DWC,TH,DTK,DSE,DPR,EPS,ALF,
```

```
   2        DT00,DT0,DT1,NP,NP1,NE,NB,NCB,IH,2)
C
    ENDIF
C
C writing head and concentration distributions
C
    IF ( ABS(TN(L)-TSUM) .LT. 1.E-5) THEN
C
      IF (IPR .LT. 3) THEN
        WRITE (6, 12) TN(L)
 12  FORMAT (/,1X,'THE OBSERVATION TIME IS:', F10.2,/)
        WRITE (6, 14)
 14  FORMAT (/, 6X,'NODE', 11X,'HEAD', 2X, 'CONCENTRATION',/)
        DO  N=1, NP
          WRITE (6, 16) N, H0(N), C0(N)
 16  FORMAT (I10, 2F15.2)
        END DO
      ENDIF
C
C writing solutions corresponding to observation wells and times
C
      DO  N=1, NP
        DO I=1, NOB
          IF (N .EQ. NOBP(I)) THEN
            HOBC(I,L)=H0(N)
            COBC(I,L)=C0(N)
          ENDIF
        END DO
      END DO
C
      IF (IPR .LT. 4) THEN
        WRITE (6, 12) TN(L)
        WRITE (6, 26)
 26  FORMAT (/, 6X, 'NODE', 11X, 'HEAD', 3X, 'CONCENTRATION',/)
        DO I=1,NOB
          WRITE (6,28) NOBP(I), HOBC(I,L), COBC(I,L)
 28  FORMAT (I10, 2F15.5)
        END DO
      ENDIF
C
C turning to the next time step
C
```

```
      L=L+1
C
   END IF
C
   IF (L .GT. NCR) GOTO  80
   IF (DT1 .LT. 0.1) DT1=DT1*1.25
   IF (DT1+TSUM.GT. TN(L)) DT1=TN(L)-TSUM
   GOTO 10
C
C generating the steady flow field
C
 45  CALL COEFF (X,Y,IJM,MA,MCPN,PS,VX,VY,DXX,DXY,DYY,TR,BR,TH,
    1       H0,DTK,DSE,DPR,DWR,CFA,CFB,PT,NP,NP1,NE,NB,IH,1)
 50  TSUM=TSUM + DT1
    CALL SNIM  (MA,MCPN,PS,IE,IEE,CFA,CFB,NHBP,H0,H1,H2,
    1       HF00,HF0,HB1,DWR,DWC,TH,DTK,DSE,DPR,EPS,ALF,
    2       DT00,DT0,DT1,NP,NP1,NE,NB,NHB,IH,1)
   DO N=1, NP
     DD=HF0(N)
     IF ( ABS(DD)/DT0 .GT. EPS(1)) THEN
       IF (DT1 .LT. 10.) DT1=DT1*1.25
       GOTO 50
     ENDIF
   END DO
C
   IF (IPR .LT. 4) THEN
     WRITE (6,52)
 52  FORMAT (/,1X,'THE STEADY STATE HEAD DISTRIBUTION',/,6X,'NODE',
    *      11X, 'HEAD',/)
     DO  N=1, NP
       WRITE (6, 54) N, H0(N)
 54  FORMAT (I10, F15.2)
     END DO
   ENDIF
C
C
   DO N=1, NP
     DO  I=1, NOB
      IF (N .EQ. NOBP(I)) THEN
        DO K=1,NCR
         HOBC(I,K)=H0(N)
        END DO
```

```
            ENDIF
          END DO
        END DO
C
C calculating mass transport in the steady folw field
C
      IF (IHC .EQ. 2) THEN
C
      CALL VD    (X,Y,IJM,DTK,DPR,DAL,DAT,DXX,DXY,DYY,VX,VY,H0,
     1           NP,NE)
      CALL COEFF (X,Y,IJM,MA,MCPN,PS,VX,VY,DXX,DXY,DYY,TR,BR,TH,
     1           H0,DTK,DSE,DPR,DWR,CFA,CFB,PT,NP,NP1,NE,NB,IH,2)
C
      L=1
      TSUM=0.
      DT00=DT
      DT0=DT
      DT1=DT
 60   TSUM=TSUM + DT1
      CALL SNIM  (MA,MCPN,PS,IE,IEE,CFA,CFB,NCBP,C0,H1,H2,
     1           CF00,CF0,CB1,DWR,DWC,TH,DTK,DSE,DPR,EPS,ALF,
     2           DT00,DT0,DT1,NP,NP1,NE,NB,NCB,IH,2)
      IF ( ABS(TN(L)-TSUM) .LT. 1.E-5) THEN
        WRITE (6,62) TN(L)
 62   FORMAT (1X,'THE OBSERVATION TIME IS:',F10.2, /)
      IF (IPR .LT. 3) THEN
        WRITE (6, 64)
 64   FORMAT (/, 6X, 'NODE',3X, 'CONCENTRATION',/)
        DO N=1, NP
          WRITE (6,66) N, C0(N)
 66   FORMAT (I10, F15.2)
        END DO
      ENDIF
C
C writing solutions corresponding to observation wells and
C times
C
        DO N=1,NP
          DO I=1, NOB
            IF (N .EQ. NOBP(I)) THEN
              COBC(I,L)=C0(N)
            ENDIF
```

```
            END DO
            END DO
C
        IF (IPR .LT. 4) THEN
            WRITE (6,70)
  70 FORMAT (/, 6X, 'NODE', 3X, 'CONCENTRATION',/)
            DO I=1, NOB
                WRITE (6, 66) NOBP(I), COBC(I,L)
            END DO
        ENDIF
C
C turning to the next time step
C
        L=L+1
C
        ENDIF
C
        IF (L .LE. NCR) THEN
            IF (DT1 .LT. 0.1) DT1=DT1*1.25
            IF (DT1+TSUM .GT. TN(L)) DT1=TN(L)-TSUM
            GOTO 60
        ENDIF
C
        ENDIF
C
C
C
  80 IF (INV .EQ. 1) THEN
        IF (IPR .LT. 5) THEN
            WRITE (6, 82)
  82 FORMAT (/,5X,'COMPARING CALCULATED HEADS WITH OBSERVED
HEADS')
            DO I=1,NOB
                WRITE (6, 84) I
  84 FORMAT (/, 15X, 'FOR OBSERVATION WELL NO.', I3)
                WRITE (6, 86)
  86 FORMAT (7X,'NO.',6X,'TIME',11X,'CAL. HEAD',11X,'OBS. HEAD')
                DO K=1,NCR
                    WRITE (6, 88) K, TN(K), HOBC(I,K), HOB(I,K)
  88 FORMAT (I10, F10.2, 2F20.5)
                END DO
            END DO
```

```
      ENDIF
C
C
      HLSQ=0.
      HTW=0.
      DO I=1, NOB
        DO K=1, NCR
          DH= HOBC(I,K)-HOB(I,K)
          DW= HSDV(I,K)
          HLSQ=HLSQ +(DH/DW)*(DH/DW)
          HTW=HTW + 1./(DW*DW)
        END DO
      END DO
      AHLSQ=HLSQ/HTW
C
C
C
      IF (IHC .EQ. 2) THEN
       IF (IPR .LT. 5) THEN
          WRITE (6, 92)
 92  FORMAT (/,5X,'COMPARING CALCULATED CONC. WITH OBSERVED
CONC.')
          DO I=1,NOB
           WRITE (6, 84) I
           WRITE (6, 94)
 94  FORMAT (7X,'NO.',6X,'TIME',10X,'CAL. CONC.',11X,'OBS. CONC.')
           DO K=1,NCR
              WRITE (6, 88) K, TN(K), COBC(I,K), COB(I,K)
           END DO
          END DO
       ENDIF
C
C
C
      CLSQ=0.
      CTW=0.
      DO I=1, NOB
        DO K=1, NCR
          DC= COBC(I,K)-COB(I,K)
          DW= CSDV(I,K)
          CLSQ=CLSQ +(DC/DW)*(DC/DW)
          CTW=CTW + 1./(DW*DW)
```

```
        END DO
        END DO
        ACLSQ=CLSQ/CTW
C
      ENDIF
C
C
      IF (IHC .EQ. 1) THEN
        CLSQ=0.
        ACLSQ=0.
      ENDIF
C
      TLSQ=WTH * HLSQ + WTC * CLSQ
      IF (IPR. LT. 5) THEN
        WRITE (6, 96) HLSQ, CLSQ, TLSQ
   96 FORMAT (//,5X,'HSLQ=',F10.3,10X,'CLSQ=',F10.3,10X,'TLSQ=',F10.3)
        WRITE (6, 98) AHLSQ, ACLSQ
   98 FORMAT (//,4X,'AHSLQ=',F10.3, 9X,'ACLSQ=',F10.3)
      ENDIF
C
      ENDIF
C
C
  100 RETURN
      END
```

C.6. Example

Figure C.2 shows a 600 m by 1200 m rectangle confined aquifer in the horizontal direction with constant **TR** = 50 m and **BR** = 0. Its bound section AB is a river with given water level **HB1** = 100 m and given concentration **CB1** = 0. Other boundary sections BC, CD, and DA are all impermeable. For numerical simulation, the flow region is divided into 40 elements with 30 nodes as shown in Fig.C.2. There are one recharge well at node #8 and one pumping well at node #23. Their rates are 500 m^3/day and 4000 m^3/day, respectively. The concentration of recharge water is 1000 mg/m^3. The aquifer has two zones. Nodes #1 to #15 belong to zone 1, other nodes belong to zone 2. The values of parameters are different in the two zones (see input file **DATAP**). Assume that the initial head distribution and the initial concentration distribution are **H00** = 100 m and **C00** =

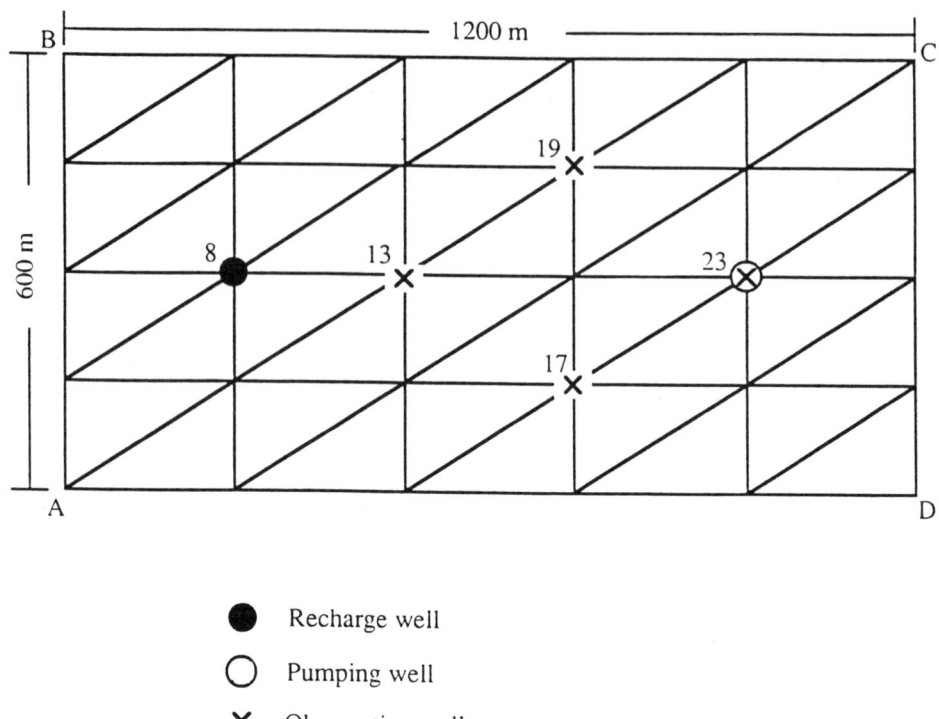

● Recharge well

○ Pumping well

✗ Observation well

Fig. C.2. Aquifer Configuration and Discretization.

0, respectively, in the whole aquifer. There are four observation wells located at nodes #13, #17, #19 and #23. Ten observation times are 0.05, 0.1, 0.2, 0.4, 0.6, 1.0, 2.0, 4.0, 6.0 and 10 (days). The inverse problem for this example involves the identification of hydraulic conductivities **ZTK(1)** and **ZTK(2)** for zone 1 and zone 2, and specific storage coefficient **ZSE(1)** for zone 1. Other parameters are assumed to be known.

Let the true values of the identified parameters be **ZTK(1)** = 5 m/day, **ZTK(2)** = 10 m/day and **ZSE(1)** = 0.001. Running the simulation model once (setting **INV** = 0 and **IPR** = 3), we can obtain all "observed" heads and concentrations. Putting these data into data file **DATAO**, we will have "observation d"ata for inverse solution. Then we change **INV** = 1 and set **PARAI(1)** = 3 m/day, **PARAI(2)** = 8 m/day and **PARAI(3)** = 0.0005 as initial estimates to solve the inverse problem. From the output file **RESUL**, we can see that the true values of unknown parameters are exactly found in only 3 iterations. Note that the above procedure implies a special assumption, i.e., there are no any model structure error, no any observation error, and no any computation error.

The following are input and output files for this example.

****

INPUT DATA FILES

****

Data File: DATAN

```
CONTROL INTEGERS
         NP       NE       NB       IHC      INV      NST
IPR
         30       40       9        2        1        0
  6
CONTROL INTEGERS
         NW       NHB      NCB      NOB      NCR      NZON
NID
         2        5        5        4        10       2
  3
//
```

Data File: DATAG

```
ZONATION, NODE COORDINATES, ELEVATIONS OF TOP AND BED ROCKS
        NODE      X COORD.00      Y COORD.00     NZONP      TR.00
BR.00
          1          0.0             0.0           1        50.         0.
          2          0.0           150.0           1        50.         0.
          3          0.0           300.0           1        50.         0.
          4          0.0           450.0           1        50.         0.
          5          0.0           600.0           1        50.         0.
          6        240.0             0.0           1        50.         0.
          7        240.0           150.0           1        50.         0.
          8        240.0           300.0           1        50.         0.
          9        240.0           450.0           1        50.         0.
         10        240.0           600.0           1        50.         0.
         11        480.0             0.0           1        50.         0.
         12        480.0           150.0           1        50.         0.
         13        480.0           300.0           1        50.         0.
         14        480.0           450.0           1        50.         0.
```

15	480.0	600.0	1	50.	0.
16	720.0	0.0	2	50.	0.
17	720.0	150.0	2	50.	0.
18	720.0	300.0	2	50.	0.
19	720.0	450.0	2	50.	0.
20	720.0	600.0	2	50.	0.
21	960.0	0.0	2	50.	0.
22	960.0	150.0	2	50.	0.
23	960.0	300.0	2	50.	0.
24	960.0	450.0	2	50.	0.
25	960.0	600.0	2	50.	0.
26	1200.0	0.0	2	50.	0.
27	1200.0	150.0	2	50.	0.
28	1200.0	300.0	2	50.	0.
29	1200.0	450.0	2	50.	0.
30	1200.0	600.0	2	50.	0.

NODE NUMBERS OF EACH ELEMENT

element	i	j	k
1	1	6	7
2	1	7	2
3	2	7	8
4	2	8	3
5	3	8	9
6	3	9	4
7	4	9	10
8	4	10	5
9	6	11	12
10	6	12	7
11	7	12	13
12	7	13	8
13	8	13	14
14	8	14	9
15	9	14	15
16	9	15	10
17	11	16	17
18	11	17	12
19	12	17	18
20	12	18	13
21	13	18	19
22	13	19	14
23	14	19	20
24	14	20	15
25	16	21	22
26	16	22	17
27	17	22	23
28	17	23	18
29	18	23	24
30	18	24	19
31	19	24	25
32	19	25	20
33	21	26	27
34	21	27	22
35	22	27	28
36	22	28	23
37	23	28	29
38	23	29	24

```
              39            24            29            30
              40            24            30            25
//
```

Data File: DATAB

INITIAL CONDITIONS OF HEAD
```
       NODE            H00.
          1            100.0
          2            100.0
          3            100.0
          4            100.0
          5            100.0
          6            100.0
          7            100.0
          8            100.0
          9            100.0
         10            100.0
         11            100.0
         12            100.0
         13            100.0
         14            100.0
         15            100.0
         16            100.0
         17            100.0
         18            100.0
         19            100.0
         20            100.0
         21            100.0
         22            100.0
         23            100.0
         24            100.0
         25            100.0
         26            100.0
         27            100.0
         28            100.0
         29            100.0
         30            100.0
```
HEAD BOUNDARY CONDITION
```
     NUMBER          NODE          HEAD.
          1            1            100.
          2            2            100.
          3            3            100.
          4            4            100.
          5            5            100.
```
INITIAL CONCENTRATIONS
```
       NODE            C00.
          1             0.0
          2             0.0
          3             0.0
          4             0.0
          5             0.0
          6             0.0
```

```
                7                    0.0
                8                    0.0
                9                    0.0
               10                    0.0
               11                    0.0
               12                    0.0
               13                    0.0
               14                    0.0
               15                    0.0
               16                    0.0
               17                    0.0
               18                    0.0
               19                    0.0
               20                    0.0
               21                    0.0
               22                    0.0
               23                    0.0
               24                    0.0
               25                    0.0
               26                    0.0
               27                    0.0
               28                    0.0
               29                    0.0
               30                    0.0
CONCENTRATION BOUNDARY CONDITION
     NUMBER        NODE          CONCEN.
          1           1            10.
          2           2            10.
          3           3            10.
          4           4            10.
          5           5            10.
//
```

Data File: DATAS

```
EXTRACTION OR INJECTION
     NUMBER        NODE            WR.0
          1          23          4000.0
          2           8          -500.0
WATER QUALITY
     NUMBER        NODE            WC.0
          1          23             0.0
          2           8          1000.0
//
```

Data File: DATAP

```
INITIAL VALUES OF PARAMETERS
   ZONE       ZTK.       ZSE.       ZPR.       ZAL.       ZAT.
      1         5.      0.001        0.1        60.        20.
```

```
  2       10.       0.002     0.2      60.       20.
//
```

Data File: DATAO

OBSERVATION WELLS
     NUMBER   NODE
          1   13
          2   17
          3   19
          4   23
OBSERVATION TIMES
     NUMBER   TIME
          1   0.05
          2   0.1
          3   0.2
          4   0.4
          5   0.6
          6   1.0
          7   2.0
          8   4.0
          9   6.0
         10   10.0
  HEAD OBSERVATIONS
        OBSERVATION WELL NO.   1
     TIME   HEAD.   HSDV.
          1   100.01   0.05
          2   100.01   0.05
          3   99.92   0.05
          4   99.56   0.05
          5   99.08   0.05
          6   98.04   0.05
          7   95.71   0.05
          8   92.52   0.05
          9   90.65   0.05
         10   88.91   0.05
        OBSERVATION WELL NO.   2
     TIME   HEAD.   HSDV.
          1   99.95   0.05
          2   99.82   0.05
          3   99.47   0.05
          4   98.72   0.05
          5   97.98   0.05
          6   96.56   0.05
          7   93.57   0.05
          8   89.51   0.05
          9   87.14   0.05
         10   84.93   0.05
        OBSERVATION WELL NO.   3
     TIME   HEAD.   HSDV.
          1   99.95   0.05
          2   99.81   0.05
          3   99.47   0.05
```
```

4	98.71	0.05
5	97.97	0.05
6	96.55	0.05
7	93.55	0.05
8	89.49	0.05
9	87.11	0.05
10	84.89	0.05

OBSERVATION WELL NO.   4

TIME	HEAD.	HSDV.
1	98.50	0.05
2	97.88	0.05
3	97.14	0.05
4	96.05	0.05
5	95.12	0.05
6	93.47	0.05
7	90.14	0.05
8	85.65	0.05
9	83.03	0.05
10	80.58	0.05

CONCENTRATION   OBSERVATIONS

OBSERVATION WELL NO.   1

TIME	CONC.	CSDV.
1	0.00	0.50
2	0.00	0.50
3	0.00	0.50
4	0.00	0.50
5	0.00	0.50
6	0.01	0.50
7	0.08	0.50
8	0.44	0.50
9	1.21	0.50
10	3.99	0.50

OBSERVATION WELL NO.   2

TIME	CONC.	CSDV.
1	0.00	0.50
2	0.00	0.50
3	0.00	0.50
4	0.00	0.50
5	0.00	0.50
6	0.00	0.50
7	0.00	0.50
8	0.00	0.50
9	0.00	0.50
10	0.00	0.50

OBSERVATION WELL NO.   3

TIME	CONC.	CSDV.
1	0.00	0.50
2	0.00	0.50
3	0.00	0.50
4	0.00	0.50
5	0.00	0.50
6	0.00	0.50
7	0.00	0.50
8	0.00	0.50
9	0.00	0.50
10	0.01	0.50

```
          OBSERVATION WELL NO.    4
   TIME                CONC.              CSDV.
     1                 0.00               0.50
     2                 0.00               0.50
     3                 0.00               0.50
     4                 0.00               0.50
     5                 0.00               0.50
     6                 0.00               0.50
     7                 0.00               0.50
     8                 0.00               0.50
     9                 0.00               0.50
    10                 0.00               0.50
```

Data File: DATAC

COMPUTATIONAL PAREAMETERS

PT.	ALF.	EPH.	EPC.	DT.	WTH.	WTC.
0.0	1.1	0.0001	0.0001	0.001	1.0	100.0

ITERATION CONTROL NUMBERS

```
       ITR       MARQ
        3          5
//
```

Data File: PARAM

********** PARAMETER IDENTIDFICATION************

NUMBER	KGRUP	KZON	P.ARAL	P.ARAI	P.ARAU
1	1	1	1.0	3.0	10.0
2	1	2	5.0	8.0	15.0
3	2	1	0.0001	0.003	0.005

//

****

OUTPUT DATA FILE

****

Data File: RESUL

```
            *****  Program runs ******

       INV= 1          IHC= 2          NST= 0
       *** BEGIN TO SOLVE THE INVERSE PROBLEM ***

       ---- Write the initial parameters ----
```

```
NO       KGUP        KZON              PARAMETER
    1         1           1                3.00000
    2         1           2                8.00000
    3         2           1                0.00300
    ---- The value of initial LS error is:      27441.7930
    ---- The modefied parameters at MARQ ----   0
NO       KGUP        KZON              PARAMETER
    1         1           1                4.50554
    2         1           2               10.13061
    3         2           1                0.00172
    ---The value of modified LS error is:        961.2468
    ---- The modefied parameters at MARQ ----   0
NO       KGUP        KZON              PARAMETER
    1         1           1                5.01130
    2         1           2               10.06022
    3         2           1                0.00098
    ---The value of modified LS error is:          3.3960
    ---- The modefied parameters at MARQ ----   0
NO       KGUP        KZON              PARAMETER
    1         1           1                4.99466
    2         1           2               10.00657
    3         2           1                0.00100
    ---The value of modified LS error is:          0.2591
    ---- The modefied parameters at MARQ ----   0
NO       KGUP        KZON              PARAMETER
    1         1           1                4.99522
    2         1           2               10.00832
    3         2           1                0.00100
    ---The value of modified LS error is:          0.2447
    ---- The modefied parameters at MARQ ----   0
NO       KGUP        KZON              PARAMETER
    1         1           1                4.99521
    2         1           2               10.00830
    3         2           1                0.00100
    ---The value of modified LS error is:          0.2416
```

References

Ahlfeld, D. P., J. M. Mulvey, G. F. Pinder, and E. F. Wood, Contaminated groundwater remediation design using simulation, optimization, and sensitivity theory, 1, Model development, *Water Resour. Res.*, *24*(d), 431-441, 1988.

Ahmed, S., and G. de Marsily, Comparison of geostatistical methods for estimating transmissivity using data on transmissivity and specific capacity, *Water Resour. Res.*, *23*(9), 1717-1737, 1987.

Ahmed, S., and G. de Marsily, Cokriged estimation of aquifer transmissivity as an indirect solution of the inverse problem: A practical approach, *Water Resour. Res.*, *29*(2), 521-530, 1993.

Andricevic, R., and P. K. Kitanidis, Optimization of the pumping schedule in aquifer remediation under uncertainty, *Water Resour. Res.*, *26*(5), 875-887, 1990.

Andricevic, R., Coupled withdrawal and sampling designs for groundwater supply models, *Water Resour. Res.*, *29*(1), 5-16, 1993.

Banks, H. T., and P. L. Daniel, Estimation of variable coefficients in parabolic distributed systems, *IEEE Trans. Auto. Control.*, AC-30, 386-398, 1985.

Bard, Y., *Nonlinear Parameter Estimation*, Academic, San Diego, Calif., 1974

Bear, J., *Dynamics of Fluids in Porous Media*, Elsevier, New York, 1972.

Bear, J., *Hydraulics of Groundwater*, McGraw-Hill, New York, 1979.

Bear, J., and A. Veruijt, *Modeling Groundwater Flow and Pollution*, D. Reidel, Hingham, Mass., 1987.

Beck, J. V., and K. J. Arnold, *Parameter Estimation in Engineering and Science*, John Wiley & Sons, pp.501, 1977.

Cacuci, D. G., C. F. Weber, E. M. Oblow, and J. H. Marable, Sensitivity theory for general systems of nonlinear equations, *Nucl. Sci. Eng.*, *75*, 88-110, 1980.

Carrera, J., E. Usnnoff, and F. Szidarovszky, A method for optimal observation network design for groundwater management, *J. Hydrol.*, *73*, 147-163, 1984.

Carrera, J., and S. P. Neuman, Estimation of aquifer parameters under transient and steady state conditions, 1, Maximum likelihood method incorporating prior information, *Water Resour. Res.*, *22*(2), 199-210, 1986*a*.

Carrera, J., and S. P. Neuman, Estimation of aquifer parameters under transient and steady state conditions, 2, Uniqueness, stability, and solution algorithms, *Water Resour. Res.*, *22*(2), 211-227, 1986*b*.

Carrera, J., and S. P. Neuman, Estimation of aquifer parameters under transient and steady state conditions, 3, Application to synthetic and field data, *Water Resour. Res.*, *22*(2), 228-242, 1986*c*.

Carrera, J., State of the art of the inverse problem applied to the flow and solute transport equations, in *Groundwater Flow and Quality Modeling, NATO ASI Ser.*, edited by E. Curtodio, A. Gurgui, and J. P. Lobo Ferreira, D. Reidel, Hingham, Mass., 1987

Carter, R. D., L. F. Kemp, A. C. Pearce, Jr., and D. L. Williams, Performance matching with constraints, *Soc. Pet. Eng. J., 14*(2), 187-196, 1974.

Chang, S., W. W-G. Yeh, A proposed algorithm for the solution of the large-scale inverse problem in groundwater, *Water Resour. Res., 12*(3), 365-374, 1976.

Chankong, V., and Y. Y. Haimes, *Multiobjective Decision Making: Theory and Methodology*, North-Holland, Amsterdam, 1983.

Chavent, G., M. Dupuy, and P. Lemonnier, History matching by use of optimal theory, *Soc. Pet. Eng. J., 15*(1), 74-86, 1975.

Chavent, G., Identification of distributed parameter system: About the output least square method, its implementation, and identifiability, in *Proceedings of the Fifth IFAC Symposium*, vol. 1, *Identification and System Parameter Estimation*, edited by R. Igermann, pp. 85-97, Pergamon, New York, 1979.

Chavent, G., Local stability of the output least squares parameter estimation technique, *Math. Apl. Comput., 2(1)*, 3-22, 1983.

Chavent, G., Identifiability of parameters in the output least square formulation, in *Identifiability of Parametric Models*, edited by E. Walter, Pergamon, New York, 1987a.

Chavent, G., A geometrical approach for the priori study of non-linear inverse problems, in *Inverse Problems: An Interdisciplinary Study*, edited by P. C. Sabatier, Academic Press, Harcourt Brace Javanovich Publishers, p.509-521, 1987b.

Chavent, G., A new sufficient condition for the well-posedness of nonlinear least squares problems arising in identification and control. In A. Bensoussan and J.-L. Lions, editors, *Lectures Notes in Control and Information Sciences*, pages 452-463, Springer, 1990.

Chen, Y. M., Generalized pulse-spectrum technique, *Geophysics, vol.50*, 1664-1675, 1985.

Chen, Y. M., and J.Q. Liu, An iterative neumerical algorithm for solving multi-parameter inverse problem of evolution partial differential equation, *J. Comput. Phys., 53*, 429-442, 1984.

Cheng, J.-M., and W. W-G. Yeh, A proposed quasi-Newton method for parameter identification in a flow and transport system, *Advances in Water Resources, 15*(4), 239-250, 1992.

Cleveland, T., W. W-G. Yeh, "Sampling Network Design for Transport Parameter Identification," *Journal of Water Resources Planning and Management, ASCE, 117*(1), 37-51, 1991.

Clifton, P. M., and S. P. Neuman, Effects of Kriging and inverse modeling on conditional simulation of the Avra Valley aquifer in southern Arizona, *Water Resour. Res., 18*(4), 1215-1234, 1982.

Cooley, R. L., A method of estimating parameters and assessing reliability for models of steady state groundwater flow, 1, Theory and numerical properties, *Water Resour. Res., 13*(2), 318-324, 1977.

Cooley, R. L., Incorporation of prior information on parameters into nonlinear regression groundwater flow models, 1, Theory, *Water Resour. Res., 18*(4), 965-976, 1982.

Cooley, R. L., and A. V. Vecchia, Calculation of nonlinear confidence and prediction intervals for groundwater flow models, *Water Resour. Bull., 23*(4), 581-599, 1987.

Cooley, R. L., Exact Scheffé-type confidence intervals of output from groundwater flow models, 1, Use of Hydrogeologic information, *Water Resour. Res., 29*(1), 17-34, 1993*a*.

Cooley, R. L., Exact Scheffé-type confidence intervals of output from groundwater flow models, 2, Combined use of Hydrogeologic information and calibration data, *Water Resour. Res., 29*(1), 35-50, 1993*b*.

Dagan, G., Stochastic modeling of groundwater flow by unconditional and conditional probabilities, 1, Conditional simulation and the direct problem, *Water Resour. Res., 18*(4), 813-833, 1982.

Dagan, G., Stochastic modeling of groundwater flow by unconditional and conditional probabilities: The inverse problem, *Water Resour. Res., 21*(1), 65-72, 1985.

Dagan, G., Statistical theory of groundwater flow and transport: pore to laboratory, laboratory to formation, formation to regional scale, *Water Resour. Res., 22*(9), 120S-134S, 1986.

Dagan, and Y. Rubin, Stochastic identification of recharge, transmissivity, and storativity in aquifer transient flow: A quasi-steady approach, *Water Resear. Res, 24(10)*, 1698-1710, 1988.

Dagan, G., *Flow and Transport in Porous Formations*, pp.461, Springer-Verlag, Berlin Heidelber, 1989.

Davis, J. C., Statistics and Data Analysis in Geology, John Wiley & Sons, pp. 646 (second edition), 1986.

Delhomme, J. P., Kriging in the hydrosciences, *Advan.Water Resour., 1*(5), 251-266, 1978.

Delhomme, J. P., Spatial variability and uncertainty in groundwater flow parameters: A geostatistical approach, *Water Resour. Res., 15*(2), 269-280, 1979.

Dennis, J. E., *Numerical methods for unconstrained optimization and non-linear equations*, Englewood Cliffs, New Jercy, Prentice-Hall, pp.378, 1983.

Devary, J. L., and P. G. Doctor, Pore velocity estimation uncertainties, *Water Resour. Res., 18*(4), 1157-1164, 1982.

Devore, J. L., *Probability and Statistics for Engineering and the Sciences*, Monterey, Calif., Brooks/Cole, Pub., Co., pp.672, 1987.

Distefano, N., and A. Rath, An identification approach to subsurface hydrological systems, *Water Resour. Res., 11*(6), 1005-1012, 1975.

Domenico, P. A., *Concepts and Models in Groundwater Hydrology*, McGraw-Hill, New York, pp.405, 1972.

Emsellem, Y., and G. de Marsily, An automatic solution for the inverse problem, *Water Resour. Res., 7*(5), 1264-1283, 1971.

Ewing, R., T. Lin, and R. Falk, Inverse and ill-posed problems in reservoir simulation, in *Inverse and Ill-Posed Problems*, edited by H. W. Engl and C. W. Groetsch, Academic, San Diego, Calif., 1987.

Ewing, R. E., and T. Lin, Parameter identification problems in single-phase and two-phase flow, in *Control and Estimation of Distributed Parameter Systems*, edited by F. Kappel, K. Kunisch, W. Schappacher, Birkhauser Verlag, 1989.

Fletcher, R., and C. M. Reeves, Function minimization by conjugate gradients, *Computer Journal*, 7, 149-154, 1964.

Fletcher, R., *Practical methods of optimization*, pp.436, John Wiley & Sons, New York, 1987.

Freeze, R. A., A stochastic-conceptual analysis of one-dimensional groundwater flow in nonuniform homogeneous media, *Water Resour. Res., 11*(5), 725-741,,1975.

Frind, E. O., and G. F. Pinder, Galerkin solution of the inverse problem for aquifer transmissivity, *Water Resour. Res., 9*(5), 1397-1410, 1973.

Gambolati, G., and G. Volpi, Groundwater contour mapping in Venice by stochastic interpolators, 1. Theory, *Water Resour. Res., 15*(2), 281-290, 1979.

Gavalas, G. R., P. C. Shah, and J. H. Seinfeld, Reservoir history matching by Bayesian estimation, *Soc. Pet. Eng. J., 16*(6), 337-350, 1976.

Gelhar, L. J., Effects of hydraulic conductivity variations on groundwater flows, paper presented at 2nd International Symposium on Stochastic Hydraulics, Int. Ass. for Hydraul. Res., Lund, Sweden, 1976.

Gelhar, L. J., Stochastic subsurface hydrology from theory to applications, *Water Resour. Res., 22*(9), 135S-145S, 1986.

Ghanem, R. G., and P. D. Spanos, Stochastic Finite Elements: A spectral Approach, pp.214, Springer-verlag, 1991.

van der Heijde, P., Y. Bahmat, J. Bredehoeft, B. Andrews, D. Holtz and S. Sebastian, *Groundwater Management: The Use of Numerical Models*, second edition, *AGU* water Resources monographs 5, 1985.

Hefez, E., V. Shamir, and J. Bear, Identifying the parameters of an aquifer cell model, *Water Resour. Res., 11*(6), 993-1004, 1975.

Hoeksema, R. J., and P. K. Kitanidis, An application of the geostatistical approach to the inverse problem in two-dimensional groundwater modeling, *Water Resour. Res., 20*(7), 1003-1020, 1984.

Hoeksema, R. J., and P. K. Kitanidis, Analysis of the spatial structure of properties of selected aquifers, *Water Resour. Res., 21*(4), 563-572, 1985*a*.

Hoeksema, R. J., and P. K. Kitanidis, Comparison of Gaussian conditional mean and Kriging estimation in the geostatistical solution of the inverse problem, *Water Resour. Res., 21*(6), 825-836, 1985*b*.

Hsu, N.-S., W. W-G. Yeh, Optimum experimental design for parameter identification in groundwater hydrology, *Water Resour. Res., 25*(5), 1025-1040, 1989.

Hudak, P. F., and H. A. Loaiciga, A location modeling approach for groundwater monitoring network augmentation, *Water Resour. Res., 28*(3), 643-649, 1992.

Hudak, P. F., and H. A. Loaiciga, An optimization method for monitoring network design in multilayered groundwater flow systems, *Water Resour. Res., 29*(8), 2835-2845, 1993.

Huyakorn, P. S., and G. F. Pinder, *Computational Methods in Subsurface Flow*, Academic Press, New York, 1983.

Jacquard, P., and C. Jain, Permeability distribution from field pressure data, *Soc. Pet. Eng. J., 5*(4), 281-294, 1965.

Journel, A. G., and Ch. J. Huijbregts, *Mining Geostatistics*, pp.600, Academic, San Diego, Calif., 1978.

Keidser, A., and D. Rosbjerg, A comparison of four inverse approaches to groundwater flow and transport parameter identification, *Water Resour. Res., 27*(9), 2219-2232, 1991.

Kiefer, J., and J. Wolfowitz, The equivalence of two extremum problems, *Canadian Journal of Mathematics*, 12, 363-366, 1960.

Kinzelbach, W., *Groundwater Modeling -- An Introduction with Sample Programs in BASIC*, Elsevier, New York, 1986.

Kitamura, S., and S. Nakagiri, Identifiability of spatially varying and constant parameters in distributed systems of parabolic type, *SIAM J. Control Optimization, 15*(5), 785-802, 1977.

Kitanidis, P. K., Parameter estimation of covariances of regionalized variables, *Water Resourc. Bull. 23(4),* 557-567, 1987.

Kleinecke, D., Use of linear programming for estimating geohydrologic parameters of groundwater basins, *Water Resour. Res., 7*(2), 367-375, 1971.

Knopman, D. S., and C. I. Voss, Further comments on sensitivities, parameter estimation, and sampling design in one-dimensional analysis of solute transport in porous media, *Water Resour. Res., 24*(2), 225-238, 1988.

Knopman, D. S., and C. I. Voss, Multiobjective sampling design for parameter estimation and model discrimination in groundwater solute transport, *Water Resour. Res., 26*( ), 2245-2258, 1989.

Knopman, D. S., and C. I. Voss, Multiobjective sampling design for parameter estimation and model discriminaton in groundwater solute transport, *Water Resour. Res., 25*(10), 2245-2258, 1989.

Knopman, D. S., C. I. Voss, and S. P. Garabedian, Sampling design for groundwater solute transport: Tests of methods and analysis of Cape Cod tracer test data, *Water Resour. Res., 27*(5), 925-950, 1991.

Kool, J. B., J. C. Parker, and M. TH. Van Genuchten, Parameter estimation for unsaturated flow and transport models -- a review, *J. Hydrol., vol.91*, 255-293, 1987.

Kravaris, C., and J. H. Seinfeld, Identification of parameters in distributed system by regularization, *SIAM J. Control Opt., vol.23*, 217-241, 1985.

Kunisch, K., and L. W. White, Regularity properties in parameter estimation of diffusion coefficients in elliptic boundary value problems, *Applicable Analysis, vol.21*, 71-88, 1986.

Kunisch, K., and L. W. White, Parameter identifiability under approximation, *Q. Appl. Math., vol.44*, 475-486, 1987*a*.

Kunisch, K., and L. W. White, Identifiability under approximation for an elliptic boundary value problem, *SIAM J. Control Opt. vol.25*, 279-297, 1987*b*.

Kunisch, K., A review of some recent results on the output least squares formulation of parameter estimation problems, *Automatica, vol.24*, No.4, 531-539, 1988.

Lapidus, L., and G. F. Pinder, *Numerical Soluton of Partial Differential Equations in Science and Engineering*, John Wiley & Sons, New York, 1982.

LaVenue, A. M., and J. F. Pickens, Application of a coupled adjoint sensitivity and Kriging approach to calibrate a groundwater flow model, *Water Resour. Res., 28*(6), 1543-1570, 1992.

Lavrent'ev, M. M., V. G. Romanov, and S. P. Shishatskii, *Ill-Posed Problems of Mathematical Physics and Analysis*, 1980, (translated from the Russian by J. R. Schulenberger), pp.275, 1986.

Lee, T., and J. H. Seinfeld, Estimation of two-phase petroleum reservoir properties by regularization, *J. Comput. Phys., vol.69*, 397-419, 1987.

Li, W.-H., *Differential Equations of Hydraulic Transients, Dispersion, and Groundwater Flow*, Prentice-Hall, Englewood Cliffs, NY, 1972.

Li, J., N.-Z. Sun, and W. W,-G. Yeh, A comparative study of sensitivity coefficient calculation methods in groundwater flow, in *Proceedings of the 6th International Conference in Finite Elements in Water Resources,* Lisboa, Portugal, June, 1986.

Loaiciga, H. A., and M. A. Mariño, The inverse problem for confined aquifer flow: Identification and estimation with extension, *Water Resour. Res., 23*(1), 92-104, 1987.

Loaiciga, H. A., An optimization approach for groundwater quality monitoring network design, *Water Resour. Res., 25*(8), 1771-1782, 1989.

Loaiciga, H. A., and M. A. Mariño, Error analysis and stochastic differentiability in subsurface flow modeling, *Water Resour. Res, 25(12),* 2897-2902, 1990.

Luenberger, D. G., *Introduction to Linear and Non-Linear Programming,* pp.356, Addison-Wesley, 1973.

Luenberger, D. G., *Linear and Non-Linear Programming,* pp.491, Addision-Wesley, 1984.

Mantoglu, A. and J.L. Wilson, The turning bands method for simulation of random fields using line generation by a spectral method, *Water Resour. Res., 18(5),* 1379-1394, 1982.

de Marsily, G., C. Lavedan, M. Boucher, and G. Fasanino, Interpretation of interference tests in a well field using geostatistical techniques to fit the permeability distribution in a reservoir model, in Geostatistics for Natural Resources Characterization, Second *NATO Advanced Study Institute, GEOSTAT 1983, Tahoe City, California,* edited by G. Verly, M. David, A. G. Journel, and A. Marechal, pp.831-849, D. Reidel, Hingham, Mass., 1984.

de Marsily, G., *Quantitative Hydrogeology,* Academic, San Diego, Calif., 1986.

Matheron, G., Principles of geostatistics, *Econ. Geol.,* 58, 1246-1266, 1963.

Matheron, G., The intrinsic random functions and their applications, *Advan. Appl. Probob.,* 5, 438-468, 1973.

McKinney, D. C., and D. P. Loucks, Network design for predicting groundwater contamination, *Water Resour. Res., 28*(1), 133-148, 1992.

McLaughlin, D. B., and E. F. Wood, A distributed parameter approach for evaluating the accuracy of groundwater model predictions, 1, Theory, *Water Resour. Res., 24*(7), 1037-1047, 1988*a.*

McLaughlin, D. B., and E. F. Wood, A distributed parameter approach for evaluating the accuracy of groundwater model predictions, 2, Application to groundwater flow, *Water Resour. Res., 24*(7), 1048-1060, 1988*b.*

Mercer, J. W., S. P. Larson, and C. R. Faust, Simulation of salt - water interface motion, *Ground Water, 18*(4), 374-385, 1980.

Mercer, J., and C.R. Faust, Groundwater Modeling: Application, *Groundwater 18(5), 1980.*

Miller, K., Least-Squares methods for ill-posed problems with a prescribed bound, *SIAM J. Math. Anal., vol.*1, pp.52-74, 1970.

Mishra, S., and S. C. Parker, Parameter estimation for coupled unsaturated flow and transport, *Water Resour. Res., 25*(3), 385-396, 1989.

Neuman, S. P., Calibration of distributed parameter groundwater flow models viewed as a multiple-objective decision process under uncertainty, *Water Resour. Res., 9*(4), 1006-1021, 1973.

Neuman, S. P., and S. Yakowitz, A statistical approach to the inverse problem of aquifer hydrology, 1, Theory, *Water Resour. Res., 15*(4), 845-860, 1979.

Neuman, S. P., G. E. Fogg, and E. A. Jacobson, A statistical approach to the inverse problem of aquifer hydrology, 2, Case Study, *Water Resour. Res., 16*(1), 33-68, 1980.

Neuman, S. P., A statistical approach to the inverse problem of aquifer hydrology, 3. Improved solution method and added perspective, *Water Resour. Res., 16*(2), 331-346, 1980.

Nishikawa, T., W. W-G. Yeh, Optimal pumping test design for the parameter identification of groundwater systems, *Water Resour. Res., 25*(7), 1737-1747, 1989.

Oblow, E. M., Sensitivity theory for reactor thermal-hydraulics problems, *Nucl. Sci. Eng., 68*, 322-337, 1978.

Pázman, A., *Foundations of optimum experimental design*, D. Reidel, Kluwer Academic publishers, Dordrecht, Holland, pp.228, 1986.

Pinder, G. F., and G. Gray, *Finite Element Simulation in Surface and Subsurface Hydrology*, Academic Press, New York, 1977.

Powell, M. J. D., An efficient method for finding the minimum of a function of several variables without calculating derivatives, *Computer Journal*, 7, 152-162, 1964.

Press, W. H., S. A. Teukoisky, W. T. Vettering and B. P. Flannery, *Numerical Recipes -- The Art of Scientific Computing*, second edition, Cambridge University Press, 1992.

Rajaram, H., and D. McLaughlin, Identification of large-scale spatial trends in hydrologic data, *Water Resour. Res., 26*(10), 2411-2424, 1990.

Remson, I., G. M. Hornberger and F. J. Molz, *Numerical Methods in Subsurface Hydrology*, Academic Press, *vol.10*, pp.1-143, 1971.

Rubin, Y., and G. Dagan, Stochastic identification of transmissivity and effective recharge in steady groundwater flow, 1, Theory, *Water Resour. Res., 23*(7), 1185-1192, 1987*a*.

Rubin, Y., and G. Dagan, Stochastic identification of transmissivity and effective recharge in steady groundwater flow, 2, Case study, *Water Resour. Res., 23*(7), 1193-1200, 1987*b*.

Sagar, B., S. Yakowitz, and L. Duckstein, A direct method for the identification of the parameters of dynamic nonhomogeneous aquifers, *Water Resour. Res., 11*(4), 563-570, 1975.

Schweppe, F. C., Model identification problems, in *Applications of Kalman Filter to Hydrology, Hydraulics and Water Resources*, edited by C.-L. Chiu, pp. 115-133, University of Pittsburgh, Pittsburgh, Pennsylvania, 1978.

Seinfeld, J. H., and W. H. Chen, Identification of petroleum reservoir properties, in *Distributed Parameter Systems, Identification, Estimation, and Control*, edited by W. H. Ray and D. G. Lainiotis, Maral Dekker, New York, 1978.

Shah, P. C., G. R. Gavalas, and J. H. Seinfeld, Error analysis in history matching: The optimum level of parameterization, *Soc. Pet. Eng. J., 18*(3), 219-228, 1978.

Silvey, S. D., *Optimal Design -- An Introduction to the Theory for Parameter Estimation*, Chapman and Hall, New York, 1980.

Steinberg, D.M., and W. G. Hunter, Experimental design: review and comment, *Technometrics, 26*(2), 71-97, 1984.

Sun, N.-Z., *Mathematical Modelings and Numerical Methods in Groundwater Flow* [in Chinese], Geological Publishing House, Beijing, 1981.

Sun, N.-Z., and W. W-G. Yeh, A proposed upstream weight numerical method for simulating pollutant transport in groundwater, *Water Resour. Res.*, *19*(6), 1489-1500, 1983.

Sun, N.-Z., and W. W-G. Yeh, Identification of parameter structure in groundwater inverse problem, *Water Resour. Res.*, *21*(6), 869-883, 1985.

Sun, N.-Z., Application of numerical methods to simulate the movement of contaminants in groundwater, *Environmental Health Perspective, 83,* 97-115, 1989a.

Sun, N.-Z., *Groundwater Pollution: Mathematical Models and Numerical Methods* (in Chinese), Geological Publishing House, Beijing, pp.361, 1989b.

Sun, N.-Z., and W. W-G. Yeh, Coupled inverse problem in groundwater modeling, 1, Sensitivity analysis and parameter identification, *Water Resour. Res.*, *26*(10), 2507-2525, 1990a.

Sun, N.-Z., and W. W-G. Yeh, Coupled inverse problem in groundwater modeling, 2, Identifiability and experimental design, *Water Resour. Res.*, *26*(10), 2527-2540, 1990b.

Sun, N.-Z., and W. W-G. Yeh, A stochastic inverse solution for transient groundwater flow: Parameter identification and reliability analysis, *Water Resour. Res.*, *28*(12), 3269-3280, 1992a.

Sun, N.-Z., and W. W-G. Yeh, On the consistency of continuous and discrete approaches in deriving the adjoint state equations, in proceedings of Computational Methods in *Water Resources IX, vol.2: Mathematical Modeling in Water Resources*, edited by T. F. Russell, R. E. Ewing, C. A. Brebbia, W. G. Gray, and G. F. Pinder, Computational Mechanics Publications, 1992b.

Sun, N.-Z., *Mathematical Modeling of Groundwater Pollution,* Springer Verlag, New York, 1994a.

Sun, N.-Z., M-C. Jeng, and W. W-G. Yeh, A proposed geological parameterization method for parameter identification in three-dimensional groundwater modeling, submitted to *Water Resour. Res*, 1994b.

Sykes, J. F., J. L. Wilson, and R. W. Andrews, Sensitivity analysis for steady state groundwater flow using adjoint operators, *Water Resour. Res.*, *21*(3), 359-371, 1985.

Tang, D. H., and G. F. Pinder, Analysis of mass transport with uncertain physical parameters, *Water Resour. Res.*, *15*(5), 1147-1155, 1979.

Tang, Y. N., Y. M. Chen, W. H. Chen, and M. L. Wasserman, Generalized pulse-spectrum technique for 2-D and 2-phase history matching, *Applied Numerical Mathematics*, North-Holland, *vol.5*, 529-539, 1989.

Tarantola, A., *Inverse Problem Theory*, Elsevier, pp.613, 1987.

Tikhonov, A. N., Solution of ill-posed problems and the regularization method, *Dokl. Akad. Nauk SSSR*, 151(163), 501-504; *Soviet Math. Dokl., vol.4*, 1035-1038, 1963.

Townley, L. R., and J. L. Wilson, Computationally efficient algorithms for parameter estimation and uncertainty propagation in numerical models of groundwater flow, *Water Resour. Res.*, *21*(12), 1851-1860, 1985.

Tsang, C. F., The modeling process and model validation, *Groundwater, 29*(6), 825-831, 1991.

Tucciarelli, T., and G. Pinder, Optimal data acquisition strategy for the development of a transport model for groundwater remediation, *Water Resour. Res.*, *27*(4), 577-588, 1991.

Usunoff, E., J. Carrera and S. F. Mousavi, An approach to the design of experiments for discriminating among alternative conceptual models, *Advances in Water Resources 15*, 199-214, 1992.

Vermuri, V., J. A. Dracup, R. C. Erdmann and N. Vermuri, Sensitivity analysis method of system identification and its potential in hydrologic research, *Water Resour. Res., 5*(2), 341-349, 1969.

Wagner, B. J., and S. M. Gorelick, Optimal groundwater quality management under parameter uncertainty, *Water Resour. Res., 23*(7), 1162-1174, 1987.

Wagner, B. J., and S. M. Gorelick, Reliable aquifer remediation in the presence of spatially variable hydraulic conductivity: From data to design, *Water Resour. Res., 25*(10), 2211-2225, 1989.

Wang, K., and A. E. Beck, An inverse approach to heat flow study in hydrologically active areas, *Geophys. J. Int., vol.98*, 69-84, 1989.

Wang, K., P. Y. Shen, and A. E. Beck, A solution to the inverse problem of the coupled hydrological and thermal regimes, in *Hydrological Regimes and Their Subsurface Thermal Effects*, eds A. E. Beck, L. Stegena, G. Garven, *Am. Geophys. Union, Monograph Series*, Washington, 1989.

Watson, A. T., J. H. Seinfeld, G. R. Gavalas, and P. T. Woo, History matching in two-phase petroleum reservoirs, *Soc. Pet. Eng. J., 21*(6), 521-530. 1980.

Weir, G. J., The direct inverse problem in aquifers, *Water Resour. Res., 25*(4), 749-753, 1989.

Willis, R., W. W-G. Yeh, *Groundwater Systems Planning and Management*, Prentice-Hall, Englewood Cliffs, New Jersey, 1987.

Wilson, J. L., and D. G. Metcalf, Illustration and verification of adjoint sensitivity theory for steady state groundwater flow, *Water Resour. Res., 21*(11), 1602-1610, 1985.

Woodbury, A. D., and L. Smith, and W. S. Dunbar, Simultaneous inversion of hydrogeologic and thermal data, 1, Theory and application using hydraulic heat data, *Water Resour. Res., 23*(8), 1586-1606, 1987.

Woodbury, A. D., and L. Smith, Simultaneous inversion of hydrogeologic and thermal data, 2, Incorporation of thermal data, *Water Resour. Res., 24*(3), 356-372, 1988.

Xiang, Y., J. F. Sykes, and N. R. Thomson, A composite L_1 parameter estimator for moodel fitting in groundwater flow and solute transprot simulation, *Water Resour. Res., 29*(6), 1661-1674, 1993.

Yakowitz, S., and L. Duckstein, Instability in aquifer identification: Theory and case study, *Water Resour. Res., 16*(6), 1045-1064, 1980.

Yeh, W. W-G., and G. W. Tauxe, Quasilinearization and the identification of aquifer parameters, *Water Resour. Res., 7*(2), 375-381, 1971.

Yeh, W. W-G., and Y. S. Yoon, Parameter identification with optimum dimension in parameterization, *Water Resour. Res., 17*(3), 664-672, 1981.

Yeh, W. W-G., Y. S. Yoon, and K. S. Lee, Aquifer parameter identification with kringing and optimum parameterization, *Water Resour. Res., 19*(1), 225-233, 1983.

Yeh, W. W-G., and N.-Z. Sun, An extended identifiability in aquifer parameter identification and optimal pumping test design, Water Resour. Res., 20(12), 1837-1847, 1984.

Yeh, W. W-G., Review of parameter identification procedure in groundwater hydrology: The inverse problem, *Water Resour. Res., 22*(2), 95-108, 1986.

Yeh, W. W-G., and C. Wang, Identification of aquifer dispersivities: methods of analysis and parameter uncertainty, *Water Resources Bulletin, 23*(4), 569-580, 1987.

Yeh, W. W.-G., and N.-Z. Sun, Variational sensitivity analysis, data requirement, and parameter identification in a leaky aquifer system, *Water Resour. Res., 26*(9), 1927-1938, 1990.

Yoon, Y. S., and W. W-G. Yeh, Parameter identification in an inhomogeneous medium with the finite-element method, *Soc. Pet. Eng. J.*, 217-226, 1976.

Index

333

Theory and Applications of Transport in Porous Media

Series Editor:
Jacob Bear, *Technion – Israel Institute of Technology, Haifa, Israel*

1. H. I. Ene and D. Poliševski: *Thermal Flow in Porous Media.* 1987
   ISBN 90-277-2225-0
2. J. Bear and A. Verruijt: *Modeling Groundwater Flow and Pollution.* With Computer Programms for Sample Cases. 1987   ISBN 1-55608-014-X; Pb 1-55608-015-8
3. G. I. Barenblatt, V. M. Entov and V. M. Ryzhik: *Theory of Fluid Flows Through Natural Rocks.* 1990   ISBN 0-7923-0167-6
4. J. Bear and Y. Bachmat: *Introduction to Modeling of Transport Phenomena in Porous Media.* 1990   ISBN 0-7923-0557-4; Pb (1991) 0-7923-1106-X
5. J. Bear and J-M. Buchlin (eds.): *Modelling and Applications of Transport Phenomena in Porous Media.* 1991   ISBN 0-7923-1443-3
6. Ne-Zheng Sun: *Inverse Problems in Groundwater Modeling.* 1994
   ISBN 0-7923-2987-2

Kluwer Academic Publishers – Dordrecht / Boston / London

Gilles Fresse

Illustrations d'Emmanuel Chaunu

L'amour
c'est tout bête

RAGEOT

ISBN : 978-2-7002-3311-7
ISSN : 1951-5758

coup de foudre

Elle est arrivée le lundi 5 janvier, juste avant la récré. Je suais sang et eau sur un exercice de grammaire. Une abomination : il fallait souligner les adjectifs qualificatifs en vert, les adverbes en bleu et les prépositions en jaune.

À peu près toutes les trois secondes, je regardais le ciel gris afin d'y dénicher la solution à mon problème grammatical. Il ne m'était d'aucun secours.

Au moment où un gros nuage noir à la forme inquiétante se mêlait à un autre nuage noir à la forme tout aussi inquiétante, trois coups secs à la porte de la classe m'ont détourné de mes observations météorologiques. Elle est apparue, derrière madame Todeschini, la directrice.

Un rayon de soleil a traversé la couche de cumulus pour venir éclairer son visage.

Elle était trop belle.

Les autres se sont levés. Pas moi. J'étais cloué sur ma chaise par une force mystérieuse.

Mademoiselle Morlot, la maîtresse, m'a jeté un regard réprobateur avant de me demander :

– Alors, Quentin, tu es trop fatigué pour te lever toi aussi?

J'ai dû prendre appui avec mes mains sur ma table pour réussir à me mettre debout.

Mon cœur creusait des trous au marteau-piqueur dans ma poitrine et je respirais aussi vite qu'après un quatre cents mètres. Je n'avais jamais connu une telle sensation et j'ai eu peur de m'évanouir.

Le rayon de soleil a disparu. J'ai entendu la voix lointaine de la directrice :

– Les enfants, je vous présente Chloé, qui nous arrive de Lorraine. Elle vient finir son année dans notre école.

Quarante-six yeux, quarante-huit avec ceux de mademoiselle Morlot, étaient fixés sur la nouvelle, la détaillant des orteils aux cheveux. Elle dansait d'un pied sur l'autre et triturait nerveusement ses doigts comme si elle avait voulu les mélanger.

– As-tu quelque chose à ajouter, Chloé ? a demandé madame Todeschini.

Le joli visage de Chloé a exprimé une grimace de panique. Elle a secoué désespérément la tête de droite à gauche tandis qu'une larme perlait dans ses yeux vert clair. J'ai eu une brusque envie de voler à son secours, de la prendre dans mes bras, de la serrer contre moi pour la cacher du regard des autres et de lui dire doucement : « Ne crains rien, je te protégerai. »

Évidemment, je n'ai pas bougé. Je me suis contenté de pousser ma règle métallique au bout de ma table. Elle est tombée en heurtant bruyamment le parquet. Cette fois, les regards se sont tournés vers moi. J'avais réussi. Je l'avais sauvée !

– Bien, je vous laisse, a conclu la directrice. J'espère que vous saurez accueillir Chloé convenablement et que vous serez gentils avec elle. Puis-je compter sur vous ?

On a tous prononcé le « oui » qu'elle attendait et elle est partie au moment où la récré sonnait. Chloé est restée en classe avec mademoiselle Morlot pendant qu'avec les copains, comme d'habitude, on footballait dans la cour.

Sans me vanter, au foot, je suis le plus fort de l'école et quand on forme les équipes, je suis toujours choisi le premier. Mais, pendant le match de ce lundi 5 janvier à dix heures, j'ai atteint les sommets de la nullité footballistique. Mes pieds ne m'obéissaient plus.

Pas un dribble correct, pas un contrôle réussi, aucune interception et, pire encore, trois échecs retentissants alors que j'étais seul face à Jules, le gardien adverse.

C'est bien simple, dès qu'on me donnait le ballon, je le perdais. Comme si j'avais deux jambes de bois avec du caramel mou à l'intérieur.

Pas question non plus d'effectuer une tête digne de ce nom puisque, justement, ma tête était ailleurs. Elle était restée en classe, aux côtés de Chloé qui rangeait ses affaires dans son casier.

Où allait-elle être placée ? Loin de moi ? Tout près ? Il n'y avait pas de table libre à proximité. Peut-être la maîtresse effectuerait-elle des changements ? Dans ce cas, tous les espoirs m'étaient permis…

– Hé, Quentin ! Tu joues ou tu rêves ? m'a demandé Alex.

Son joli visage flottait dans mon esprit comme un cerf-volant sans ficelle. Ses cheveux d'un brun doré, le bandeau mauve qui les tirait en arrière laissant des mèches s'échapper et tomber en cascade sur ses oreilles.

Je me rappelais du moindre détail. Pourtant, je ne l'avais regardée que quelques secondes.

– Quentin ! Vas-y bon sang !

Ses yeux vert clair, son teint légèrement bronzé…

– Quentin ! Tu vas réagir ? On en a déjà pris trois !

Mais ces quelques secondes m'avaient chamboulé. Chamboulé à tel point que je n'ai pas vu le poteau des buts sur lequel je me suis écrasé lamentablement.

J'ai senti un liquide chaud couler sur la partie droite de mon visage. J'ai vu la tête affolée de Jules et, juste avant de tomber dans les pommes, la belle Chloé m'est apparue dans un mirage. Je pouvais mourir tranquille.

réveil en douleur

Je n'étais pas mort. Non.

Je me suis réveillé doucement et quand j'ai ouvert les yeux, ce n'est pas la jolie figure de Chloé qui est apparue au-dessus de moi mais la grosse moustache rousse et le regard paniqué de monsieur Ravet, le maître des CP.

Plus haut, se découpant dans le ciel, les têtes des copains dessinaient un cercle presque parfait. Alex a crié :

– Il est vivant !

– Ne bouge pas ! m'a intimé monsieur Ravet d'une voix douce mais ferme. Les pompiers arrivent.

C'est en entendant le mot pompier que je me suis rendu compte que j'avais très mal et qu'une main invisible me déchirait méchamment les chairs au niveau du sourcil.

Quelques minutes plus tard, deux pompiers m'ont embarqué dans une ambulance...

Je me suis retrouvé aux urgences devant une interne qui, en observant les clichés de la radio, m'a déclaré :

– Il n'y a apparemment rien de bien grave. Pas de fracture du crâne…

J'ai poussé un ouf de soulagement. Maman allait venir me chercher et on rentrerait à la maison. Je retournerais à l'école et je reverrais Chloé. J'ai vite déchanté quand l'interne m'a annoncé :

– Cependant, nous allons te garder vingt-quatre heures en observation. C'est plus prudent.

J'ai tenté de protester :

– Mais si je n'ai rien, ça ne…

– Tu as l'arcade sourcilière ouverte et tu as reçu un gros choc. Si tout va bien, tu sortiras demain. La prochaine fois que tu joueras au foot, méfie-toi des poteaux ! C'est dangereux, ces machins-là !

Elle m'a fait deux points de suture puis m'a posé une grosse compresse sur l'œil droit. On m'a poussé dans un fauteuil roulant jusqu'à une chambre et on m'a allongé sur un lit. Maman est arrivée, le teint pâle. Elle m'a serré contre elle en chuchotant :

– Mon chéri, mon chéri !

Je lui ai raconté l'accident. Elle a été soulagée de constater que je me rappelais ce qui s'était passé et que ma tête semblait fonctionner normalement. Elle a extirpé une petite glace de son sac et me l'a tendue :

– Tu t'es vu en pirate de l'hôpital ?

J'ai jeté un œil – forcément je ne pouvais plus me servir de l'autre – sur le miroir. Pas vraiment réjouissant. Heureusement que Chloé ne me voyait pas dans cet état.

J'ai tout enduré sans broncher : les repas dégoûtants, l'infirmière qui passait chaque heure et me plongeait le rayon lumineux de sa lampe dans l'œil pour y déceler je ne sais quoi. La douleur, les secondes, les minutes, les heures d'ennui…

Une chose et une seule me broyait le cœur : je n'avais pas revu Chloé. C'était ça le vrai supplice. Le reste n'avait aucune importance.

Maman est venue me chercher le lendemain en fin de matinée.

Elle avait pris un après-midi de congé pour rester avec moi. J'ai tué le temps devant la télévision et la console vidéo mais avec un seul œil, ce n'est pas très confortable.

Vers cinq heures, Alex et Jules sont arrivés. On s'est installés dans ma chambre avec le paquet de bonbons qu'ils avaient apporté.

– Tu nous as fichu une sacrée trouille, a commencé Jules, la bouche pleine.

– Ouais, a rigolé Alex. J'ai bien cru que t'avais cassé le poteau !

J'ai ri aussi, prudemment car ça tirait sur mon œil. Il a renchéri :

– T'imagines un peu ? Si on n'avait plus de but, comment on ferait ? Un joueur, ça se remplace facilement mais un but ?

Je brûlais de leur demander des nouvelles de la nouvelle mais une question trop directe leur aurait mis la puce à l'oreille. Je ne voulais surtout pas qu'ils se doutent que j'étais tombé amoureux.

– Comment ça va en classe ? ai-je tenté innocemment.

Jules a eu l'air un peu surpris de la question, mais il m'a répondu :

– Ça va.

Je n'apprendrais rien comme ça. Je me suis jeté à l'eau :

– Et la nouvelle ? Comment elle s'appelle déjà ?

– Chloé ?

– Oui, c'est ça, Chloé !

– Ça va, a répété Jules, toujours aussi laconique.

– Si tu veux mon avis, a continué Alex, c'est une chochotte.

– Une chochotte ? Qu'est-ce que c'est une chochotte ?

Alex s'est gratté le nez pour mieux trouver ses mots.

– Ben… Euh… Elle fait des tas de manières et on a l'impression qu'elle est toujours prête à pleurer. Pourtant personne ne lui veut de mal.

– Elle est comme les autres filles, quoi ! a conclu Jules en rigolant. En fait, chochotte est un synonyme de fille !

Je me suis forcé à sourire, mais au fond de moi j'ai senti un pincement de colère. J'ai failli leur dire qu'ils se trompaient, que Chloé était peut-être tout simplement timide et que ça ne devait pas être évident de débarquer en milieu d'année dans une classe peuplée de spécimens comme eux.

Mais Alex et Jules n'auraient pas compris que je la défende. C'est normal. D'ailleurs, moi jusqu'à hier je pensais exactement la même chose qu'eux à propos des filles.

– Et en plus, a ajouté Alex en tendant la main vers le sachet de bonbons, elle est devenue copine avec Marthe. Ce n'est pas bon signe. Je les ai vues échanger des images d'animaux. La nouvelle les collectionne. Elle en a deux gros classeurs.

Là, il marquait un point. Marthe est la peste de la classe. Je me suis raisonné en me disant que Chloé s'était contentée de Marthe parce qu'elle était un peu perdue à son arrivée, mais qu'elle comprendrait vite. Ou bien c'était Marthe qui lui avait mis le grappin dessus et, pleine de gentillesse, Chloé n'avait pas eu le cœur de la rembarrer.

– Tu reviens en classe jeudi? m'a demandé Jules.

– Non. Je dois rester à la maison toute la semaine.

– Waouh ! T'as trop de chance ! s'est exclamé Alex qui pense qu'à l'école, le seul moment supportable c'est la récréation.

Tu parles d'une chance ! J'allais encore me morfondre cinq jours à la maison avec un œil gros comme un œuf d'autruche tandis que Chloé…

– Elle est assise où, la nouvelle ? J'espère qu'elle n'est pas à côté de moi ?

– Non, t'inquiète, a cru me soulager Jules. Elle est carrément à l'opposé. Mais pourquoi tu nous questionnes sans arrêt sur elle ? T'es amoureux ou quoi ?

J'ai haussé les épaules d'un air offusqué :

– Ça va pas, non !

Alex a pouffé en déclarant :

– Si, si ! À mon avis, elle t'a tapé dans l'œil !

la chloïte aiguë

Cinq jours. Cinq longues journées à me morfondre loin de Chloé, à ressasser les images de son arrivée à l'école.

J'essayais de me raisonner, de me dire que, comme l'avait prétendu Alex, cette fille était peut-être une chochotte, que je ne la connaissais pas et que je n'avais même pas entendu le son de sa voix, que je ne l'avais vue qu'un bref instant! Oui mais un bref instant qui tournait en boucle dans ma tête comme un film que l'on passe et repasse sans jamais se lasser. Une véritable maladie! Tout le reste me semblait sans intérêt.

J'ai pris mon *Foot plus* sans réussir à terminer le premier article. Forcément, lorsque j'attaquais la deuxième phrase j'avais déjà oublié de quoi parlait la première.

Mes jeux vidéo? Insipides. Mes séries télé préférées? De la guimauve. Mes bandes dessinées? Assommantes. Les repas avec maman? Je n'y touchais que du bout des lèvres, parce qu'elle insistait et que je voulais lui faire plaisir.

Mais le pire, je crois, c'était ces multiples pertes de mémoire. J'avais besoin d'aller chercher quelque chose dans la cuisine, je m'y rendais et, une fois sur place, impossible de me rappeler ce que j'étais venu y faire. Le pull enfilé devant derrière, les chaussons à l'envers étaient d'autres symptômes fréquents.

Si ce n'est pas une maladie ça, je ne m'appelle plus Quentin! La chloïte aiguë, j'avais attrapé la chloïte aiguë! Est-ce que c'est grave, docteur? Est-ce qu'il existe un médicament qui me guérira?

Durant des heures, je m'allongeais sur mon lit et je rêvais. Je rêvais à nous deux. Chloé et moi. Moi et Chloé. Plus rien d'autre n'existait.

Voilà. Voilà à quoi je passais mes journées. Mais j'avais beau penser à elle à longueur de temps, je n'avais toujours pas trouvé la réponse à la seule question qui comptait : comment la séduire ? Je n'avais aucune expérience dans ce domaine et ça me semblait encore plus compliqué que les exercices de grammaire de mademoiselle Morlot. Les filles, je n'y connaissais rien. Forcément, je ne m'y étais jamais intéressé.

Une bonne nouvelle : mon coquart avait désenflé. Une mauvaise : ce n'était pas demain la veille que mon œil retrouverait son aspect normal.

Quand maman a enlevé les compresses, elle s'est exclamée :

– Oh! Quel joli œil au beurre noir!

Au beurre noir. Je trouvais l'expression comique. Mais lorsque je me suis vu dans la glace de la salle de bains, je n'ai pas eu envie de rire. Pas le moins du monde. Du noir, du violet, du bleu...

Si j'avais voulu jouer Zorro, je n'aurais eu besoin que de la moitié de son masque.

J'ai imaginé Chloé me regardant et j'ai paniqué :

– Il va pas rester comme ça, hein ?

– Mais non, ne t'inquiète pas, a répondu l'image de maman dans le miroir. Il va passer par les couleurs de l'arc-en-ciel avant de redevenir normal. Dans une quinzaine de jours, tu n'y penseras plus.

Une quinzaine de jours avec ce maquillage multicolore ? Autant dire une éternité de ridicule !

– Mais qu'est-ce qu'on peut faire ?

– Pas grand-chose, à part attendre que ça passe.

J'ai senti la colère monter en moi. Maman en a fait les frais :

– Que ça passe ! Que ça passe ! On voit que c'est pas toi qui trimballes un truc pareil !

Son visage s'est assombri. J'ai eu des remords.

– Excuse-moi, m'man. Tu n'y es pour rien.

Elle m'a souri et a pris ma main :

– Viens.

Elle m'a entraîné jusqu'au cagibi et a farfouillé dans mes affaires de ski avant d'y dénicher ma paire de lunettes de soleil.

– Tiens. Elles cacheront un peu les dégâts.

Bon, je n'avais pas vraiment le choix : OK pour les lunettes de soleil. Tant pis si on était en plein mois de janvier et qu'il n'y avait eu qu'un rayon de soleil depuis une semaine, celui qui avait éclairé le visage de Chloé à sa première apparition en classe.

Pour me remonter le moral, maman a proposé :

– Et si on préparait des crêpes ? Tu me donnes un coup de main ?

Pendant que je cassais les œufs, une pensée m'a brusquement effleuré.

– Dis, m'man. Je peux te poser une question ?

– Tu peux.

J'ai pris ma respiration.

– Mon père, tu l'aimais ?

Je connaissais les circonstances de ma naissance, nous avions déjà abordé le sujet. Mais jamais aussi directement. Elle s'est assise sur la table de la cuisine, a réfléchi quelques secondes, comme si elle triait les mots qu'elle s'apprêtait à prononcer.

– Oui, je l'ai aimé. Très fort mais pas longtemps car il est parti trop vite et sans laisser d'adresse. Tu sais, Quentin, quelquefois, l'amour grandit et devient indestructible. D'autres fois, il se rabougrit et disparaît. Avec ton père, nous n'avons pas eu le temps…

Elle parlait sans tristesse. Moi, je me disais que c'était dommage, qu'elle aurait mérité d'être aimée comme moi j'aimais Chloé.

– Mais ton père m'a fait le plus beau cadeau du monde !

Je me suis creusé la cervelle sans trouver.

– Ah oui ? C'était quoi ?

– Qui. Pas quoi. C'est toi mon plus beau cadeau, idiot !

On a dévoré les crêpes en écoutant les informations à la radio. Comme d'habitude, dans le monde, tout allait mal : des guerres, des attentats, des famines, des tremblements de terre, des ouragans, le prix du pétrole qui monte, les salaires qui stagnent, les impôts qui augmentent… Bref, le contraire de ce qu'on aimerait entendre. Heureusement que, une fois la radio coupée, on oublie tout.

Une nouvelle toutefois sortait de l'ordinaire : une bande de voleurs déguisés en clowns avait encore frappé.

Ils venaient de braquer leur neuvième banque de la région en deux mois. Leur butin s'élevait maintenant à plusieurs dizaines de milliers d'euros. Pas des rigolos, ces clowns-là !

La météo, quant à elle, ne prévoyait que du mauvais temps. J'aurais aimé du soleil pour justifier mes lunettes noires. Un instant, je me suis pris à imaginer une info sensationnelle :

Bonne nouvelle ! La jolie Chloé aime Quentin malgré son œil au beurre noir.

Mais bon, ne rêvons pas.

Maman s'est décidée à changer de station et la musique classique a remplacé la voix d'employé des pompes funèbres de l'animateur.

Pendant qu'on débarrassait la table, elle m'a annoncé :

– Samedi, je sors avec mon fils préféré !

– Ah oui ?

– Je l'emmène voir une exposition !

– Sur quoi ?

– Sur… prise !

drôles de bêtes !

L'expo était géniale. Elle s'appe-
lait *Le comportement amoureux des
animaux*. Vrai de vrai.

Dans la vie, il y a des hasards
extraordinaires. Le lundi, j'ai le
coup de foudre pour Chloé et, le
samedi suivant, maman m'entraîne
à une expo expliquant justement

les méthodes de séduction des différentes espèces !

Je l'ai dévorée des yeux, enfin de l'œil. Et comme maman s'est rendu compte que le sujet me passionnait, elle m'a offert le livre qui en reprenait les principaux thèmes. Tandis qu'elle le payait, j'ai repensé à ce qu'Alex m'avait confié : Chloé était copine avec Marthe. Elles s'échangeaient des images d'animaux.

J'ai dû me retenir pour ne pas crier de joie devant la caisse. Ce bouquin représentait un moyen inespéré d'entrer en relation avec Chloé.

Maman et moi, nous sommes revenus à la maison. Après le repas du soir, je me suis isolé dans ma chambre pour me plonger dans mon livre.

Il était aussi formidable que l'exposition, sûr qu'il intéresserait Chloé.

En le refermant, je me suis demandé comment j'allais m'y prendre, moi, pour la séduire. Le livre était formel sur ce point : dans 99 % des cas, le mâle fait le premier pas !

Une chose était certaine, jamais je n'oserais lui avouer mon amour. Pas par crainte du ridicule, non. Plutôt par peur d'un refus de sa part qui briserait net toutes mes espérances.

Comment procéder alors ? Faire comme le paon ? J'attendrais Chloé à la sortie de l'école et je déploierais ma splendide roue de plumes pour lui montrer que je suis le plus fort et le plus beau ?

Le hic, c'est que je ne voyais pas bien comment je pourrais me procurer le stock de plumes nécessaires. Et puis il faudrait que j'attende le carnaval.

Comme le microcèbe, alors ? Ce petit mammifère imprègne les arbres d'urine pour indiquer aux femelles qu'il est disponible. Il paraît que cette méthode est très efficace car l'odeur est transportée par l'air sur de grandes distances.

– Ce pipi-là est pour vous, ma chère !

Le souci était que je m'imaginais mal en train d'uriner sur le territoire de Chloé pour deux raisons. Une : j'ignorais où elle habitait et s'il y avait des arbres, deux : elle aurait du mal à comprendre mon comportement. Il me fallait quelque chose de plus romantique.

Le brame du cerf était une autre solution mais un peu trop sonore à mon goût. De plus, je ne tenais pas à être poursuivi par une bande de chasseurs.

Le moustique battant des ailes était une possibilité moins bruyante. Malheureusement, cette option était impossible car c'était la femelle qui le produisait.

J'aurais aussi pu choisir la façon de procéder du mandrill, ce singe dont le museau devient bleu et rouge et dont les fesses sans poils prennent une jolie couleur rose. J'y ai vite renoncé. Trop voyant et me promener à moitié nu, non merci ! La seule couleur qui ornait ma figure était l'arc-en-ciel qui me servait d'œil et il n'avait rien d'attirant.

J'ai dû me rendre à l'évidence. Les techniques de séduction du monde animal n'étaient pas applicables à l'être humain. Je devrais me débrouiller autrement…

Je me suis endormi sur les images de deux dauphins pendant leur grande parade nuptiale. J'ai rêvé que Chloé me regardait dormir, veillant sur mon sommeil.

Elle se penche sur moi, remonte tendrement les mèches de cheveux qui barrent mon front. Elle y dépose un tendre baiser. Je soupire :

– Ah, Chloé…

– Chloé ? Quelle Chloé ?

J'ai ouvert les yeux. Maman était juste au-dessus de moi, assise sur mon lit.

Elle était venue m'embrasser. Elle a répété sa question :

– Qui est Chloé ? Je ne la connais pas. Une copine de l'école ?

J'ai senti mon visage qui s'empourprait et je me suis retourné.

– Humm ! J'ai sommeil.

Elle n'a pas insisté et s'est éloignée. Juste avant d'éteindre la lumière, elle a quand même dit :

– Tu sais, mon petit Quentin, le rouge de ton visage va très bien avec le noir de ton œil !

école, le retour

Le lundi, mon sac dans le dos et mes lunettes de soleil sur le nez, je suis arrivé à l'école avec une demi-heure d'avance, le cœur battant.

Je ne voulais pas rater Chloé. En observant des oiseaux planer au-dessus du toit du groupe scolaire, j'ai pensé au milan royal et à sa technique de séduction géniale.

Pour épater son amoureuse, il monte très haut dans le ciel en décrivant des cercles puis il se laisse tomber en piqué pour ne reprendre son vol qu'à quelques mètres au-dessus du sol. Sauf que moi, si je sautais depuis le toit de l'école, je m'écraserais lamentablement aux pieds de Chloé. J'ai préféré l'attendre sur mes deux jambes. Mais, quand madame Todeschini a ouvert le portail, elle n'était toujours pas là. Les copains, eux, ont fêté mon retour à grands coups de calembours :

– Ouvre l'œil, Quentin ! Et le bon !

– Hé ben, toi, t'as pas les yeux dans ta poche !

– T'as pas les yeux en face des trous, ce matin. Tu vas pas tourner de l'œil, hein ?

Je m'en fichais. Seule l'arrivée de Chloé m'intéressait. Pas de chance, elle était absente. À croire que, justement, le mauvais œil s'acharnait sur moi.

Je n'ai pas osé demander de ses nouvelles à Marthe parce que je ne lui parlais que lorsque j'y étais obligé. Elle aurait trouvé ça louche. Alors, j'ai menti à la maîtresse en lui affirmant que le docteur avait dit qu'il serait plus prudent que je reste en classe pendant la récréation. Elle a gobé mon invention sans problème.

Quand j'ai été seul et tranquille, après m'être assuré que personne ne traînait dans les couloirs, je me suis dirigé vers le bureau de la maîtresse. Deuxième tiroir gauche, le registre d'appel, je savais qu'elle le rangeait là. Je l'ai ouvert, la peur au ventre, et j'y ai découvert ce que je cherchais : l'adresse de Chloé.

Chloé CHAMBERTIN
10, allée des Acacias.

En lisant le mot « acacias », j'ai repensé au microcèbe. Est-ce que cet arbre-là l'inspirait, question urinoir ?

Je n'ai rien avalé à la cantine. Ce qui, naturellement, a permis à Alex de s'esclaffer :

– On peut pas dire que t'as les yeux plus gros que le ventre !

Lorsque enfin, à quatre heures et demie, la sonnerie nous a libérés, j'ai filé vers l'allée des Acacias. La nuit commençait à tomber et le ciel gris foncé semblait vouloir écraser la ville. J'ai enlevé mes lunettes.

Le numéro 10 était une minuscule maison plantée au milieu d'un jardinet. Une lumière brillait par une fenêtre du rez-de-chaussée, une autre au premier étage. Peut-être la chambre de Chloé ?

Je ne connaissais pas les raisons de son absence en classe mais je l'ai imaginée, blottie sous les couvertures, le front en sueur à cause de la fièvre, serrant contre elle le nounours qu'on lui avait offert lorsqu'elle était petite. Tiens, les ours, comment se débrouillaient-ils, eux?

Je ne voulais pas rester planté au beau milieu du trottoir. J'avais peur qu'on me remarque. Alors, j'ai effectué de petits allers et retours dans la rue. Au bout du quatrième, un vieux break cabossé et rouillé est apparu. Il a ralenti, s'est garé devant le 10.

Malgré l'obscurité, j'ai pu détailler l'homme qui en est descendu. Il valait le coup d'œil! Le genre de type à endosser le rôle de la méchante brute dans un film à suspense. Frissons garantis à chaque apparition sur l'écran!

Sa taille avoisinait les deux mètres, il avait des épaules de déménageur, des bras de monsieur Muscle et le crâne rasé. Il portait un tee-shirt. En plein mois de janvier! Il a verrouillé sa voiture avant d'entrer dans la maison. Ce devait être le père de Chloé.

Un peu refroidi par l'irruption de ce géant, j'ai décidé de regagner notre appartement. En passant devant la voiture, j'ai aperçu deux gros sacs de voyage en toile dans le coffre.

De retour à l'appartement, j'ai expédié mes devoirs pour me replonger dans mon super bouquin. Je n'ai rien trouvé sur les ours. Vers six heures et demie, maman a appelé pour m'annoncer en vrac qu'elle rentrerait tard du journal, que je pouvais réchauffer le reste des lasagnes de la veille au micro-ondes, que je ne devais pas oublier de m'enduire l'œil avec ma pommade, qu'elle passerait me donner un bisou dans ma chambre dès son retour et que, bien sûr, si j'avais un souci quelconque, son portable était allumé. La liste entière sans respirer.

Je suis habitué à ce genre d'exploit qui se produit une fois par mois, lorsqu'il s'agit pour l'équipe de son mensuel de boucler le prochain numéro.

J'ai avalé les lasagnes en regardant la télé. J'avais repéré un documentaire animalier sur le programme. Mais comme il était encore trop tôt, j'ai écouté la fin du journal de vingt heures. Maman me l'interdit. Elle prétend que certaines images peuvent choquer la sensibilité des jeunes. Les horreurs à la radio, passe encore, mais sur le petit écran, pas question. Allez comprendre !

Parmi le flot de nouvelles déprimantes que martelait la présentatrice, une seule a réellement éveillé mon intérêt.

Les enquêteurs n'avaient pas la moindre piste à propos du gang des clowns. J'ignore pourquoi, mais cette bande-là me plaisait bien et j'étais content qu'ils échappent à la police.

Le documentaire parlait des insectes. J'ai éteint la télévision juste après les images d'une mante religieuse dévorant le pauvre mâle avec lequel elle venait de s'accoupler. Brrrr! Heureusement Chloé n'avait rien d'une mante religieuse. Elle n'avait rien de l'insecte tout court. Dommage, elle aurait fait le premier pas et ça m'aurait arrangé. À condition bien sûr qu'elle me choisisse!

pisaura mirabilis

Le lendemain, mardi, toujours pas de Chloé. Mais, à la récré, j'ai appris un truc extraordinaire qui m'a chamboulé pour le reste de la journée. Comme je ne pouvais pas jouer au foot à cause de mon œil, je me suis assis sous le préau avec un bouquin.

Juste à côté, cette peste de Marthe discutait avec Marie et Élise, deux autres filles de la classe. Si j'étais handicapé côté vision, je n'avais aucun problème avec mon audition et mes oreilles ont grandi de quelques centimètres lorsque j'ai entendu Marthe dire :

– Je suis allée chez Chloé, hier soir. La maîtresse m'avait priée de lui apporter les devoirs. Vous ne devinerez jamais ce qu'elle m'a confié !

Priée, confié. Elle parle comme ça, Marthe, avec des mots de marquise et un air pincé et supérieur qui la rend encore plus antipathique.

– Quoi ? ont questionné en chœur Marie et Élise.

– Eh bien, figurez-vous que...

Elle s'est interrompue avant de poursuivre :

– Non. Je ne vais pas vous le dire aussi facilement. Vous m'interrogez et je réponds par oui ou par non. Ce sera plus drôle.

Les questions ont commencé, un torrent de questions auxquelles, je l'ai senti, Marthe prenait plus de plaisir à répondre par non que par oui. Je brûlais d'envie d'en poser moi aussi, mais j'avais encore plus envie de me lever, d'attraper Marthe par les épaules et de la secouer comme un prunier pour qu'elle recrache tout ce qu'elle savait.

Évidemment, je n'ai pas bougé : je ne suis pas un sauvage. Je me suis contenté d'écouter.

Au bout d'une dizaine de minutes de ce jeu idiot, j'ai fini par connaître le secret de Chloé. Son père avait été en prison. C'est la raison pour laquelle sa famille avait emménagé ici, pour recommencer une nouvelle vie. Par contre, Marthe ignorait la raison pour laquelle il avait été condamné.

Elle a fait jurer cracher à Marie et Élise de ne jamais rien révéler de ce qu'elle venait de leur avouer. Elles ont promis.

Je n'aurais pas parié un centime sur leur silence. Mais que faire ? Je ne pouvais quand même pas leur couper la langue !

J'étais dans une rogne noire contre Marthe en qui Chloé avait mis sa confiance et qui s'était empressée de la trahir. J'ai dû lutter de toutes mes forces afin de m'empêcher de lui crocheter la jambe dans les escaliers lorsque nous sommes remontés en classe.

Après le choc de cette nouvelle, je me suis remémoré le père de Chloé, ses allures menaçantes de brute épaisse. Un frisson m'a de nouveau parcouru la colonne vertébrale.

Je me suis également rappelé les paroles d'Alex à propos de Chloé : « On a l'impression qu'elle est toujours prête à pleurer. Pourtant personne ne lui veut de mal. »

Pas étonnant. Si j'avais été à sa place, j'aurais moi aussi été inquiet que mon père commette une autre bêtise.

Le soir, à table, j'ai demandé à maman :

– Tu crois que quelqu'un qui est libéré de prison a plus de chance d'y retourner que quelqu'un qui n'y est jamais allé ?

– C'est une question délicate. Les cas de récidive existent mais certains réussissent à s'en sortir.

– Récidive ! Qu'est-ce que ça veut dire ?

– Récidiver, ça veut dire recommencer. En fait, a continué maman, j'imagine que le comportement des anciens prisonniers après leur peine dépend beaucoup de la situation dans laquelle ils se trouvent à leur sortie. Ont-ils un logement, un travail, une famille stable ?

Maman a continué son exposé sans que je sois plus avancé.

Le jeudi, Chloé est enfin revenue. En classe, elle était au premier rang à gauche, alors que j'étais placé dans le fond à droite. Cette disposition me permettait de l'observer sans qu'elle le remarque.

Et je ne m'en suis pas privé! J'ai passé ma journée à la détailler sous toutes les coutures. J'ai vu chacun de ses gestes, j'ai observé chacune de ses moues, j'ai chaviré de plaisir devant son profil magnifique, j'ai été aux anges quand j'ai remarqué que, lorsqu'elle écrivait, elle penchait légèrement sa tête de côté.

Elle était belle, si belle et il était tout à fait naturel que je sois tombé raide amoureux. Je me suis même dit que tous les garçons de la classe étaient peut-être comme moi.

Cette idée ne m'a pas plu du tout. Et si un autre que moi réussissait à la séduire? Je ne m'en remettrais jamais…

Alex avait prétendu qu'elle était sans arrêt sur le point de pleurer. C'était complètement faux. Elle n'avait pas l'air fragile bien que, de temps en temps, un voile de tristesse se peigne sur son doux visage. Peut-être qu'à ces moments-là, elle repensait à son père quand il était en prison? Ça avait dû être drôlement difficile.

J'aurais tellement voulu lui faire oublier ces mauvais souvenirs. On serait seuls, rien que nous deux. N'importe où mais seuls. Je lui imiterais le paon, le lion. Elle rirait de mes blagues. Tous ses sourires seraient pour moi. Adieu tristesse ! Elle serait si gaie qu'elle se laisserait aller à imiter la libellule ou la si jolie biche.

Inutile de m'interroger sur ce qui s'est passé en classe ce jeudi matin, je l'ignore. Heureusement, par miracle, mademoiselle Morlot m'a fichu une paix royale.

Je n'ai pas osé parler à Chloé. Je ne savais pas quoi lui dire et, même si j'avais su, j'aurais été incapable d'aligner deux mots. Je ne tenais pas à passer pour l'idiot du village planté devant elle la bouche ouverte.

Et puis, avec mon coquart et mes lunettes de soleil, je ressemblais à un boxeur après un combat perdu par K.-O. au dernier round.

Pourtant, si je voulais rentrer en contact avec elle, il fallait que je tente quelque chose !

L'idée, je l'ai finalement trouvée à la cantine, alors que je dessinais avec ma fourchette la silhouette d'une *pisaura mirabilis* dans ma purée. Cette araignée a une parade nuptiale vraiment très originale. Le mâle capture une proie et l'emmaillote d'un filet de soie. Ensuite, il offre son joli cadeau à la femelle. Si elle l'accepte et le dévore, c'est qu'elle est d'accord ! Et moi, mon cadeau était tout trouvé.

parade nuptiale

Dans la cour, à la récréation, je suis descendu avec ma proie, enfin avec mon livre sur les animaux.

Je me suis posté sous le préau, à quelques mètres du groupe de filles où se trouvait Chloé. Marthe n'y était pas. Chloé n'avait eu besoin que de quelques jours pour s'apercevoir que c'était une peste. Encore un bon point pour elle.

J'ai ouvert le livre et j'ai fait semblant de lire. En réalité, j'attendais le bon moment pour passer enfin à l'action.

Lorsque Chloé a été seule, je me suis levé et, le plus naturellement possible, je me suis dirigé vers elle en lui demandant :

– Est-ce que tu pourrais garder mon bouquin ? Je ne veux pas l'emmener sur le terrain de foot.

Sans attendre sa réponse, je lui ai collé le livre dans les mains et j'ai rejoint les copains. En jetant un coup d'œil vers l'arrière, j'ai vu qu'elle regardait la couverture.

– Tu joues! s'est étonné Alex.

– Enlevez vite les poteaux! a crié Jules.

– Non, non. Il vaut mieux pas. Je vais arbitrer.

– Wouah! Génial! a conclu Alex avant que la partie ne reprenne.

Je me suis placé le long de la ligne de touche opposée au préau, de manière à observer les moindres gestes de Chloé. Elle avait ouvert le livre et le parcourait.

Lorsque la sonnerie a retenti, elle est venue me le rendre. Nous avancions l'un vers l'autre.

Dix mètres, neuf, huit.

J'ai fourré mes mains dans mes poches afin qu'elle ne remarque pas qu'elles tremblaient.

Sept, six, cinq, quatre.

Je me suis raclé la gorge afin de pouvoir prononcer la phrase que j'avais préparée. C'était quoi déjà, cette phrase ? Pas moyen de me la rappeler. J'ai paniqué.

Trois, deux, un.

C'était fichu, j'allais être ridicule...

– Ça a l'air super ! a-t-elle dit.

Sa voix douce a fait voler mon trac en éclats. J'ai failli hurler de joie. J'avais réussi la première phase de mon plan ! Vite, passer à la deuxième. D'un ton plein d'assurance, je lui ai demandé hypocritement :

– Tu t'intéresses aux animaux ?

– J'adore !

Bien que cette passion ne m'ait envahi que depuis quelques jours, je lui ai affirmé haut et clair :

– Moi aussi.

Elle a souri, alors j'ai ajouté :

– Si tu veux, je te le prête.

J'ai lu de l'hésitation dans ses yeux. De l'hésitation et de la timidité. C'était l'instant de vérité. Je devais la persuader d'accepter ma proie.

– Tu n'auras qu'à me prêter un de tes livres en échange, ai-je tenté.

Son sourire est revenu.

– D'accord ! T'es gentil, Quentin !

C'était la première fois que j'entendais mon prénom dans sa bouche et ça m'a réchauffé tout le corps, mais elle avait eu du mal à prendre mon cadeau et cela ternissait quelque peu mon succès.

N'empêche, le plus difficile était accompli, j'avais noué le contact avec elle et je ne semblais pas lui déplaire. La preuve, elle avait dit que j'étais gentil.

Ensuite, comme chaque jeudi en fin de journée, nous sommes allés au gymnase.

Toujours à cause de mon œil, je me suis assis sur un banc et j'ai assisté à l'échauffement des autres. Enfin, à l'échauffement de Chloé. Elle était si gracieuse. C'est simple, elle dansait et elle courait avec la grâce d'une gazelle. Que dis-je, elle volait tel un superbe papillon de jour dans toute sa beauté. Je serais resté de longues heures à la regarder s'ébattre, si aérienne.

Après l'échauffement, mademoiselle Morlot a demandé aux garçons d'installer les tapis de gymnastique. Lors d'une roulade avant, le tee-shirt de Chloé s'est soulevé. J'ai alors aperçu deux vilaines traces bleu-violet qui barraient le bas de son dos. Des traces de la même couleur que celles de mon œil.

Je n'ai pas réagi tout de suite mais, sur le chemin du retour, j'y ai pensé sans arrêt et je suis arrivé à une conclusion qui m'a retourné l'estomac.

J'ai attendu Alex et Jules et je leur ai annoncé :

– Il faut que je vous parle, les gars.

Alex a été surpris par la gravité de ma voix.

– Ça a l'air sérieux ! Ben, vas-y, on t'écoute.

– Non, pas ici. Marchons jusqu'au parc. Je vous expliquerai là-bas.

le courage du lion

Le parc Nicaud se trouve à environ trois cents mètres de l'école. Pendant le trajet, personne n'a desserré les dents.

Même pas Alex qui, habituellement, est plus bavard qu'une assemblée de merles réunis sur un arbre.

À peine étions-nous installés sur un banc entouré d'arbustes qu'Alex, n'y tenant plus, m'a demandé :

– Alors, qu'est-ce qui se passe ?

J'ai jeté un regard aux alentours avant d'attaquer :

– Je crois que Chloé se fait frapper par son père.

Le silence s'est abattu sur nous. Les rares oiseaux présents dans les arbres à cette saison se sont arrêtés de chanter et la nuit, tout à coup, a semblé tomber plus vite.

– Comment tu sais ça ? m'a interrogé Jules.

Je leur ai raconté la conversation que j'avais surprise entre Marthe, Élise et Marie. Je leur ai expliqué les traces violacées dans le bas de son dos. Je leur ai décrit les muscles de son père sans mentionner les circonstances dans lesquelles je l'avais rencontré. Je n'ai rien dit non plus du vif intérêt que je portais à Chloé.

– Et alors ! s'est exclamé Alex. Ça ne prouve rien. Tu t'emballes un peu trop vite…

– C'est vrai, a renchéri Jules. Ce n'est pas parce que son père est baraqué et qu'il a fait de la prison qu'il frappe sa fille !

– Tout ça, a continué Alex, c'est des préjugés !

– Ouais, a confirmé Jules. Chloé s'est peut-être juste blessée…

Ils avaient raison. Je n'avais aucune preuve de ce que j'avançais. Et comme disait mademoiselle Morlot, on n'accuse pas sans preuves. J'ai shooté du bout de ma chaussure dans un caillou avant de leur proposer :

– Le seul moyen d'être sûr, c'est de mener une enquête.

– Comment veux-tu qu'on mène une enquête ? m'a objecté Alex.

– Ben… J'sais pas. On observe, on guette, on épie. Je ne vous oblige pas à m'aider. Vous êtes libres, je ne vous en voudrai pas si…

– Je marche avec toi, m'a interrompu Jules.

– Moi aussi ! a affirmé Alex.

C'était un grand moment. J'étais fier d'eux et j'ai déclaré tel le général parlant à ses troupes avant la bataille :

– Je savais que je pouvais compter sur vous. Vous êtes des vrais copains. Voilà ce que je vous propose. Demain, on observe Chloé chacun de notre côté. À la sortie, on fait le point et on va surveiller ce qui se passe chez elle.

Je leur ai tendu la paume de ma main. Ils l'ont énergiquement tapée du bout de leurs doigts en signe de pacte.

Alex m'a demandé :

– Tu connais son adresse ?

J'ai menti en annonçant avec l'assurance que se doit de manifester un
vrai chef :

– Non. Mais t'inquiète ! Je me
débrouillerai pour la dénicher.

Le lendemain, vendredi, notre
espionnage ne nous a rien appris de
nouveau. Nous nous sommes donc
retrouvés au parc, au même endroit
que la veille, à six heures.

Le parc est situé à environ deux cents mètres de l'allée des Acacias et constitue donc un excellent lieu de ralliement et de repli. J'avais pensé à cet instant pendant la journée entière et le moment était enfin venu de passer à l'action. J'ai donné mes dernières instructions :

– Bon, on est d'accord ? En cas de coup dur, on se sépare et on file dans des directions différentes. O.K. ?

– O.K. ! ont répondu Alex et Jules dans un ensemble parfait.

J'ai tendu ma main et, comme la veille, ils l'ont frappée. Nous nous sommes dirigés vers la maison de Chloé sans prononcer la moindre parole.

Nos bouches envoyaient des panaches de vapeur blanche dans la nuit qui s'annonçait glaciale. Bizarrement, je n'avais pas froid. Au contraire, la sueur perlait sur mon front et descendait en grosses gouttes le long de mon dos.

Au coin de la rue des Acacias, Jules s'est brusquement arrêté. Il tremblait de tous ses membres. Il s'est assis sur le premier muret venu et, en bégayant, il nous a avoué :

– J'peux pas, les gars. J'peux pas. J'ai trop la trouille.

Ses dents du haut claquaient sur celles du bas dans un rythme effréné de castagnettes.

Je l'ai encouragé :

– Moi aussi, je meurs de trouille. C'est normal. Allez, viens !

Jules n'a pas bougé. Comme s'il était collé par le gel sur son muret.

– Tu te rends compte? il a dit. Si ce type nous attrape et s'il est aussi méchant que tu le crois? Il va nous découper en petits morceaux et nous cacher dans son congélateur. Il sortira un bout de temps en temps pour le donner à son chien. Je veux pas finir comme ça.

J'ai failli lui dire qu'il n'y avait pas de chien mais j'ai renoncé, cela n'aurait servi à rien. Impressionné par les derniers mots de Jules, Alex s'est mis à trembler à son tour. Il a bégayé :

– Moimoimoi non pluplus. Jejeje veux papapas finir coco comme çaça!

J'ai encore tenté de les motiver :

– Et Chloé? Vous y pensez à Chloé?

J'ai lu dans leur regard qu'à ce moment précis ils s'en fichaient éperdument. Ils n'ont pas osé me l'avouer.

– On n'a qu'à appeler le 119! a proposé Jules.

– Ouais, a renchéri Alex, en commençant à faire demi-tour. Bonne idée. Allez, viens, Quentin. On laisse tomber.

– Le 119? C'est quoi le 119?

– Le numéro de téléphone pour les enfants maltraités, m'a expliqué Jules.

Un instant, j'ai failli craquer. Mais les yeux de Chloé sont apparus devant moi, des larmes y brillaient. Je ne pouvais pas reculer.

– J'y vais tout seul.

Penauds, ils ont baissé la tête avant de s'éloigner. Après quelques pas, Alex s'est retourné et m'a dit :

– Fais gaffe, Quentin.

Ils s'étaient dégonflés et normalement, j'aurais dû leur en vouloir. Ce n'était pas le cas. En fait, j'étais presque content d'agir seul. Si j'accomplissais ma mission en solitaire, ce serait moi et moi seul qui en récolterais la gloire auprès de Chloé. Je l'aimais, je devais la sauver ! C'était entre elle et moi.

Quand la lionne est en danger, c'est au lion, le roi des animaux, de voler à son secours. J'étais le lion. J'ai senti ma poitrine se gonfler de courage.

un chat sur les talons

Mes pieds m'ont amené dans le jardin du numéro 10 sans que je leur demande rien.

Je me suis glissé le long de la façade jusqu'à une fenêtre illuminée. Collé contre le volet, je jetais un regard de temps à autre à l'intérieur.

Ils étaient quatre. Chloé, son père et un garçon d'environ cinq ans, étaient assis à la table. La mère de Chloé, quant à elle, tournait le contenu d'une cocotte avec une cuillère en bois.

Le petit s'amusait à assembler les pièces d'un puzzle, Chloé lisait une bande dessinée et le mastodonte écrivait dans un cahier. Le stylo paraissait minuscule dans ses gros doigts puissants. Un instant, j'ai été étonné que de telles mains puissent tracer des lettres. Il s'est mis à lire à voix haute ce qu'il venait de noter.

Je n'ai évidemment rien entendu, mais Chloé et sa mère ont éclaté de rire. Que pouvait-il bien gribouiller de si drôle dans son cahier ?

Il l'a posé, s'est levé, a ouvert la porte de la cuisine.

Au moment même où sa large silhouette s'engouffrait dans l'encadrement, j'ai senti un frôlement dans mes jambes.

J'ai eu si peur qu'un grand vide s'est creusé dans mon ventre. Ma respiration s'est arrêtée quelques secondes. Un chat. C'était un chat noir qui caressait le bas de mon pantalon de sa queue dressée.

J'ai laissé échapper un rire nerveux tandis que mon cœur éclatait en mille morceaux.

Je n'ai pas eu le temps de me remettre de mes émotions. J'ai entendu la porte du garage s'ouvrir. J'ai aussitôt quitté mon emplacement pour filer derrière la maison. Bien m'en a pris car, à cet instant, la grosse voix grave du père de Chloé a empli les ténèbres :

– Belzébuth ! Belzébuth !

Le Belzébuth en question, qui n'était autre que le chat, m'a suivi, prenant mon départ pour un jeu. L'imbécile !

J'ai effectué le tour de la maison le chat sur les talons et le père de Chloé sur les talons du chat.

– Belzébuth ! Belzébuth !

Arrivé à la fin de mon périple, j'avais deux solutions. Jouer au recordman du monde sur cent mètres en traversant le jardinet pour disparaître dans la rue, ou entrer dans le garage ouvert et m'y cacher. J'ai réfléchi à une vitesse sidérale et j'ai opté pour le garage. Impossible de traverser le jardinet sans être repéré par la brute. Et, à la course, je n'avais aucune chance de lui échapper.

Je me suis jeté sous le break cabossé. Belzébuth est apparu entre les roues arrière, s'apprêtant à me rejoindre. J'étais cuit. J'ai pensé au congélateur de Jules en me disant que j'allais finir mes jours entre les dents d'un chat.

Heureusement, la grosse main du colosse est venue cueillir cette idiote de bestiole avant qu'elle ne me rejoigne.

– Je t'ai eu, Belzébuth. Qu'est-ce que tu croyais? Que j'étais moins rapide que toi?

Depuis le dessous de la voiture, j'ai entendu la porte se refermer. La lumière s'est éteinte et je suis resté seul, éclairé par la faible lueur du lampadaire de la rue filtrant à travers les hublots de la porte.

J'ignore combien de temps j'ai passé ainsi, couché, le nez sous le pot d'échappement, n'osant pas bouger un orteil, paralysé par la peur. La brute devait réparer sa voiture car des outils étaient éparpillés sous le châssis.

Le sol était glacé. Il fallait que je me décide à bouger. Centimètre après centimètre, j'ai fini par me dégager du dessous du break.

Je me suis redressé doucement, les articulations endolories. Le plus dur restait à accomplir : sortir du garage et franchir la distance me séparant de la rue. Chaque geste me demandait un courage énorme tant mon angoisse était forte.

J'ai fait un pas, deux, trois. Au quatrième, mon pied droit est venu buter contre quelque chose.

Dans la pénombre, j'ai reconnu un des sacs que j'avais vus quelques jours auparavant dans le coffre du break.

La curiosité l'a emporté sur la peur et, après m'être agenouillé, j'ai tiré le plus silencieusement possible sur la fermeture. J'ai vidé son contenu sans bien voir de quoi il s'agissait.

Lorsque, brusquement, j'ai identifié ce qui s'étalait en vrac sur le sol, la panique est revenue au grand galop. Une panoplie de clown. Le sac était rempli d'une panoplie de clown !

La tête m'a tourné. Tout s'y mélangeait : les clowns, les braquages de banques, le père de Chloé ancien prisonnier. C'était trop pour moi !

Je me suis rué vers la porte. Trop vite. Je me suis étalé de tout mon long, provoquant un boucan infernal. Je me suis redressé et, après avoir manipulé le système de fermeture du garage pendant de longues secondes, les lumières de la rue sont enfin apparues. J'étais sauvé ! J'ai couru vers la liberté.

Une ombre s'est dessinée sur le gravier devant moi. Ma course a été stoppée net comme si j'étais rentré dans un mur invisible.

Une main a agrippé mon blouson et m'a soulevé telle une petite plume insignifiante.

Mes pieds ont décollé et je me suis retrouvé à soixante-dix centimètres du sol, nez à nez avec le père de Chloé. J'ai eu une pensée pour le congélateur, alors j'ai fermé les yeux pour ne pas voir la suite…

Je les ai rouverts en même temps que mes pieds reprenaient contact avec le sol du garage.

– Qui es-tu? Que fais-tu ici?

Je lui aurais bien répondu mais aucun son ne voulait sortir de ma bouche. Une des mains du mastodonte a actionné l'interrupteur tandis que l'autre m'immobilisait toujours par le col.

J'aurais volontiers récité une prière mais je n'en connaissais pas. Il a vu le contenu du sac éparpillé par terre. Une grimace de mécontentement a rendu son visage encore plus menaçant.

– Ramasse ! m'a-t-il ordonné sèchement.

Il m'a lâché comme un vulgaire crapaud et je me suis agenouillé. Il était juste au-dessus de moi et me dominait de toute sa hauteur, les jambes écartées. Il était en chaussons.

J'ai aperçu une grosse clé à molette sous la voiture. Si je ne voulais pas terminer mon existence dans des sachets de congélation, c'était le moment ou jamais.

J'ai rangé un masque et une perruque bleue. Puis, rapide comme l'éclair, je me suis emparé de la clé et j'ai tapé de toutes mes forces sur le chausson droit. J'ai entendu un craquement juste avant qu'un vilain grognement de douleur ne sorte de la bouche du gorille.

Je ne suis pas resté à l'écouter hurler. Je me suis glissé entre ses jambes et, prenant les miennes à mon cou, j'ai filé dans la rue. Je pensais à la peur abominable que devaient ressentir les chevreuils traqués par les chasseurs et par les chiens. Je me suis juré de ne jamais participer à une pareille horreur.

Je suis repassé par le parc où, dissimulé dans un bouquet d'arbustes, je me suis arrêté quelques minutes pour reprendre mon souffle et vérifier qu'il ne m'avait pas suivi. Tout semblait calme.

Lorsque ma respiration a repris son rythme presque normal, je suis sorti de ma cachette et j'ai déguerpi vers la maison.

J'ai trouvé maman dans le hall d'entrée, le téléphone sur l'oreille, le visage inquiet et les yeux rouges. Lorsqu'elle m'a vu, elle a dit dans le combiné :

– Le voilà. C'est lui. Il est enfin rentré. Excusez-moi de vous avoir dérangé.

Elle a reposé l'appareil et m'a serré dans ses bras, longtemps, sans prononcer la moindre parole. J'étais heureux d'être là, si heureux après ce que je venais d'endurer que je me suis mis à pleurer, pleurer, pleurer. Un véritable torrent de montagne.

Les mains sur mes épaules, maman m'a demandé doucement :

– Mais enfin, Quentin, où étais-tu passé? J'étais folle d'inquiétude. J'ai téléphoné à tous tes copains. Je viens juste de parler avec la police!

un vrai clown

L'idée de lui mentir ne m'a même pas effleuré. Je lui ai raconté l'histoire de A jusqu'à Z.

Elle m'a écouté sans m'interrompre. Au moment où je terminais le récit de mes exploits, la sonnette de la porte d'entrée a carillonné. Nous sommes allés ouvrir.

J'ai frôlé la crise cardiaque quand la silhouette menaçante du père de Chloé s'est dessinée dans l'encadrement. Maman a eu un mouvement de repli. Moi, je n'ai pas bougé, j'étais comme collé à la super glu sur les lattes du parquet.

La brute épaisse a grimacé méchamment. Il allait nous massacrer, nous écrabouiller comme de minuscules moustiques.

– Salut, Quentin! Je crois que tu as quelques explications à me fournir. Non?

J'aurais bien voulu lui répondre mais ma langue était comme mes pieds, scotchée dans ma bouche desséchée. Pleine de sang-froid, maman, qui avait identifié le colosse, a rétorqué fermement :

– Je pense que c'est plutôt à vous de nous en fournir !

Le mastodonte a aussitôt haussé le ton :

– Vous plaisantez, j'espère, madame. Votre fils s'introduit chez moi, fouille dans mes affaires, me donne un coup de clé à molette avant de s'enfuir et c'est moi qui devrais m'expliquer ?

Bluffer, je devais bluffer pour l'empêcher de nous broyer maman et moi entre ses grosses pattes menaçantes.

J'ai enfin réussi à articuler :

— La police est prévenue. Ils vont arriver d'une minute à l'autre…

Maman m'a gratifié d'un regard admiratif avant de poursuivre :

— Vous ne feriez qu'aggraver votre cas en nous agressant…

Il nous a lancé un regard étonné :

— Aggraver mon cas ? Vous agresser ? Mais c'est une histoire de fous ! Je suis venu ici pour que tu m'expliques ce que tu faisais dans mon garage. J'ai tout de même le droit de le savoir, non ?

— Quand on braque des banques, on n'a le droit de rien.

Je m'attendais à ce qu'il devienne furieux après mes dernières paroles, mais je me trompais. Il a répété plusieurs fois :

– Braquer des banques, braquer des banques, braquer des banques… Mais qu'est-ce que c'est que cette histoire de banques?

Victime d'une soudaine faiblesse, il s'est adossé au chambranle de la porte, le regard vide et lointain.

– Je ne m'en sortirai donc jamais.

Il paraissait anéanti, sa grande carcasse ratatinée semblait à présent totalement inoffensive. On était sauvés.

Il s'est redressé et, avant de s'éloigner, il nous a annoncé :

– Vous pourrez dire aux policiers que je rentre chez moi. Tu connais l'adresse, n'est-ce pas ? Tu leur indiqueras.

Il a accompli quelques pas dans le couloir, les épaules basses et la démarche claudicante. Je n'y étais pas allé de main morte avec la clé à molette !

J'ai repensé à la scène que j'avais épiée dans la cuisine, quand il avait lu dans son cahier. J'ai revu le visage souriant de Chloé. Et, soudain, j'ai été pris d'un doute. Je m'étais peut-être trompé. Bien sûr, les apparences étaient contre lui mais si j'accusais son père à tort, jamais Chloé ne me pardonnerait. Je n'aurais plus qu'à faire une croix sur son amour.

J'ai presque crié :

– Attendez ! Ne partez pas !

Maman m'a observé d'un air curieux.

J'ai ajouté :

– C'est pas vrai ! On n'a pas prévenu la police !

M. Muscles s'est arrêté, s'est retourné et a planté son regard dans le mien :

– Faudrait savoir, petit. Il y a deux minutes tu m'accusais d'être un braqueur de banques et maintenant tu m'empêches de partir ! Tu veux quoi, à la fin ?

Je n'ai pas répondu, me contentant de l'inviter à entrer et à s'asseoir sur le canapé du salon. Il a émis un gémissement de douleur sous son poids. Le père de Chloé nous regardait tour à tour, attendant des éclaircissements.

Je lui ai retracé les événements de ces derniers jours en faisant bien attention de ne rien oublier. Lorsque j'ai parlé des traces violacées sur le dos de Chloé et de mes soupçons, son visage s'est crispé. Il a serré les dents pour ne pas m'interrompre.

Quand j'ai eu terminé mon récit, ses joues avaient repris des couleurs. Il a dit dans un sourire :

– Bien, à mon tour de raconter.

Il a avalé sa salive et s'est lancé :

– Tu as raison, j'ai bien effectué une peine de prison. De six mois. Avant, j'étais charpentier. Je travaillais dans une grosse entreprise. Avec les collègues, on avait terminé un important chantier, la charpente d'un gymnase. On a décidé de fêter ça. Un verre, deux verres, trois, et puis quatre. Je n'avais pas l'impression d'être saoul et j'ai pris ma voiture. Je n'ai pas réussi à éviter un jeune cycliste d'à peu près ton âge qui a fait un écart sur la route...

Je n'ai pas pu m'empêcher de lui demander :

– Il... il... il est mort ?

– Non. Il est dans un centre spé-
cialisé pour les accidentés de la
route depuis plus d'un an. J'ai toujours l'espoir qu'il reprendra une
existence normale.

Il s'est tu à nouveau avant d'ajouter en soupirant :

– Vous imaginez la suite... l'ambulance, les gendarmes, prise de sang,
jugement et condamnation. Les
pires moments de ma vie. Non pas
parce que j'étais condamné mais
parce que j'étais coupable. J'avais
failli tuer quelqu'un. J'ai passé cinq
mois horribles en prison.

– Cinq mois ? a répété maman.
Vous nous avez confié il y a un instant que vous aviez été condamné à
une peine de six mois !

– Exact. Mais ce sont les cinq premiers qui ont été horribles car, au début du sixième, des clowns sont venus donner un spectacle. Ça a été une révélation pour moi. J'ai ri. Oui, j'ai ri. Oubliant mes ennuis et ma culpabilité. En sortant de prison, j'ai endossé une panoplie de clown puis suis allé jouer mes numéros devant des enfants malades. Et pas braquer des banques comme tu l'as imaginé. Je n'ai rien à voir avec cette bande qui sévit depuis quelque temps. Voilà, vous savez tout.

J'ai hésité avant de le contrarier :

– Euh… Pas vraiment. Les traces noires dans le dos de Chloé…

– Le judo. Chloé pratique le judo depuis qu'elle est petite. Elle a pris un vilain coup lors de sa dernière compétition, dimanche. C'est d'ailleurs pour cette raison qu'elle était absente lundi et mardi. Et toi, Quentin, tu boxes ?

Je n'ai pas compris alors il a pointé un doigt vers mon œil.

– Ah ! Euh… Non, non. J'ai testé la solidité des poteaux dans la cour.

Il a ri de bon cœur. Maman et moi aussi.

– Tu sais que c'est grâce à ton œil que je t'ai retrouvé ! m'a-t-il annoncé. Après le coup de clé à molette, j'ai inspecté le garage : rien n'avait disparu. Je me suis dit que tu n'avais peut-être pas l'intention de voler quoi que ce soit. Comme j'avais constaté que tu avais à peu près le même âge que Chloé, je lui ai demandé si, par hasard, elle ne connaissait pas un jeune garçon avec un œil au beurre noir. Elle n'a pas hésité une seconde. J'ai pris un annuaire téléphonique et j'ai trouvé votre adresse.

J'ai bredouillé :
– Je suis désolé. Excusez-moi. Je ne pensais pas que...

Il a tendu sa main pour m'intimer le silence :

— Ne t'excuse pas. J'aurais agi comme toi. Tu es drôlement courageux et généreux. T'intéresser ainsi à une fille que tu connais à peine, chapeau bas, Quentin !

Je n'ai pas bronché. Je ne lui ai surtout pas avoué mon coup de foudre pour Chloé. Il s'est levé et nous a salués avant de disparaître dans l'escalier en boitant.

bêtes de scène !

C'est à mon tour. Je m'élance dans mon costume de paon. Je débouche sur la scène improvisée sous les oh et les ah. L'immense Grattouille me poursuit avec un long bâton en criant :

– Va-t'en ! Va-t'en ! Maudite bestiole !

Je disparais de l'autre côté du rideau tandis qu'il se prend les pieds dans le tapis et chute en se cognant la tête contre une échelle. Rires ! Je réapparais à l'opposé et, tandis qu'il a le plus grand mal à se relever, j'en profite pour dévoiler ma plus belle roue à Chloé, elle aussi déguisée en paon. Grattouille nous tourne le dos. Les enfants lui crient :

– Derrière, Grattouille ! Là-bas, regarde !

Il fait semblant de ne pas entendre et sort un immense entonnoir de sous son chapeau. Il le colle sur son oreille. Il comprend enfin et reprend la poursuite. Cette fois, il se sert de son bâton comme si c'était un fusil en ponctuant chaque coup de feu :

– Paon ! Paon ! Paon !

Le jeune public se régale. Moi aussi.

C'est Pascal alias Grattouille – le père de Chloé – qui a eu l'idée d'écrire une scène d'après le livre que j'avais prêté à Chloé et qui a décidé de l'incorporer dans son grand spectacle intitulé *Pas si bêtes!*

J'avais l'air un peu idiot travesti en paon mais je m'en fichais. L'important était de passer le maximum de secondes et de minutes auprès de Chloé à qui, d'ailleurs, je n'avais toujours pas osé avouer mon amour. Je laissais faire le temps.

Peut-être qu'un jour je me déciderais. Pour l'instant, mes rêves, mes espoirs et sa présence auprès de moi me suffisaient.

Le lendemain de ma virée nocturne au 10 allée des Acacias, j'ai attendu Chloé devant l'école en compagnie d'Alex et de Jules à qui j'avais raconté mes aventures une bonne dizaine de fois au moins. J'ai cru comprendre qu'ils regrettaient de ne pas avoir eu le courage de m'accompagner. Tant pis pour eux ! N'est pas un aventurier qui veut !

Lorsqu'elle a pointé le bout de son joli nez, j'ai laissé mes deux copains pour la rejoindre. J'ai juste dit :

– Chloé, j'espère que tu pourras me pardonner.

Elle m'a répondu du tac au tac, d'un air pincé et sévère :

– Non. Impossible.

Le monde s'est écroulé autour de moi. La journée allait être très mauvaise. Pas seulement la journée, les lendemains aussi.

J'ai baissé la tête, piteux. Je m'apprêtais à la quitter lorsqu'elle a répété, dans un sourire cette fois :

– Non. Impossible, Quentin, car il n'y a rien à pardonner.

Le monde s'est reconstruit. La journée serait bonne.

J'enlève mon costume de fier vola-
tile derrière le rideau qui sépare
la scène des coulisses. Chloé me
rejoint, s'approche de moi et dépose
un baiser sur ma joue. Je me sens
rougir comme un coq.

Voyons, Quentin, réfléchis.
Comment se débrouillent-ils,
les coqs, pour faire la cour aux
poules?

TABLE DES MATIÈRES

Gilles Fresse est né dans une petite ville des Vosges en 1960. Il y a passé son enfance parmi les sapins.

Instituteur et comédien amateur, il crée voici quelques années un atelier théâtral pour enfants. Il écrit alors des pièces qu'il met en scène. Touché par le virus de l'écriture, il aime écrire des romans drôles et haletants.

Marié et père de deux enfants, il vit dans un petit village lorrain riche en vieilles pierres. Il partage son temps entre l'écriture, le théâtre et… l'école.

✎ L'ILLUSTRATEUR

Emmanuel Chaunu est né à Caen en 1966. Dessinateur infatigable, il dirige une agence de communication et illustre l'actualité dans une vingtaine de titres de presse.

Pour la littérature jeunesse, il imagine des univers peuplés d'animaux intelligents et drôles. Caricaturiste, il aime rencontrer ses lecteurs au gré des Salons du livre. Il met régulièrement son coup de crayon et son sens de l'observation au service des autres, comme pour le concours international de plaidoiries pour les Droits de l'Homme, au Mémorial de Caen.

Retrouvez la collection
Rageot Romans
sur le site www.rageot.fr

Achevé d'imprimer en France en novembre 2008
par CPI – Hérissey à Évreux (Eure).
Dépôt légal : décembre 2008
N° d'édition : 4852 - 02
N° d'impression : 109999